Advanced Electrocatalysis Materials Design: Innovations and Applications

Advanced Electrocatalysis Materials Design: Innovations and Applications

Guest Editor

Junxian Liu

MDPI

Basel • Beijing • Wuhan • Barcelona • Belgrade • Novi Sad • Cluj • Manchester

Guest Editor
Junxian Liu
Faculty of Engineering
Queensland University of Technology
Brisbane
Australia

Editorial Office
MDPI AG
Grosspeteranlage 5
4052 Basel, Switzerland

This is a reprint of the Special Issue, published open access by the journal *Inorganics* (ISSN 2304-6740), freely accessible at: https://www.mdpi.com/journal/inorganics/special_issues/3EQ2S8W085.

For citation purposes, cite each article independently as indicated on the article page online and as indicated below:

Lastname, A.A.; Lastname, B.B. Article Title. *Journal Name* **Year**, *Volume Number*, Page Range.

ISBN 978-3-7258-7336-4 (Hbk)
ISBN 978-3-7258-7337-1 (PDF)
https://doi.org/10.3390/books978-3-7258-7337-1

Contents

inorganics

MDPI

Review

Recent Progress Using Graphene Oxide and Its Composites for Supercapacitor Applications: A Review

Ganesan Sriram [1,*], Muthuraj Arunpandian [1], Karmegam Dhanabalan [1], Vishwanath Rudregowda Sarojamma [2], Selvaraj David [3], Mahaveer D. Kurkuri [2] and Tae Hwan Oh [1,*]

[1] School of Chemical Engineering, Yeungnam University, Gyeongsan 38541, Republic of Korea
[2] Centre for Research in Functional Materials (CRFM), Jain (Deemed-to-be University), Jain Global Campus, Bengaluru 562112, Karnataka, India; mahaveer.kurkuri@jainuniversity.ac.in (M.D.K.)
[3] Department of Physics Education, Chonnam National University, Gwangju 61186, Republic of Korea
* Correspondence: sriramyu@yu.ac.kr (G.S.); taehwanoh@ynu.ac.kr (T.H.O.)

Abstract: Supercapacitors are prospective energy storage devices for electronic devices due to their high power density, rapid charging and discharging, and extended cycle life. Materials with limited conductivity could have low charge-transfer ions, low rate capability, and low cycle stability, resulting in poor electrochemical performance. Enhancement of the device's functionality can be achieved by controlling and designing the electrode materials. Graphene oxide (GO) has emerged as a promising material for the fabrication of supercapacitor devices on account of its remarkable physiochemical characteristics. The mechanical strength, surface area, and conductivity of GO are all quite excellent. These characteristics make it a promising material for use as electrodes, as they allow for the rapid storage and release of charges. To enhance the overall electrochemical performance, including conductivity, specific capacitance (C_s), cyclic stability, and capacitance retention, researchers concentrated their efforts on composite materials containing GO. Therefore, this review discusses the structural, morphological, and surface area characteristics of GO in composites with metal oxides, metal sulfides, metal chalcogenides, layered double hydroxides, metal–organic frameworks, and MXene for supercapacitor application. Furthermore, the organic and bacterial functionalization of GO is discussed. The electrochemical properties of GO and its composite structures are discussed according to the performance of three- and two-electrode systems. Finally, this review compares the performance of several composite types of GO to identify which is ideal. The development of these composite devices holds potential for use in energy storage applications. Because GO-modified materials embrace both electric double-layer capacitive and pseudocapacitive mechanisms, they often perform better than pristine by offering increased surface area, conductivity, and high rate capability. Additionally, the density functional theory (DFT) of GO-based electrode materials with geometrical structures and their characteristics for supercapacitors are addressed.

Keywords: energy storage; supercapacitor; symmetric; asymmetric; cycle stability; DFT

1. Introduction

High levels of energy consumption and the depletion of fossil fuel sources are all a consequence of fast global development and a significant increase in the rate of population growth around the globe. Consequently, there is a growing need for energy daily to power essential operations such as operating electronic gadgets and motors and supplying distant places with electricity. Currently, supercapacitors (SCs) are considered cutting-edge technology for storing energy because of their fast charging and discharging capabilities, high power density, safety, and extended cycle performances [1,2]. Nevertheless, the energy density (ED) of supercapacitors is comparatively lower than that of batteries, and a key challenge encountered by SCs is the sluggish ion transfer rate, which can be attributed to the comparatively low electrical conductivity of their electrode materials [3]. Batteries are believed to store energy by ion intercalation or deintercalation, resulting in slow ion migration into active electrode materials and having the capacity to keep high ED [4–6]. However, batteries also have some limitations, including diminished performance at elevated temperatures, lower cycle performance, poor power density, and safety issues. Owing to SCs' safety, electrochemical researchers are consistently striving to advance the study to boost the ED in supercapacitors for commercial use [7], given that their ED is not competitive with batteries and fuel cells. Accordingly, Figure 1a shows the Ragone plot of fuel cells, which are observed as high-energy storage materials, while supercapacitors are regarded as high-power devices, and batteries exist between them. Recently, the application of high-charge storage in SCs has been the primary focus of electrochemistry-based scientists [4,8]. The reason SCs are being explored as battery substitutes is because they store energy electrostatically, which takes place promptly at the electrode/electrolyte interface or via redox processes at the electrode interface [9–11].

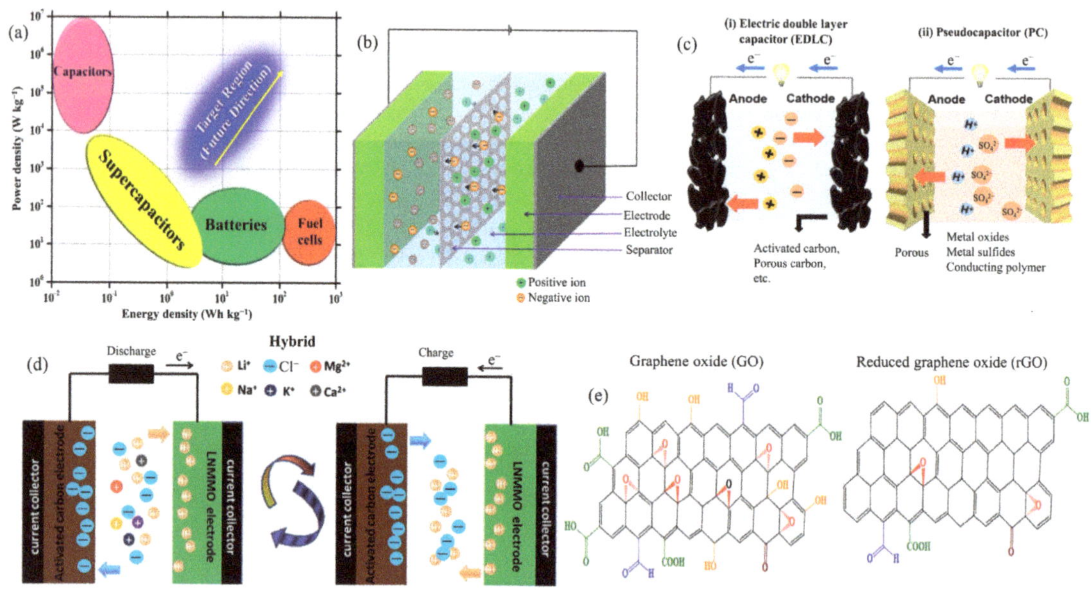

Figure 1. (**a**) Ragone plot for the capabilities of various energy storage devices (reproduced with permission from [8]). Copyright 2023, Elsevier. (**b**) Schematic diagram of supercapacitor working mechanism (reproduced with permission from [12]). Copyright 2022, John Wiley & Sons, Inc. (**c**) Schematic diagram of working of (i) EDLC and (ii) PC supercapacitor (reproduced with permission from [13]). Copyright 2023, Royal Society of Chemistry. (**d**) The charge storage mechanism of the hybrid supercapacitor (reproduced from [14]). Copyright 2020, Elsevier. (**e**) GO and rGO chemical structures (reproduced with permission from [15]). Copyright 2023, Elsevier.

Figure 1b schematically shows that SCs consist of two electrodes separated by an ion-permeable membrane. Current collectors and an electrolyte are also used [12]. SCs can be divided into two types based on their configuration, i.e., symmetric and asymmetric. Symmetric SCs contain an identical electrode (both the positive and negative electrodes are made of the same materials), whereas asymmetric SCs contain dissimilar electrodes (positive and negative electrodes of different materials). The charge storage properties of SCs are significantly impacted by the type of electrolyte, which can be classified as either aqueous or non-aqueous. Charge storage is achieved through the active electrodes, which are facilitated by electrolytes. Aqueous electrolytes are capable of attaining high capacitance and conductivity [16]. Aqueous electrolytes, which are inexpensive and environmentally benign, are ideally suited for SCs on account of their high conductivity and capacitance. The electrode material is the most important component in supercapacitors and has received substantial attention. The design of electrode materials, which constitute the most crucial component of energy storage systems, plays a pivotal role in deciding electrochemical performance [17]. Hence, it is essential to fabricate electrode materials with the requisite composition to create high-performance devices for supercapacitor applications [18,19]. On the other hand, the separator is a component that is used in supercapacitors to separate electrode materials (cathode and anode) against a short circuit. It is present in the form of a porous structured membrane to facilitate the transfer of ions [20,21]. Based on the electrochemical reaction, SCs can be classified as an electrochemical double-layer capacitor (EDLC) or a pseudocapacitor (PC) [22–24]. An EDLC derives its capacitance from charge separation at the interface between the electrode and electrolyte. In contrast, a PC (redox-type capacitor) utilizes charge-transfer pseudocapacitance generated by reversible faradaic redox reactions that take place at the surface of the electrode material and electrolyte [4,25]. Materials for electrodes of the EDLC types include graphene, carbon aerosol, activated carbon, porous carbon, carbon fibers, carbon nanotubes (CNTs), etc. [26–31]. Electrode materials of the PC type are classified as conductive polymers, metal oxides, metal carbides, metal chalcogenides, metal sulfides, conducting MOFs, etc. [32–38].

Based on the differences in the energy storage process, SCs may be classified as EDLC, PC, and hybrid SCs [39–42]. To store charges, the EDLC mechanism depends on electrolyte ion adsorption–desorption at the interfaces or internal pores of materials, as shown in Figure 1c (i). The storing charges in the EDLC mechanism are produced by a physical adsorption process free of any redox reaction [13]. PC goes through a fast and reversible faradaic process on electrodes to store energy, and the charge storage mechanism is caused by a shift in the valence state of a component in the electrode caused by electron transport, as shown in Figure 1c (ii) [43]. On the other hand, hybrid SCs exhibit both charge storage mechanisms [44]. Hybrid supercapacitors are divided into asymmetric and symmetric SCs. To this degree, the asymmetric SCs' two-electrode system has a distinct electrode material that may take advantage of multiple voltage windows to optimize the device's operating voltage and attain a higher ED [38,45]. The two-electrode system of symmetrical SCs uses an identical electrode material, is simple to build, and is inexpensive [30,46]. Figure 1d depicts an asymmetric hybrid electrode system with battery-type cathode electrodes and double-layer anode electrode materials. The activated carbon (AC) electrode's high cycle stability and the capacitance of $LiNi_{0.03}Mo_{0.01}Mn_{1.96}O_4$ (LNMMO) contribute to this [14]. However, the conductivity of the electrode may vary depending on whether the anode and cathode materials possess comparable characteristics. Table 1 shows a comparison of EDLC, PC, and hybrid SCs.

Table 1. Comparison of EDLC, PC, and hybrid capacitors [24,44].

EDLC	PC	Hybrid
Carbon-based materials are utilized as electrode materials.	Metal oxides and conducting polymers are utilized as electrode materials	Electrode materials are used as both carbon and metal oxides/conducting polymers.
The charge storage mechanism is via the non-Faradaic process (electrochemical double-layer formation).	The charge storage mechanism is via the Faradaic process (redox reactions).	The charge storage mechanism is via both Faradaic and non-Faradaic processes.
It has low energy density, good rate capability, cyclic stability, and low specific capacitance (C_s).	High energy density, high power density, low rate capability, and high C_s.	With high energy and power density and good cyclability, the system could have a low or high cost, which depends on the design of electrode materials and modest stability.

Electrodes for supercapacitors have not exhibited significant efficiency despite recent developments in the synthesis of metal oxide materials due to their exhibiting poor efficiency, low capacitance, and low cycle stability throughout the redox reaction as a result of their low electrical conductivity and surface area [47,48]. This may restrict their applicability to PCs in the industrial sector. Hence, the configuration of an innovative composite electrode material, characterized by exceptional electrochemical properties and a synergistic effect among various electrode materials, has garnered significant attention from an extensive array of scientific investigations. Consequently, it is pragmatic to accept that the combination of metal oxides and carbonaceous materials in SCs can produce favorable ED and PD. Due to its high surface area (2630 $m^2 g^{-1}$), mechanical stability, zero band gap, thermal conductivity, and electrical conductivity, graphene is an effective active material [49–51]. Due to their superior chemical stability and electrical conductivity in comparison to metal oxide electrodes, their popularity has been progressively increasing. It is a two-dimensional carbon allotrope family with a hexagonal lattice of atomic size. Chemical decomposition of graphite, epitaxial growth, chemical vapor deposition, and other processes have been utilized to produce graphene. Among these techniques, mechanical cleavage holds promise for the synthesis of high-quality graphene, while chemical methods have the potential to enhance graphene productivity. Graphene oxide (GO), which is a member of the graphene family, has a lower surface area than graphene. GO is a honeycomb-structured carbon material formed by the single-layered arrangement of carbon atoms with sp^2 bonds [52]. GO demonstrates a high specific capacitance (C_s) as a result of the abundance of oxygen-functional groups on its surface. This phenomenon results in an enhanced attraction of water towards GO, whereas graphene exhibits the opposite effect due to its oxygen functional group deficiency [53,54]. GO is more hydrophilic in water than graphene, making it easier to composite with other materials. This improves the electrochemical performance and makes it more useful in supercapacitor applications [55]. In SCs, GO is regarded as a more viable alternative to graphene due to its lower cost and superior performance. Graphene possesses remarkable conductivity and mechanical stability, rendering it a viable material for active materials. However, achieving high ED and power density (PD) with pure carbon materials for SCs is challenging. Nevertheless, their excellent surface area and superior conductivity continue to be crucial factors in developing high-performance SCs.

Recently, interest in graphene oxide-based electrode materials for supercapacitors has gradually increased [15,56,57]. GO was synthesized in the 19th century by the methods of Brodie, Staundenmaier, Offeman, and Hummers [15]. GO is a graphene derivative that can be directly synthesized from graphite material. Chemical modification is used to add oxygen-containing functionalities to GO, including hydroxyl, carboxyl, epoxy, and carbonyl groups. rGO is produced by reducing GO employing chemical, mechanical, thermal, and green procedures to eliminate the oxygen functional groups. A significant distinction between GO and rGO is brought about by the concentration of the oxygen functional groups [15]. GO's and rGO's chemical structure are illustrated in Figure 1e. GO is hydrophilic because oxygen functional groups are present on its surface. Due to the intolerable nature of complete reduction, rGO demonstrates a partially hydrophilic nature. GO is an outstanding material for energy storage due to its high conductivity, mechanical properties, specific surface area, and stability in basic and acidic environments [58]. By utilizing the structural and chemical variety of oxygenated functional groups, the physicochemical properties of GO can be effectively modified to achieve the desired properties for energy storage applications [59].

Recently, reviews on using graphene and GO-based composites in supercapacitor applications, including GO, AC/graphene, graphene/vanadium, graphene-based gels, graphene/nano cellulose, graphene/metal oxide, GO/MXenes, metal oxide/rGO, rGO/cellulose, iron oxide/graphene, rGO/polypyrrole, GO/MOFs, LDHs/rGO, GO-based aerogels, and poly(cyclotriphosphazene)/GO, have been published [15,49,60–75]. In contrast, this review focuses on several types of materials, including metal oxides, metal sulfides, metal chalcogenides, MOFs, and MXene composites with GO, as well as fluorine, bacteria, and other organic functionalization with GO. It discusses their synthesis, as well as their surface, structural, and morphological properties for supercapacitor applications.

GO-based composites' synthesis conditions, time, and cost were thoroughly provided. In addition, their C_s, resistances, rate capability, cycle stability, energy, and power density for the application of supercapacitors were discussed, and the electrochemical results of the various GO composites utilizing three- and two-electrode setups were tabulated for the reader's ease of access. To assist researchers in examining the properties of electrode materials for use in supercapacitors, earlier work using DFT calculations is also reviewed. Figure 2 shows the maximum C_s of GO and its composite made of various materials for application in supercapacitors, as measured by GCD in recent studies.

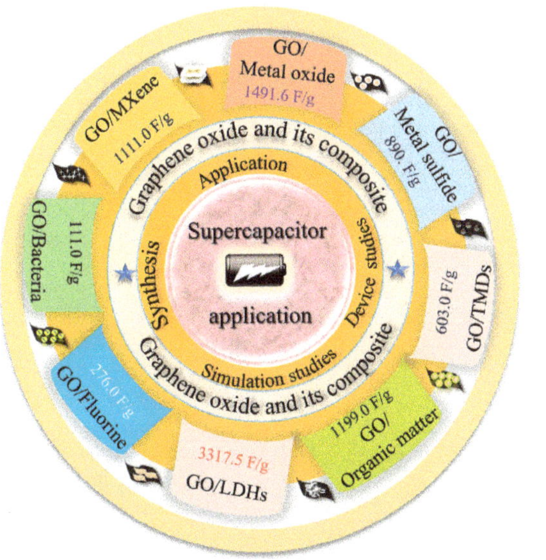

Figure 2. Schematic diagram of GO and its composite, showing the highest specific capacitance (C_s) measured by galvanostatic charge–discharge (GCD) for the supercapacitor application.

2. Synthesis

2.1. Fluorine-Functionalized GO

Fluorine's (F^-) strong electronegativity makes it possible to tweak and adapt graphene's characteristics via fluorination. The fluorinated graphene-based materials exhibit increased electron transport kinetics to this extent, and since this functionalization creates defects on the graphitic surface, it may change the electrical properties of the graphene surface [76,77]. Thus, for the application of SCs, Sim et al. synthesized fluorine-functionalized graphene oxide (FGO) via plasma treatment [77]. In this procedure, GO was synthesized using chopped graphite fibers using Hummer's method. The chemical reaction occurred in the presence of potassium permanganate ($KMnO_4$), sulfuric acid (H_2SO_4), and sodium nitrate ($NaNO_3$) in a cold bath. The reaction required five days to complete after the ice bath was removed and replaced with water. Hydrogen peroxide (H_2O_2) was employed to terminate the reaction, and GO was obtained through multiple washes with water. GO was subsequently functionalized with fluorine using a plasma reactor developed in-house. Plasma was generated in this procedure utilizing nitrogen (N_2), nitrogen trifluoride (NF_3), and hydrogen (H_2) as plasma gases via a dielectric barrier discharge reactor. NF_3 was utilized as the F-radical-generating gas in the reactor, and by acting as a catalyst, H_2 caused NF_3 to dissociate. When the reactor was consistently supplied with GO and plasma gases, FGO was generated. The morphology, surface area, and crystalline characteristics are significant factors that may greatly influence the electrochemical performances of electrode materials. Hence, it is important to thoroughly analyze and study the characteristics of electrode materials to determine their suitability in supercapacitor applications. Accordingly, the

FGO morphology revealed very thin layers with interspaces, while GO seemed to be ruffled and flake layered structures. In addition, it could be either due to the morphology of GO or due to its having very few pores that it has a low surface area of 3.1 $m^2\ g^{-1}$, and accordingly, no significant surface area was found on FGO (3.7 $m^2\ g^{-1}$). The use of an in-house plasma reactor for the functionalization of fluorine on GO may require highly skilled workers and entail significant expenses due to using gases, electrodes, and Teflon coating in the reactor for corrosion prevention by NF_3, and FGO may be produced on large scales using this plasma reactor. Figure 3a (i and ii) illustrates the schematic design of the FGO production process and a photo image of the fluorine plasma treatment reactor.

2.2. GO-Metal Oxides

TiO_2 may create oxygen vacancies, which function as redox centers and improve the capacitive properties. Additionally, it is environmentally friendly, easily synthesized, electrically conductive, and has high pseudocapacitance. Further, the capacitance, high conductivity, and redox activity of TiO_2 can be enhanced by doping of other metal oxides. In light of this, El-Shafai et al. recently created GO@TiO_2 and doped it with several metal oxides (MgO, SiO_2, Co_3O_4, and V_4O_7) for use in supercapacitors [78]. Initially, TiO_2 was doped on GO using the reflux technique. After obtaining GO@TiO_2, it was combined with other nanoparticles (GO@TiO_2-NPs) to produce GO@TiO_2-MgO, GO@TiO_2-SiO_2, GO@TiO_2-Co_3O_4, and GO@TiO_2-V_4O_7 by grinding and a two-hour ultrasonication process. Figure 3b schematically depicts the reflux process used to produce the GO composite with TiO_2, followed by the composite of different nanoparticles to the GO@TiO_2. The produced composite particles have an aggregated form and a spherical shape on the GO sheet. All composite particles were highly crystalline except for GO@TiO_2, which showed low crystallinity. The introduction of additional nanoparticles with GO@TiO_2 may have been influenced by the high crystallinity. However, the composite's preparation through physical grinding led to morphological damage and raised the possibility of bulk structure. This approach may seem straightforward. Still, the synthesis of different metal oxides and subsequent composites makes this kind of synthesis quite expensive.

The energy storage sector has focused much attention on activated carbon (AC) because of its unique surface area, porous structure, and surface functional groups. Also, supercapacitors have considered transition metal oxides (TMOs) as potential electrode materials; their charge storage mechanisms are predicated on pseudocapacitance. Therefore, Veeresh et al. designed two types of nanocomposites for the use of supercapacitor application: AC-GO-SnO_2 and AC-GO-TiO_2-Zn, which incorporate AC-GO with SnO_2 and TiO_2-Zn, respectively [79]. By combining carbon materials with conducting graphene and TMOs, the electrochemical properties of GO can be enhanced. Initially, AC was made in a muffle furnace using charcoal under 800 °C for 2 h. Then, GO was produced using a modified Hummer's process using graphite powder. Later, nanocomposites AC-GO-SnO_2 and AC-GO-TiO_2-Zn were produced in water and thiourea reaction media using the hydrothermal technique at 180 °C. A schematic representation of the synthesis of AC from charcoal via chemical activation at 400 °C and the preparation of an AC-GO-SnO_2 composite via hydrothermal methods is depicted in Figure 3c. To produce AC-GO-TiO_2-Zn, the same hydrothermal method may be used. Typically, the morphologies of nanoparticles or composites produced via hydrothermal techniques are well-structured. However, in this procedure, the synthesized AC-GO-SnO_2 and AC-GO-TiO_2-Zn morphologies were found to be aggregated and irregular spherical morphologies solely due to AC showing an aggregated and planar structure. A hydrothermal process is used to prepare these composites' highly hybrid structures. However, the purity of the composites was found to be quite poor, as shown by a large number of dust particles on the composite's morphology. A poor crystalline structure was also seen in the composites that were formed, which may be attributed to the low crystallinity of AC. The composite may have a relatively low surface area as a result of the aggregated morphologies and poor crystallinity. This synthesized nanocomposite may be a simple and modest cost, but there are multiple processes, and a time-consuming technique may be seen.

Recently, Abdallah et al. created a ternary composite form of activated carbon-GO-TiO_2 (AC-GO-TiO_2) for supercapacitor use [80]. According to the synthesis, AC was originally produced by calcining H_2SO_4-treated corncob powder at 600 °C in a muffle furnace. Then, using corncob powder, GO was made via a modified Tour method. To make charcoal, corncob powder that had been diluted and treated with HCl, along with 0.1 N NaOH, was combined with H_2SO_4 and autoclaved for 8 h at 90 °C. The resulting charcoal was subsequently treated in an ice bath with H_2SO_4 and H_3PO_4, after which $KMnO_4$ was added gradually. After forming a reddish-brown hue, water was introduced into the solution, followed by the dropwise addition of 30% H_2O_2 to obtain a GO. Finally, the ternary composite was created by combining AC, GO, and TiO_2 with PVDF as an electrode for the supercapacitor. The morphologies of AC-TiO_2-GO were observed to be the GO sheet, and TiO_2 particles were deposited on the surfaces of AC, which blocked most of the pores. Following composites, the morphology of AC was found to be fluffy. During this technique, the level of crystallinity was reduced, but the surface area of the composites with AC increased at each stage. AC, AC-TiO_2, and AC-TiO_2-GO exhibited surface areas of 212.5, 344.5, and 439.2 m^2 g^{-1}, respectively. The decrease in crystallinity might be attributed to the amorphous nature of GO or perhaps its low quality. The AC-TiO_2-GO composites obtained were inexpensive. However, the synthesis of AC from corncob using chemical activation and the synthesis of GO from corncob using the modified Tour method can be costly due to the multi-step process, the use of multiple organic compounds, and the lack of an environmentally friendly approach.

Copper oxide (CuO) has shown that low conductivity and rapid capacitance drop cannot be used in a supercapacitor application as a consequence of the chemical ion insertion or extraction process causing CuO crystal structure degradation [81,82]. To address this concern, composites of CuO with other metal oxides and carbon materials may be able to solve the CuO issue in supercapacitor applications. Accordingly, El-Shafai et al. developed a composite of GO with CuO, CuO/ZnO, CuO/TiO_2, and CuO/ZnO/TiO_2 using a simple procedure [82]. To create GO-CuO, CuO nanorods were produced by reacting $CuSO_4 \cdot 5H_2O$ with 1.0 M KOH in ethanol. Hummer's technique was used to synthesize GO, which was then composited with CuO (GO-CuO) by reacting it with $CuSO_4 \cdot 5H_2O$ and adding 1.0 M KOH at 90 °C for 3 h. GO-CuO/ZnO nanorods were produced by reacting $Zn(CH_3COOH)_2 \cdot 2H_2O$ with 1.0 M NaOH in ethanol. The resulting ZnO was then composited with GO-CuO under ultrasonic agitation for three hours to create GO-CuO/ZnO. For GO-CuO/TiO_2, TiO_2 was first produced using the reflux technique with titanium (IV) isobutoxide. The produced TiO_2 was then ultrasonically mixed with GO-CuO for 2 h to produce GO-CuO/TiO_2. Finally, a mixture of GO and all nanoparticles was subjected to a straightforward ultrasonication process to generate GO-CuO/ZnO/TiO_2. This approach included the observation of nanorod morphologies of CuO, ZnO, and aggregated TiO_2 irregular particles on a GO sheet. The structure of GO was found to be amorphous. However, the composite of CuO, ZnO, and TiO_2 resulted in an enhanced crystallinity of GO compared to CuO alone, as well as CuO/ZnO composites on GO. Therefore, the number of distinct nanoparticles has the potential to enhance crystallinity. Ternary compounds with structured morphologies and crystalline structures have the potential to be materials with a significant surface area. These synthesis processes seem relatively straightforward; nonetheless, synthesizing numerous metal oxides and composites with GO may be time-consuming.

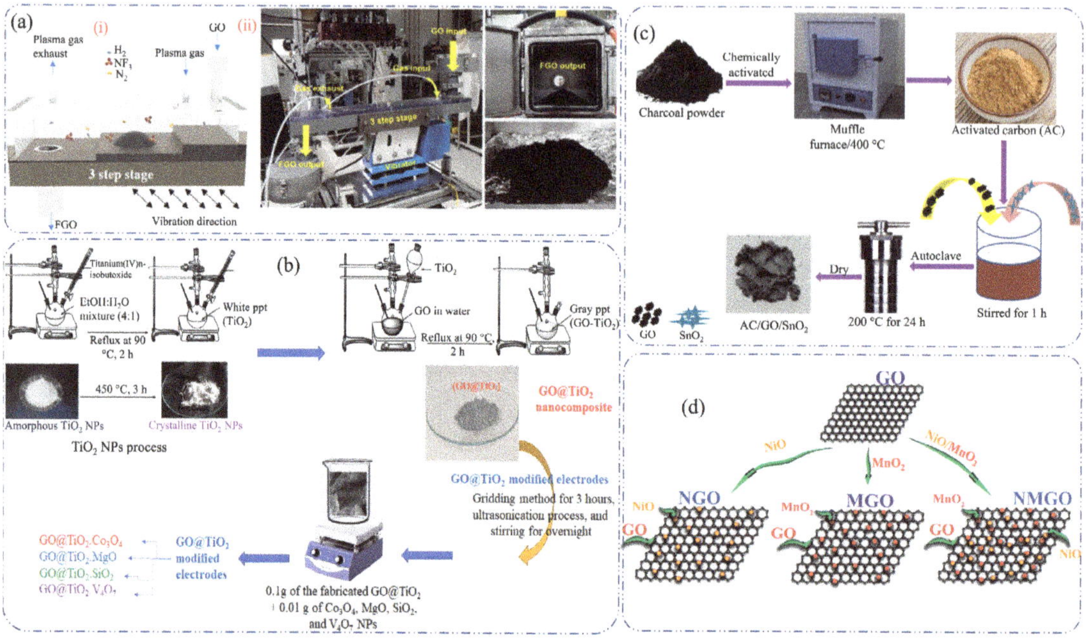

Figure 3. (**a**) Synthesis procedure of (i) FGO and (ii) Photograph image of fluorine plasma treatment system for FGO synthesis (reproduced with permission from [77]). Copyright 2022, Elsevier. (**b**) GO@TiO_2-x composite (reproduced with permission from [78]). Copyright 2023, Elsevier. (**c**) AC/GO/SnO_2 composite (reproduced with permission from [79]). Copyright 2023, Elsevier. (**d**) NMGO composite (reproduced from [83]). Copyright 2023, Elsevier.

The $CuFe_2O_4$ nanoparticles (anode material) have been attracted to the design of supercapacitor devices due to their ease of availability and environmental friendliness [84]. However, $CuFe_2O_4$ particles are hardly used for supercapacitor applications due to their low conductivity, which quickly weakens capacitance. Furthermore, heteroatoms like nitrogen (N), sulfur (S), phosphorus (P), and boron (B) may significantly influence the electrical characteristics of GO, resulting in rapid ion diffusion at the electrolyte–electrode interface and a considerable amount of charge storage. Therefore, the best approach could be to use N-doped GO to enhance the capacitance characteristics of $CuFe_2O_4$. Accordingly, Lashkenari et al. studied NGO composited with $CuFe_2O_4$ nanocomposite by hydrothermal methods [85]. For this approach, they first created GO using Hummer's method. Later, N was doped into GO (NGO) and employed by thermal breakdown procedure under the reaction of urea at 350 °C for 30 min in a tubular furnace. Furthermore, NGO/$CuFe_2O_4$ was created using the hydrothermal technique. As a result, the produced NGO was reacted with ferric chloride hexahydrate and copper (II) chloride dihydrate in ethanol while stirring. A total of 6.0 M of NaOH was used to adjust the pH of the solution to 12, and the solution was then autoclaved at 180 °C for 24 h to produce a composite NGO/$CuFe_2O_4$. Due to thermal expansion during the thermal decomposition process, the GO's sheet shape became wrinkly after N doping, and rough surfaces were found. For structural stability, it could be preferable to dope N on GO surfaces using low-temperature processing. $CuFe_2O_4$ is produced using a hydrothermal method, which results in the formation of nanocubes on the NGO sheet. Upon a blend with $CuFe_2O_4$, it was found that the crystallinity of the NGO was improved. The surface area of this composite may be substantial owing to the nanocube structure of $CoFe_2O_4$ and the NGO sheet. This approach might be expensive because of the several steps involved, such as the synthesis of GO using different organic compounds, thermal breakdown to dope a heteroatom into GO, and, finally, the hydrothermal procedure for composite formation.

Fe_3O_4 is one of the metal oxides used for supercapacitors due to its low toxicity, simple preparation, cost-effectiveness, high capacitance, and wide range of operational windows [86]. However, its conductivity is low, which hinders the rapid electron transfer in the electrode materials. On the other hand, sodium tungstate (Na_2WO_4) has a high electrochemical redox property and a high conductivity, and it has the potential to be used in the enhancement of the characteristics of Fe_3O_4. In addition, ionic liquids (ILs) are used as an efficient electrode material for supercapacitors. These liquids possess a variety of distinctive characteristics, including high electrical conductivity and a broad range of electrochemical window functioning. Due to this, Dalvand et al. studied GO-modified Fe_3O_4 on IL@tungstate (GO/Fe_3O_4-IL@W) to develop a supercapacitor [58]. Consequently, IL was produced through a reaction involving 1,4-diazabicyclo [2.2.2] octane (DABCO), 1.0 mmol of 3-(chloropropyl) trimethoxy silane, and sodium iodide in methyl formamide at 120 °C under N_2 atmosphere for 48 h. Then, GO/Fe_3O_4@IL was produced by combining ammonia, $FeCl_2{\cdot}4H_2O$ (0.75 g), and $FeCl_3{\cdot}6H_2O$ in a toluene solution at 80 °C. GO/Fe_3O_4 was acquired via aqueous rinsing after the reaction. IL and GO@Fe_3O_4 were additionally combined with TEOS and DMF/H_2O at 80 °C while agitating to produce GO/Fe_3O_4@IL. A product of GO/Fe_3O_4-IL@W was subsequently produced by combining Na_2WO_4 with GO/Fe_3O_4@IL while agitating for 24 h. After functionalization of Fe_3O_4-IL-W, the GO sheet was folded, indicating that the sp^3 defect was caused by the introduction of oxygenated groups into the GO surfaces, and the spherical shape of Fe_3O_4 and W particles may be adsorbed randomly on GO surfaces. However, the cloud of particles was found to be low on the GO surfaces, whereas the composite exhibits a highly crystalline structure owing to the crystalline properties of Fe_3O_4 and W. Based on the morphologies and crystalline properties, this composite may have a moderate surface area. This process incurs significant costs due to the observation of numerous processes and the use of numerous compounds in the synthesis of a composite.

Due to its strong electrochemical stability, pseudocapacitive behavior, and outstanding electrical conductivity, bismuth oxide (Bi_2O_3) has received a lot of attention as a material for supercapacitors [87]. It may also be used instead of traditional carbon materials to increase the C_s and high energy density of supercapacitors. The electrochemical reaction and electrochemical capacity are hampered, nonetheless, by the particle agglomeration. One potential solution to address the issue is to include highly electrically conductive elements in Bi_2O_3. Thus, Hounkanrin et al. recently created a Bi_2O_3-GO composite by using a microwave process [88]. This compound is very easy to synthesize from raw materials, such as sugar and pentahydrate bismuth nitrate. The mixture was thereafter transferred to a glass beaker and heated for 120 s at 600 W in a microwave oven. Following this procedure, the powder was crushed and heated to 450 °C to produce the yellow Bi_2O_3-GO composite product. In this study, sugar was used as a source for GO. This method yields high-purity composite particles. GO morphology revealed sheet and layered structures but collapsed following composite with Bi_2O_3 at 600 W for 120 s. Increasing Watts (800 W) and time (90 s and 120 s) strongly impacted morphologies. Thus, a microwave under 600 W with a short processing time is best. Elevated crystallinity was observed as a result of the composite's high purity. As a result of the dominance of Bi_2O_3 on GO surfaces, no GO peaks were witnessed. Additionally, the highly crystalline and purity levels of this study might account for its high surface area. Using a microwave to create a composite or nanoparticle might be a relatively straightforward process.

A composite material consisting of NiO, MnO_2, and GO (NiO/MnO_2/GO) was recently created by ultrasonication for the supercapacitor application [83]. To this degree, Tour's approach was first used to produce GO for the composite preparation. Secondly, NiO was subsequently synthesized via co-precipitation using nickel chloride hexahydrate at a pH of 10. Thirdly, $MnCl_2{\cdot}4H_2O$ was employed in the co-precipitation process to produce MnO_2. The various ratios of prepared materials, including GO, NiO, and MnO_2, were mixed with water and ultrasonicated for two hours to make a $NiO/MnO_2/GO$ (NMGO) hybrid. Figure 3d shows a schematic diagram that illustrates the NMGO composite formation using sonication. The shape of GO was found to be a thin nanosheet layer, and the very dense particles were completely adsorbed on the GO surfaces after they were composited with NiO and MnO_2. However, nanoparticles on the GO sheet did not exhibit any noticeable shape. Electrostatic interaction or other chemical bonds might be the mechanism by which the adsorbing nanoparticles bind to GO. When compared to other chemical techniques, this study might be advantageous to producing GO-based composites without the use of solvents or organic media. The shape of the particles showed that the study was unable to produce particles of high purity on GO surfaces. A low crystalline NiO/MnO_2 peak and a strong crystalline GO peak were noticed since GO was the dominant component [89]. It was determined that the produced GO and nanoparticle may have high and poor purity, respectively, by analyzing their crystalline structure. Following the composite with NiO/MnO_2, the surface area of GO increased due to the high surface-to-volume ratio of nanoparticles. Accordingly, the surface areas of GO and NMGO were observed to be 94 and 102 $m^2\ g^{-1}$, respectively. This ultrasonication process for making composites is very easy, quick, and cheap.

Abdolahi et al. recently synthesized a hierarchical $Ni_{0.5}Co_{0.5}WO_4$ ($NCWO_4$)/f-MWCNTs/GO via a hydrothermal technique [90]. $NCWO_4$ was produced using a hydrothermal process at 220 °C for 12 h using trisodium citrate, water, cobalt chloride, and nickel chloride precursors. Following the synthesis, the procured GO and functionalized multi-walled carbon nanotubes (f-MWCNTs) were disseminated in water using sonication. The prepared $NCWO_4$ was added to the solution while continuously sonicating for 2 h to create the composite $NCWO_4$/f-MWCNTs/GO. Through the use of this procedure, $NCWO_4$ was shown to have an urchin-like morphology and to be heavily adsorbed randomly on a composite made of GO thin sheet and f-MWCNTs in the form of wires. The electrostatic interaction between $NCWO_4$ and f-MWCNTs/GO may be the cause of the adsorption. However, the clouds of urchin-like structure were low on the f-MWCNTs/GO. To enhance clouds, the $NCWO_4$ ratio must be optimized. Furthermore, it was noted that the composite $NCWO_4$/f-MWCNTs/GO had low crystallinity, perhaps as a result of low $NCWO_4$ purity. As a result, high-purity composite particles could not be produced via this synthesis method. The predominance of $NCWO_4$ resulted in the appearance of the majority of crystalline peaks, while f-MWCNT and GO exhibited only two peaks. To increase the surface area of NCWO4/f-MWCNTs, GO is introduced. As a result, $NCWO_4$/f-MWCNTs/GO had more surface area (172.2 $m^2\ g^{-1}$) than $NCWO_4$@f-MWCNTs (146.4 $m^2\ g^{-1}$). The surface area of the composite may be enhanced due to the high porosity of GO. Lastly, procuring GO and f-MWCNTs may be costly, and the many steps involved make the process time-consuming.

Similarly, a manganese ferrite ($MnFe_2O_4$)/GO composite was produced using a modified Hummer technique [91]. In this process, $MnFe_2O_4$ was initially produced via co-precipitation utilizing precursors $Fe(NO_3)_3$ and $Mn(NO_3)_2$ in the presence of KOH to modify the pH to 12 at 70 °C, after which the sample was calcinated at 600 °C. Furthermore, synthesized $MnFe_2O_4$ was mixed with GO, which was made using Hummer's method to create a $MnFe_2O_4$/GO hybrid. The modified Hummer's approach demonstrates heavily adsorbed fine spherical $MnFe_2O_4$ on GO's stacked sheet. The stacked morphology may improve the electrochemical reaction rate by allowing ions access to high electroactive sites in the electrode materials. It is noteworthy that the co-precipitation-synthesized morphology of $MnFe_2O_4$ showed high stability since the synthesized nanoparticle composites with GO by Hummer's technique did not significantly alter their morphology. However, $MnFe_2O_4$ and GO were found to have low and high purity, respectively. Therefore, the $MnFe_2O_4$ and GO were found to have a low and a highly crystalline structure, respectively. After compositing with GO, $MnFe_2O_4$ crystallinity enhanced significantly. The composite synthesis using a modified Hummer's method may increase $MnFe_2O_4$ powder purity. The combination of dense $MnFe_2O_4$ nanoparticles and thin GO sheets may have a high surface area. Making this composite seems like a simple, low-cost way to make an electrode for a supercapacitor. In another investigation, GO-modified graphite felt (GF) was hydrothermally combined with MnO_2 to form GO/GF-MnO_2 [92]. In this procedure, GF that had been treated with sulfuric acid, water, sodium nitrate, and nitrate was functionalized with GO (GO/GF) using Hummer's method at a reaction temperature of 180 °C for 12 h. The hydrothermal method was employed in the synthesis of GO-GF/MnO_2. In this step, $KMnO_4$, $MnSO_4$, and GO/GF were added to water and agitated for five minutes. They were subsequently transferred to an autoclave for a 12 h reaction at 140 °C to acquire the composite. The high-purity GF was found to have a cylindrical tube structure and hydrothermally produced a sheet-like structure with uneven development on its surface after MnO_2 modification was found. On the other hand, MnO_2 with GO composite on GF using the same technique exhibits rod forms that are almost covered on the GF surfaces. Consequently, the shape of MnO_2 changes when it is combined with other materials. As a result of GF's predominance, the composites exhibit a single peak corresponding to GF/GO; no peaks of significance were detected for MnO_2. The broad appearance of the crystalline peaks suggests a low degree of crystallinity. The synthesized composites, MnO_2-GO/GF, may have a greater surface area than MnO_2/GF because of the high clouds of MnO_2/GO nanorods produced on GF when compared to the uneven sheet shape of MnO_2. Due to the procurement of GF and the multiple processes required to synthesize the composite and its application in batteries, the synthesis procedure may be moderately expensive. Nevertheless, the synthesized composite has the potential to be utilized in supercapacitors.

The heterojunction structure of transition metal oxides (TMOs) has the potential to enhance the material's overall conductivity through favorable electronic effects [93]. As a result of its advantages, chemical stability, low cost, and simplicity of synthesis, Cu_2O has been the subject of extensive research as a supercapacitor material [94]. CoO has a favorable theoretical capacitance and is made from cost-effective materials. To further optimize the electrochemical characteristics, metal oxide composites may incorporate carbon-based materials recognized for their exceptional electrical conductivity. For this reason, Yu et al. utilized hydrothermal methods to produce Cu_2O-CoO composited with GO (Cu_2O-CoO/GO) to investigate supercapacitors [95]. Therefore, the precursors $CoCl_2$ and $CuCl_2$ were added to water together with the synthesized GO solution, and the mixture was agitated for 10 min before KOH was added. The combination was then placed in an autoclave to undergo a reaction for eight hours at 160 °C, producing black powder. Moreover, they were calcined for two hours at 350 °C in an N_2 environment to produce Cu_2O-CoO/GO. Formed like a folded sheet, the Cu_2O/GO shape was seen, but the Cu_2O on or with GO could not be recognized. In contrast, irregular CoO particles can be found on GO. Additionally, particles of spherical-like Cu_2O-CoO are visible on GO. Consequently, the hydrothermal method exhibits particle formation on GO exclusively under conditions involving CoO and Cu_2O-CoO. In comparison to CoO/GO, the samples of Cu_2O/GO and Cu_2O-CoO/GO may have a higher purity because CoO/GO lacks a smooth morphology. Therefore, Cu_2O/GO and Cu_2O-CoO/GO were found to have a high crystallinity. These materials may also have a high surface area. The process for producing this kind of composite via hydrothermal means is simple and inexpensive.

Tungsten oxide (WO_3) has drawn a lot of interest because of its benefits, which include cheap cost, many oxidation states, fast ion insertion characteristics that enable electrochemical activity, significant absorption in different electrolytes, and good ion transport properties. Therefore, using hydrothermal methods, Shembade et al. investigated the use of supercapacitors on the GO/WO_3 composite [96]. GO was first produced via this process using a modified version of Hummer's technique from graphite flakes. The various percentages (5, 10, and 15%) of GO were then hydrothermally synthesized to form a composite with WO_3. Thus, 0.1 M sodium tungstate was dissolved in water. HCl and H_2O_2 were added to the aforementioned solution and served as pH adjusters and reducing agents, respectively. GO was subsequently added to the solution, which was sonicated for a few minutes before being transferred to an autoclave for a 12 h reaction at 180 °C; after drying and rinsing the solution several times with ethanol and water, a GO/WO_3 composite was obtained, and finally, it was annealed at 500 °C. GO/WO_3 nanorod hybrids were observed via this method; as the ratio of GO to WO_3 was increased, thus were the particle sizes and minor morphological changes. The high degree of hybrid formation hinders the identification of WO_3 particles and GO sheets based on their morphologies. We found that after increasing the GO ratio from 5% to 10%, the crystallinity of WO_3 significantly decreased due to the dominance of GO. As a result, optimization studies on crystalline and morphological properties are of critical importance. According to the morphology and crystalline characteristics of GO/WO_3, all composites have the potential to be very pure. Because 10%GO/WO_3 has a strong crystalline quality, the surface area was calculated and found to be low at 33.9 $m^2\ g^{-1}$. The low surface area of the composite may be due to the lack of pores on its surfaces. This procedure appears to have a moderate cost and allows for the synthesis of fine-sized particles with high purity, as well as the synthesis of highly hybrid particles through systematic investigation.

Carbon-based materials (GO or CNT) typically have problems with self-aggregation, which results in poor electrochemical performances [97]. Consequently, using carbon-based materials that can be composited with metal oxides has improved supercapacitor performance. Consequently, because of its exceptional electrochemical reversibility, high conductivity, capacitance value, and cycle efficacy, cobalt ferrite ($CoFe_2O_4$) is primarily used in supercapacitors. To enhance the electrochemical performance in supercapacitor applications, Verma et al. investigated a composite $CoFe_2O_4$ with GO/CNT ($GO/CNT/CoFe_2O_4$) [97]. Initially, Hummer's technique was adapted to synthesize GO for this procedure. Second, cobalt nitrate hexahydrate and ferric nitrate nonahydrate were combined with glycine to produce $CoFe_2O_4$ using the solution-combustion technique. A precipitate of black cobalt ferrite was produced by undergoing a reaction in water at 80 °C; this precipitate was subsequently calcined at 800 °C to yield $CoFe_2O_4$. Using this solution-combustion method to produce nanoparticles could be an inexpensive and straightforward process. Composites were synthesized utilizing a simple stirring method. Further pure composites were obtained by combining varying weight percent (wt.%) of GO, CNT, and $CoFe_2O_4$ with water while agitating for six hours. The mixture was then washed with 10% HCl and dried at 100 to 120 °C to obtain $GO/CNT/CoFe_2O_4$. The morphologies of GO and CNT were found to be tubes and sheets, respectively. In contrast, $GO/CNT/CoFe_2O_4$ shows the tubular shape of the CNT and the spherical shape of the $CoFe_2O_4$ on the GO sheets. On GO sheets, however, very few $CoFe_2O_4$ particulates were observed in comparison to CNTs. Further optimization of the composite's ratio is required to achieve a more balanced formation of all materials. In addition to several crystalline peaks associated with $CoFe_2O_4$, the presence of GO and CNTs is distinguished by a single crystalline peak each. The crystalline quality of composites serves as an indicator of their purity. The presence of CNTs and $CoFe_2O_4$ nanoparticles on GO could potentially account for its considerable surface area. In this study, a low-cost method was used to prepare the composite. However, it might be more expensive to use commercial graphite and CNTs, as well as to synthesize GO using the modified Hummer's method and $CoFe_2O_4$ by the solution-combustion method. Simply stirring different materials in water may not effectively produce hybrid particles.

2.3. GO-Metal Sulfide

Because of their high van der Waals forces and pi–pi interactions, GO sheets are bound and layered. By functionalizing nanoparticles with GO surfaces, the degree of GO layer restacking can be substantially reduced through a blend of GO and transition metal compounds. In consequence, transition metal sulfides (TMSs) are extensively employed as potential materials in the field of supercapacitor technology owing to their favorable electrochemical characteristics. Furthermore, because their electrochemical properties are superior to those of TMOs, TMSs are employed in pseudocapacitors. As a result, Chandraraj et al. have recently investigated the use of a composite of CoS/GO and NiS/GO in a supercapacitor application [98]. For CoS/GO, GO that had been synthesized hydrothermally was composited with CoS. To acquire a CoS/GO composite, they combined GO with $CoCl_2$

and Na_2SO_3 in a solution of water and ethanol while agitating. The resulting mixture was then autoclaved at 190 °C for one night. GO was likewise incorporated into NiS/GO via a hydrothermal process involving $NiSO_4$, NaOH, triacetonamine, and water, followed by its autoclaving at 175 °C for three days. Furthermore, this study did not describe the synthesis of CoS/NiS/GO. The NiS/GO and CoS/GO composites, made using hydrothermal methods, had a very hybrid structure. The particles formed a uniform thin coating on the GO sheets, thereby preventing the nanoparticles from agglomerating. GO sheets retain their exceptional stability even after composites. However, no discernible nanoparticle form was seen on GO. In addition, GO's crystallinity increased after composites with CoS and NiS. As a result, the produced nanoparticles are very pure and crystalline. These composite particles may have a modest surface area because they appear as coatings on GO, which is related to their low porosity rather than their actual particle form. This procedure may be straightforward, but it may be moderately priced due to the need for multiple syntheses and the impossibility of mass production.

Similarly, lanthanum (La_2S_3) and manganese (MnS) sulfides make excellent pseudocapacitive materials because of their cheap cost, strong C_s, and simplicity of production [99]. Additionally, these materials possess high power density, a broad voltage window operation range, and high redox properties due to their diminished electrical conductivity, low energy density, and poor kinetics during cycle testing. For these reasons, it may be possible to effectively enhance the voltage window, capacitance, electrical conductivity, and cyclic stability by combining MnS and La_2S_3 with carbon-based conducting materials in the form of a composite. As a result, Mane et al. investigated the composite of MnS and La_2S_3 with GO for use in supercapacitors [100]. Consequently, films can be fabricated using GO dispersed in a water solution in an ultrasonic chamber. By performing 100 SILAR (Successive Ionic Layer Adsorption and Reaction) cycles on a stainless steel (SS) substrate, GO film was produced. A MnS film was subsequently deposited onto another SS substrate using the 90 SILAR cycles method and a solution of $MnCl_2$ and Na_2S. For the MnS-La_2S_3 film, $MnCl_2$ and $LaCl_3$ were dissolved in water following the addition of Na_2S. The film was subsequently coated by dipping the SS substrate in the solution for 90 SILAR cycles. The 90 SILAR cycle method was employed to fabricate a MnS-La_2S_3/GO film on an SS substrate using the identical procedure. Figure 4a illustrates the fabrication process of composite film using the SILAR method. The application of this film manufactured using the SILAR method may be straightforward and efficient. The candy-like morphology of MnS on SS was observed, in contrast to the cotton-like morphology of densely aggregated particles in MnS-La_2S_3 composites, which may have a high porosity. The composite of these particles with GO does not exhibit a discernible MnS-La_2S_3 morphology. When the porous cotton-shaped particles interact with GO, they transform into a film or sheet morphology. Despite this, the morphology of the layered structure of GO following composites appears unstable. This type of porous morphology may facilitate the adsorption and desorption of electrolyte ions on electrode materials by providing a large number of active sites, thereby enhancing electrochemical performance. The considerable enhancement of the surface area of MnS (47.0 $m^2\ g^{-1}$) following composite with La_2S_3 (82.0 $m^2\ g^{-1}$) and GO (156.0 $m^2\ g^{-1}$) is intriguing. It may account for the increase in porosity between composite stages; in particular, GO, which is a highly porous material, may be the cause. Nevertheless, the synthesized composites exhibited a decreased degree of crystallinity, which may indicate a lower degree of purity. The process of anion and cation coating on an SS substrate may appear straightforward and inexpensive; however, 100 SILAR cycles are required to achieve a clean coating.

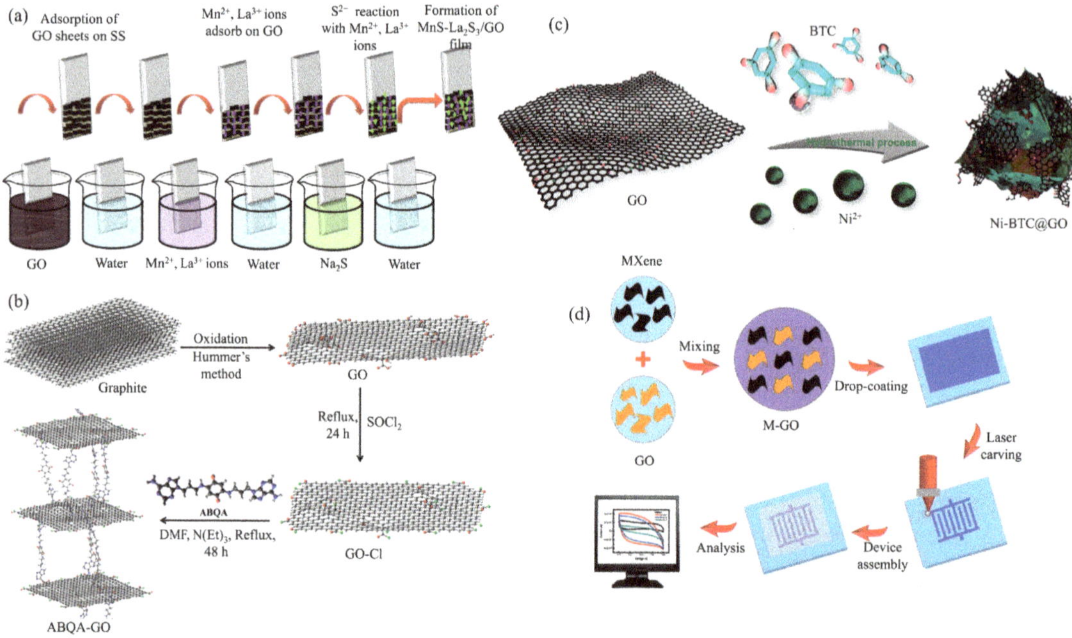

Figure 4. (**a**) Fabrication process of MnS-La_2S_3/GO film (reproduced from [100]). Copyright 2023, Elsevier. (**b**) Synthesis scheme for ABQA-GO composite (reproduced with permission from [101]). Copyright 2023, Elsevier. (**c**) The schematic diagram for the synthesis of Ni-BTC@GO composite (reproduced from [102]). Copyright 2023, American Chemical Society. (**d**) Fabrication of GO-MXene composite electrode (reproduced with permission from [103]). Copyright 2023, American Institute of Physics.

2.4. GO-Transition Metal Chalcogenide

Because of its excellent redox characteristics, large surface area, crystal structure, and electrical conductivity, transition metal chalcogenides, or TMDs, have drawn interest in the field of energy storage. The benefit of including sulfur and selenium is that they enhance the metal-to-carbon-based material ion exchange. Thus, carbon materials may be useful for supercapacitor applications because of the TMDs' excellent ion exchange capability. As a result, Yasoda et al. used a straightforward mixing procedure to create a manganese sulfoselenide (MnSSe) composite with GO (GO-MnSSe) [104]. Consequently, manganese anodium sulfide was incorporated into GO (GO-SW) using graphite flakes in a modified Hummer's process. By agitating a GO-SW in water for a whole night, repeatedly washing it with ethanol, and drying it at 60 °C, the GO-MnS was produced using GO-SW. To prepare GO-MnSSe, GO-SW and Na_2S were dissolved in water, and then a Se compound was added while continuously stirring. Following the reaction, it was repeatedly washed in ethanol and dried at 60 °C to obtain the composite. The wrinkled appearance of GO-MnS can be attributed to the potential film-like formation of MnS. On the other hand, finely spherical MnSSe nanoparticles were observed to form on GO surfaces. The interaction between Se and MnS to form particles may be the cause of the uneven particle formation on GO surfaces and the lack of particles on some GO surfaces. Therefore, the ratio of GO to MnSSe should be optimized to ensure the particle's uniform formation. Enhancement of ion transport through energy storage in the structure's attributes may be contingent upon the number of nanoparticles present on the GO surfaces, thereby leading to improved electrochemical performances. Furthermore, the nanoparticles that form between the GO layers potentially contribute to a favorable conductivity, resulting in enhanced electrochemical performance.

Additionally, the low crystallinity observed in both composites may be attributed to the possibly low purity of the particles acquired in this investigation. Both composites may have poor surface area due to irregular GO particles and low crystallinity. The synthesis of this composite might follow a relatively straightforward and quick approach, but since it involves many steps and takes a lot of time, the process could be quite expensive.

2.5. GO-Layered Double Hydroxide

Layered double hydroxides (LDHs) are synthetic two-dimensional (2D) nanostructured anionic clays [105,106]. It is an ionic, positively charged brucite structure with interlayer portions that contain charge compensation anions. It features a multilayer structure with a large surface area, many active sites, and rich redox characteristics [107]. The spacing between layers is more favorable to ion deintercalation, and insertion contributes to illuminating its C_s. Thus, the C_s of GO may be increased by combining it with Co_2-Ni_1 LDH (GO/Co_2-Ni_1 LDH) for supercapacitor applications. Thus, Zhao et al. produced GO/Co_2-Ni_1 LDH using the reflux technique at 90 °C [106]. The mixture of water, hexamethylenetetramine, GO, 2.0 M $NiCl_2{\cdot}6H_2O$, and 1.0 M $CoCl_2{\cdot}6H_2O$ was refluxed at 90 °C. After a 3 h reaction, the mixture was washed with water and ethanol to yield a GO/Co_2-Ni_1 LDH. The sample exhibited a sheet morphology with a wrinkly structure, which may account for the solid sheet structure of LDHs; it may have exerted a contraction force on the GO surfaces. On the other hand, the type of composite particles may possess high conductivity and active sites, allowing for rapid ion movement and efficient adsorption of electrolyte ions, which is advantageous for high-energy storage. This composite showed low crystallinity, and the GO/Co_2-Ni_1 LDH composite showed no GO crystalline peak. It can be a highly thin sheet of LDH and govern in terms of quantity as compared to GO. In general, the high-porosity nature of LDHs results in high surface area [108–110]. Accordingly, GO/Co_2-Ni_1 LDH was attributed to a high surface area (84.6 $m^2\ g^{-1}$). The approach for this composite seems to be low-cost and simple.

2.6. GO-Organic Material

Graphene sheets are experiencing aggregation via the process of stacking two graphene sheets on top of one other, which results in a large surface area that is not achievable for high electrical conductivity [111]. On the other hand, GO that has been generated by the sp^3 hybridization of carbon atoms is not conductive and exhibits poor charge transfer, which contributes to its unstable cycling [112]. Therefore, to solve this problem, one active method for supercapacitor applications is to increase the performance of GO materials by including organic molecules as an interlayer in graphene sheets. In light of this, Biradar et al. produced a composite material called 2,5-bis((3-(6-amino-9H-purin-9-yl)propyl)amino)cyclohexa-2,5-diene-1,4-dione (ABQA)-GO by using Hummer's technique, which was then followed by a chemical approach [101]. The reaction of 1,4-benzoquinone, 3-bromopropylamine hydrobromide, and adenine in dimethyl sulfoxide (DMF) under the N_2 environment was the process that was followed to make ABQA. They were able to obtain the ABQA using the process of low-pressure distillation. To generate acyl chlorides of GO surface (G-COCl), the synthesized GO was then combined with $SOCl_2$ and DMF in a reflux procedure while the environment was controlled to be N_2. To produce an ABQA-GO, G-COCl and ABQA were first reacted into DMF under reflux conditions. This was then followed by washing in methanol and freeze-drying. Figure 4b depicts the schematic synthesis of ABQA-GO, which began with graphite and subsequently reacted with $SoCl_2$ to form GO-Cl under reflux. GO-Cl was then organically reacted with ABQA under reflux to obtain the final product. The synthesized GO was sheet-structured and unsmooth. However, GO became wrinkled and weakened after ABQA functionalization. It may be because organic solvents damage GO surfaces during ABQA functionalization. Thus, this method affects the GO structure when modified or functionalized. Functionalized GO has poor crystallinity due to bio-inspired organic molecules' low crystalline characteristics. But, it slightly enhances the GO's crystallinity. The strong redox property, quick ion transfer, and superior conductivity of ABQA allow the GO to store more charges. In this study, the use of organic molecules to composite GO in this technique may be both expensive and not environmentally friendly.

MOFs perform poorly in supercapacitors due to their weak conductivity and electrolyte instability. Trimesic acid, which is readily accessible, stable, and inexpensive, has been used recently to synthesize neodymium (Nd)-MOF. This material was then combined with GO for use in supercapacitors [113]. In the media of water and ethanol mixtures, a GO was composited with Nd-MOF using the hydrothermal technique. Consequently, trimesic acid and neodymium nitrate hexahydrate were added to the GO-containing mixture above, and the mixture was sonicated to disperse the particles further. The solution was placed in an autoclave set at 200 °C for 12 h. Following the reaction, the precipitate was repeatedly cleaned with ethanol and water to create composite GO/Nd-MOF. For Nd-MOF, smooth hexagonal-rod-shaped crystals were seen in this process; these crystals adhere to GO after compositing with it. However, the MOF crystal is slightly damaged and unsmooth after the composite. Furthermore, the GO reveals wrinkled and broken sheet morphology. The composite Nd-MOF/GO demonstrates a strong crystalline property because of the crystallinity of Nd-MOFs. Fast ion transport to the pores of MOFs/GO and highly active sites for the adsorption of high electrolyte ions may be made possible by the combination of the sheet structure of GO and rod-shaped MOFs to allow high energy storage applications. The high purity of the materials synthesized via this method for MOFs and their composites may facilitate superior electrochemical performance. Additionally, MOFs containing GO composites may possess a substantial surface area. The procedure of synthesizing GO and its composites seems to be low-cost, yields a well-structured MOF morphology, and produces particles with highly active sites and purity, and it could be a time-consuming process.

An easy chemical procedure was employed by Chen et al. to fabricate a Ni-BTC MOF/GO composite [102]. Firstly, GO was synthesized via a modified Hummer's method. A subsequent solvothermal reaction was used to produce Ni-BTC MOFs. This was accomplished by sonicating a solution of $Ni(NO_3)_2{\cdot}6H_2O$, 1,3,5-trimesic acid (BTC), and 2-methylimidazole in DMF. The resulting solution was then transferred to an autoclave at 170 °C for a duration of 48 h. For subsequent use, the obtained particulates were rinsed with DMF and ethanol. Subsequently, composite GO/Ni-BTC MOFs were produced by sonicating the reaction of GO in DMF. Following that, 2-methylimidazole, BTC, and $Ni(NO_3)_2{\cdot}6H_2O$ were introduced into the GO solution. Additionally, the solution was autoclaved at 170 °C for 48 h to obtain composite. Furthermore, the quantity of GO varied from 0 to 40 mg when producing a composite for comparison reasons. The schematic representation of the Ni-BTC@GO process, which involves the hydrothermal reaction of GO, BTC, and nickel nitrate precursor, is illustrated in Figure 4c. The morphology of this Ni-BTC MOF is octahedral, with smooth surfaces. As the amount of GO on Ni-BTC MOF increased from 10 to 40 mg, the morphology was observed to be unevenly adherent GO on a smooth MOF surface for 10 mg GO. Furthermore, rough surfaces, GO aggregation, and unevenness on MOF surfaces were found for the 20 and 40 mg doped GO. As a result, increasing the GO caused the MOF surfaces to become rougher. However, partly covered GO on the MOF surface will enable Ni^{2+} to gain certain advantages, such as interaction with electrolyte ions, shortening the ion diffusion route, and protecting the active site of Ni^{2+} in the electrochemical process. Interestingly, the crystalline characteristics of Ni-BTC MOF grew when the GO dosage was raised from 10 to 20 mg, and after 40 mg of GO, the crystalline peaks of MOF were severely reduced owing to excess GO, which terminated the crystallinity of MOF. As a result, combining higher-dosage GO with Ni-BTC MOF may be ineffective for supercapacitor applications. Based on the crystallinity, the resulting particles may be of high purity in the investigations. Inevitably, the prepared composite may possess a higher surface area as a result of its elevated crystallinity, high purity, and potentially porous characteristics. The synthesis process appears to be straightforward and economical. However, it is not an environmentally friendly procedure because of the use of organic solvents in the hydrothermal reaction, and it is a time-consuming process due to the 48 h reaction and drying particles maintained for 24 h. Similarly, Khakpour et al. produced sulfur-doped GO/ZIF-8 MOFs (SGZ) through a straightforward chemical process [114]. As a result, a modified version of Hummer's method was employed to prepare the GO. Sulfur (S) was introduced into GO (S-GO) via the hydrothermal process. This included sonicating Na_2S and GO in water and stirring for an hour. After that, the mixture was put in an autoclave to undergo a hydrothermal reaction for 10 h at 200 °C, yielding S-GO. For the SCZ composite, an hour was spent sonicating a mixture of S-doped-GO and 2-methylimidazole in methanol. After that, Zn salt and methanol were combined separately and left for 10 min under stirring to create another solution. To create the SCZ composite, the two solutions mentioned above were combined, agitated for two hours, and then centrifuged. In this work, no significant morphology or irregular shape was found for the ZIF-8 MOF, and the purity of the MOF seems poor. Probably, the MOF synthesis carried out in an aqueous solvent with mechanical stirring for 1 h in an open environment affected the morphology and maybe contaminated dust particles from the air. For GO, the morphology was severely wrecked after S doping. The broken structure might be because the GO was sonicated and mechanically stirred with Na_2S before the hydrothermal reaction, which could affect the shape of the GO. Similarly, no notable morphology or irregular shape was seen in composite SCZ. As a result, this approach may not be successful in producing morphology for MOFs and/or composite particles. The crystallinity of the SGO composite with ZIF-8 MOF decreased dramatically as a result of the presence of GO, which reduced the crystalline quality of the MOF. The process for SCZ composite might be inexpensive, straightforward, and time-saving. However, in the process of preparing composites, the hydrothermal method should produce GO and sulfur-doped GO (SGO). This method may have a modest overall cost, but it results in particles with low purity and no distinct shape.

2.7. GO-MXene

MXene is a two-dimensional material extensively used in several applications, including energy storage, water treatment, and sensors. MXene has strong electrical conductivity, high hydrophilicity, and a multilayer structure, enabling it to function as a current collector [115,116]. Compared to stacked layers of GO or MXene materials, the conductivity of MXene-GO composites may significantly enhance ion transit and charge storage for supercapacitor applications. The MXene-GO composite has considerable promise for development in energy storage applications. Fu et al. created an MXene-GO film using drop casting [103]. GO and MXene were combined in a specific ratio while being continuously stirred. The solution was applied onto a glass plate using the drop-casting method and then dried in a vacuum. Subsequently, the undesired section of the film was eliminated using a CO_2 laser to serve as an electrode for supercapacitor use. Figure 4d shows a schematic diagram of the process of making the MXene-GO composite film using drop casting and further electrode assembly for electrochemical testing. The drop-coated morphology of the MXene-GO reveals rough surfaces, and MXene particles were discovered to form islands on the GO. Because of the electrostatic interaction, GO and MXene may bond strongly together. The flat composite thin film surfaces may provide high interaction with electrolyte ions and quicker ion motions into the MXene-GO electrode, allowing for efficient energy storage. Though the synthesis method appears straightforward, developing the electrode design using a CO_2 laser could be an expensive process.

Conducting polymers were investigated for their potential application as energy storage materials due to their high C_s, rapid redox protonation/deprotonation processes, and cheap cost of operation. Following this, poly (1,5-diaminoanthraquinone) (PDAAQ) is classified as a conducting polymer. It has favorable structural properties, including a π-conjugated structure, making it an ideal candidate for use as novel electrode material in supercapacitors. As a result of the exfoliation of MXene ($Ti_3C_2T_X$), nanosheets tend to self-restack. This is because of the intermolecular force, which impedes the movement of electrons and reduces the surface area. An et al. designed cetyltrimethylammonium bromide (CTAB)-modified graphene oxide (GO)-coated PDAAQ nanotubes@MXene for use in supercapacitors to circumvent this issue [117]. CTAB was initially incorporated into the GO (CGO) surface via ultrasonication in water for 15 min. Then, using an initiator, 1,5-diaminoanthraquinone monomer (DAAQ) was dissolved in a mixture of dichloromethane and acetonitrile. Following the solution's dissolution of ammonium persulfate and sodium dodecyl benzene sulfonate, CGO was added to the mixture. CGO/PDDAQ was produced by combining the DAAQ solution with the CGO solution and observing the reaction for 48 h. Before undergoing modification with CGO/PDDAQ, an organic process was employed to etch MXene utilizing lithium fluoride and hydrochloric acid. Following the etching process, CGO/PDDAQ and MXene sheets were combined using ultrasonication in water to produce CGO/PDAAQ/MXene by the force of electrostatic interaction. The synthesis of CGO/PDAAQ@MXene film-type electrode is illustrated schematically in Figure 5a. This is achieved by coating CTAB-functionalized graphene oxide with poly(1, 5-diaminoanthraquinone) nanotubes via interfacial polymerization and then connecting the GO and MXene through vacuum-assisted filtration and electrostatic self-assembly. The produced CGO/PDDA shows the morphology of CGO adsorbed on PDDA nanotubes in the form of a sheet. After modification with an MXene solution, it exhibits an uneven morphology, thick, wrinkled textures, and a rough morphology. However, this morphology may have a highly active site, high purity, strong structural stability, offer high electrons to attract electrolyte ions, and allow for much quicker ion transportation to effectively store charges. The crystallinity of the CGO/PDAAQ/MXene was discovered to have a weak crystalline structure owing to the polymer nature of PDAAQ, which may be an amorphous structure. According to the synthesis, it may be time-consuming, expensive, and environmentally unfriendly due to the use of organic compounds and the involvement of multiple processes.

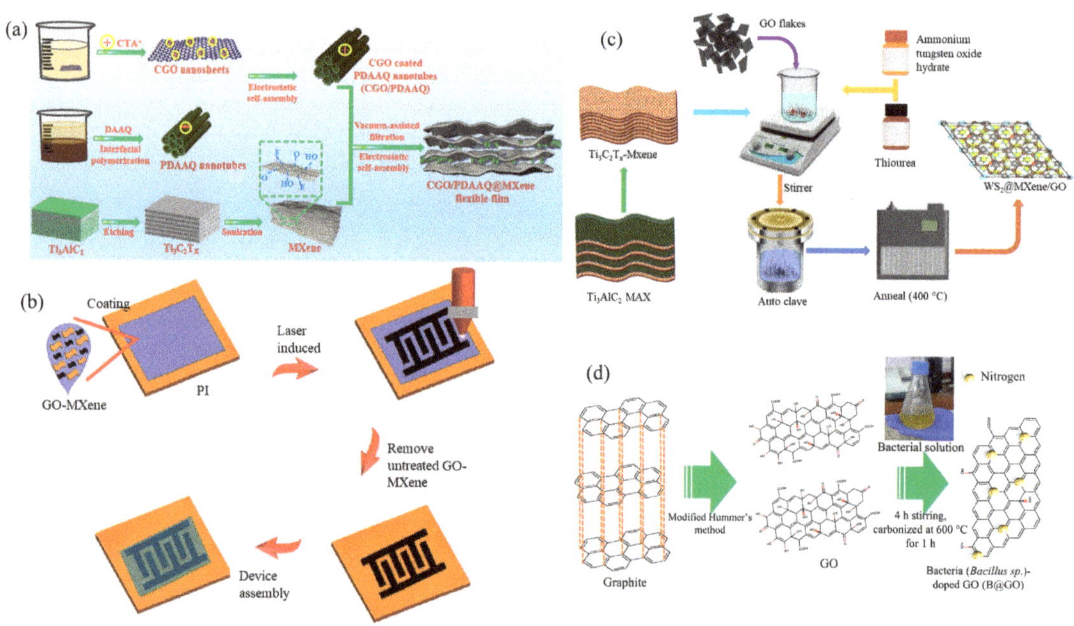

Figure 5. (**a**) Schematic diagram of the synthesis of CGO/PDAAQ/MXene (reproduced with permission from [117]). Copyright 2023, Elsevier. (**b**) GO-MXene composite electrode (reproduced with permission from [118]). Copyright 2023, Elsevier. (**c**) WS_2@MXene/GO composite (reproduced with permission from [119]). Copyright 2023, Elsevier. (**d**) Bacteria/GO (reproduced with permission from [120]). Copyright 2023, Elsevier.

Recently, dip coating was utilized to implement an MXene-GO composite film onto polyamide material, followed by laser-induced graphene (LIG) to etch the film [118]. GO and MXene, both of which were obtained in solution form, were initially combined while agitating. The composite material was subsequently coated onto a polyamide substrate via dip coating. The interdigital design or pattern was subsequently etched into the film via a single laser-induced graphene process. The resulting film was subsequently utilized for studies of supercapacitors, and the pattern generated through this method appeared to be effective. For comparison, they also developed LIG on bare polyamide surfaces for electrochemical studies. The goal of this study was to demonstrate how to combine GO and MXene to create films with improved microstructure modification and ordered MXene particle distribution. They also wanted to demonstrate how to use laser-induced graphene to create patterned porous graphene structures on MXene/GO-polyamide surface electrodes that can make good contact with electrolyte ions for high electrochemical performance. The schematic representation of Figure 5b shows the film fabrication process using a single laser-induced graphene structure on the MXene-GO-polyamide film composite for the supercapacitor application. According to the morphology of the composite film, the polyamide film coated with a highly porous MXene/GO composite had a uniform distribution of MXene fine particles with twinkled GO. On the other hand, the absence of GO results in the observation of MXene particles aggregating on the polyamide film. Following the laser-induced graphene process, the composite film acquires a fibrous structure that is densely layered. The porous structure of this structured electrode film may facilitate contact with electrolyte ions, while the presence of graphene structure (after the LIG process) in the composite film contributes to its high conductivity. As a result, this electrode film has the potential to be significantly effective in supercapacitors. Because the polyamide material contains MXene and GO, it may have a low-crystalline property. This procedure for making films and fabricating electrodes with laser-induced graphene could provide highly pure film electrodes. Nevertheless, the procedure of procuring MXene and GO compounds and employing laser techniques might be quite costly.

Because of its vast surface area, high crystallinity, and quick ion diffusion, tungsten sulfide (WS_2) has high electrochemical activity. On the other hand, WS_2 exhibited inadequate electrical conductivity, specific capacitance, and cycling stability. To enhance the electrochemical characteristics of WS_2, Hussain et al. fabricated a composite material consisting of MXene-GO and WS_2 (WS_2@MXene-GO) intended for use in supercapacitors [119]. Initially, MXene was produced for this composite by reacting commercially available Ti_2AlC_2 powder with hydrofluoric acid to remove Al content. To produce WS_2/MXene-GO, commercially available GO and the as-synthesized MXene were combined with water, sonicated, and stirred at 60 °C for 1 h. The mixture was then supplemented with ammonium tungsten oxide hydrate and CH_4N_2S for two hours. They were then transferred to an autoclave set to 200 °C overnight to obtain WS_2@MXene-GO. The acquired particles were subsequently annealed at 400 °C in an H_2/Ar atmosphere to eliminate any remaining impurities. Figure 5c depicts a schematic layout of the process of converting Ti_3AlC_2 MAX to Ti_3C_2TX MXene, followed by a hydrothermal reaction of MXene, GO, and ammonium tungsten oxide hydrate with thiourea to produce WS_2@MXene-GO, which is subsequently annealed at 400 °C. The treated MXene was found to have stacked sheet structures, while WS_2 showed layered sheet structures. After compositing MXene/WS_2, the morphology reveals agglomerated sheet-shaped grains. The GO composite with WS_2@MXene morphology showed that the grains were distributed throughout the GO surfaces. The morphology of the composite potentially offers an abundance of active sites for electrolyte ion interaction, thereby improving electrochemical performance. Furthermore, this composite exhibits a high degree of crystallinity due to the crystalline properties of the individual particles, such as MXene, WS_2, and GO. The prepared composite (WS_2@MXene/GO (12.9 $m^2\ g^{-1}$) reveals a very low surface area due to the low surface area of the individual particles (MXene (2.7 $m^2\ g^{-1}$), WS_2 (5.2 $m^2\ g^{-1}$) and GO (2.9 $m^2\ g^{-1}$)). The acquisition of MXene precursors, GO, and other chemicals may contribute to the high cost of the synthesis process. Additionally, the involvement of multiple processes can be time-consuming.

2.8. GO-Bacteria

To save energy and safeguard the environment, most synthesis processes for electrode materials use poisonous substances or unfavorable reaction conditions, making them expensive and toxic. Finding a low-cost, environmentally friendly method that uses precursors that might be found in nature would be beneficial. As a result of their robust cell walls, bacteria belong to the class of prokaryotic organisms that can endure a variety of harsh environments. Bacteria are inexpensive, widely distributed, and safe for the environment. By converting the material in the electrolyte, bacteria can raise C_s and electrical conductivity. Thus, to develop supercapacitors, Verma et al. investigated the interaction of bacteria and graphene oxide (B@GO). They, therefore, utilized Bacillus subtilis, a Gram-positive bacterium [120]. For four hours, the produced Bacillus subtilis solution and the synthesized GO solution were combined by stirring. After that, the solutions were carbonized at 600 °C in an N_2 environment and cleaned with water and ethanol to produce B@GO. Figure 5d depicts a schematic representation of bacteria functionalization on GO utilizing GO produced from graphite powder and stirring with bacterial solution for 4 h, followed by calcination at 600 °C under N_2 gas for heteroatom doping. The resulting GO has an unsmooth and sheet-like morphology, while B@GO has an irregular and wrecked sheet shape with a larger size. This might be the reason why bacterial solution produces highly wettable GO, which is attributed to agglomeration. Furthermore, it was mechanically stirred for 4 h; thus, the morphology was affected. This B@GO composite was found to have semi-crystalline properties owing to structural deformation and interlayer gap caused by the bacterial solution connection with a heteroatom (N)-doped GO. The bacterial component with the GO materials may have good electrochemical performance because of their abundant porosity and plenty of foreign atoms. The composite method may be easy to use; however, further optimization is needed when utilizing bacterial cell culture in conjunction with carbon materials.

2.9. Discussion

Supercapacitor electrode materials must be low-cost, be easy to synthesize, porous, have highly active sites, be conducive, pure, have a well-structured particle shape, chemical stability, structural stability, and reproducibility. Based on the reported studies, Table 2 summarizes the advantages and disadvantages of synthesis methods for GO-based composite electrode materials and their characteristics. This review shows that the hydrothermal technique was mostly used for producing GO-based composites. Because of its low cost, ease of processing, ease of the composite, environmentally friendly process, autoclave reaction at 120–220 °C, high purity and crystalline particles, structural stability, packed hybrid particles, well-structured shape particles, that water is mostly used for the reaction media, low energy consumption, and rich active sites' nanostructured materials, it enables high electrochemical performance. However, the hydrothermal reaction could happen for at least 12 h and up to 48 h for particles to form, and the produced particles may have a low surface area. In this technique, the particles might not form if the autoclave is not tightly locked. Using a mortar and pestle to grind two particles together to make a composite could also be very cheap and easy with highly crystalline particles and a large-scale production method that does not use any energy. However, there is a chance of serious structural and morphological damage, high particle contamination, particle agglomeration, and blockage of the pores on the particle surface, which could affect the electrochemical performance and particle contact with the eyes and nose during grinding, which can cause health problems. Utilizing direct plasma reactive treatment to produce doping materials enables large-scale production and uniform distribution and is ideally suited for doping powdery substances due to its high stability and the ability of the numerous functional groups on the doped GO materials to facilitate a rapid electrochemical reaction and improve performance. However, the global applicability of this technique may be hindered by its proficient complexity, costly equipment, demand for highly experienced personnel, high maintenance expenses, a doping process that could affect the surface area of the composites, and substantial energy consumption. The simple reaction employed for the synthesis of composite particles may be inexpensive, straightforward, energy-free, large-scale, operate at room temperature, and produce a high number of active sites with substantial surface area particles and high or semi-crystalline particles. However, this approach to composite preparation proves to be ineffective due to its time-consuming procedures, inability to produce well-structured morphology as a result of reaction under a mechanical stirrer, and there is a possibility for low interaction between materials, which leads to poorly hybridized particles, and a high risk of particle contamination that blocks surface pores, thereby resulting in low electrochemical performance. Ultrasonication may not be the optimal method for preparing composite particles for supercapacitors due to the possibility for high-frequency ultrasound to break the interfacial bond, introduce surface defects and contamination, damage morphology and low crystalline particles, and particles with a moderate surface area, all of which could contribute to suboptimal electrochemical performance.

Table 2. Advantages and disadvantages of synthesis technique for various GO-based composite electrode materials, including their morphology, surface area, and structural properties.

Electrode Materials	Methods	Advantages	Disadvantages	Morphology	Surface Area ($m^2\ g^{-1}$)	Structure	Ref.
GO-Fluorine							
FGO	Direct plasma reactive treatment	Achieving uniform coating of fluorine on GO and rich active sites, large-scale preparation, and operation under atmospheric pressure could be the best surface coating method for powdery materials.	High-cost process and methods may not be environmentally friendly due to the use of NF_3 gas, which is toxic. Skilled labor is required to handle the reactor, possible corrosion in the reactor through NF_3 gas, the plasma generates high electron temperature affects the sensitive material, maintenance of the reactor, low surface area of material, and high energy consumption.	Thin layers with large interspaces	3.7	-	[77]
GO-Metal oxides							
GO@TiO_2-NPs	Grinding	Low-cost, simple, easy to composite, rich active sites, no energy consumption, large-scale production, and environmentally friendly procedure.	High aggregation, particle sizes could be very high, poor quality, the interaction between nanoparticles and GO may be weak, morphological shape damages during grinding, possibly low surface area, and different techniques used to synthesize different nanoparticles consume the cost of the process and time.	Aggregation and spherical shape	-	High crystalline	[78]
AC-GO-SnO_2	Hydrothermal	Low-cost, simple, safe, environmentally friendly, doping, high-purity particles, highly active sites, easy to composite, low-temperature process, and well-structured morphologies.	Small-scale production, time-consuming process, moderate surface area, moderate energy consumption, and this method requires a vacuum or unvacuumed oven to synthesize particles.	Aggregated and spherical shape	-	Low crystalline	[79]
AC-TiO_2-GO	Simple reaction	Cheap, easy-to-obtain composite using a binder, large-scale production, rich active sites, high hybrid structure, and no energy consumption.	Synthesizing AC, TiO_2, and GO separately and then combining them may be a costly, time-consuming process; there is poor composite quality owing to dust particles during outdoor mixing; and it is not environmentally friendly since organic chemicals are added during particle mixing.	Micro-TiO_2 and GO sheet hybrid AC (Fluffy)	439.2	High crystalline	[80]

Table 2. *Cont.*

Electrode Materials	Methods	Advantages	Disadvantages	Morphology	Surface Area (m^2 g^{-1})	Structure	Ref.
GO-CuO/ZnO/TiO_2	Ultrasonication	Cheap, very simple process, easy to operate instrument, has no energy consumption, has highly active sites, time-saving and room-temperature process, applies to all common materials for composites, the composite surfaces may exhibit high surface functional groups, low particle aggregation, and homogeneity of particles.	Mechanical forces cause morphological damage to particles and produce heat while performing small-scale production, and long-term operation may affect the instrument; high-time consuming procedures; this technique may produce low bonding between particles, and synthesizing many nanoparticles using chemical procedures might be expensive and expensive.	Nanorods of CuO and ZnO and irregular shape TiO_2 on GO sheet	-	High crystalline	[82]
NGO-$CuFe_2O_4$	Hydrothermal	Cheap, simple, environmentally friendly, high-purity particles, easy to hybrid particles, rich active sites, doping, particles could be of high surface area, low-temperature process, well-structured morphologies, and particles produced by this method are nanometer in size.	Commonly used to prepare metal oxides, temperatures below 220 °C could be used for particle synthesis, particles cannot be produced if the autoclave is not tightly closed, small-scale production, time-consuming process, doping N on GO by thermal decomposition in the tubular furnace could be high energy consumption and expensive, and high-temperature thermal decomposition can affect the stability of the material.	$CuFe_2O_4$ nanocubes on GO sheet	-	High crystalline	[85]
GO/Fe_3O_4-IL@W	Simple reaction	Low-cost procedure, highly hybrid particles, high-purity particles, high structural stability, highly active sites, large-scale production, the composite surfaces may exhibit high surface functional groups, low-temperature process, and no energy consumption.	Synthesis of GO and its functionalization with Fe_3O_4-IL may be time-consuming and energy consumption due to the large number of organic compounds and processes involved, and GO-Fe_3O_4-IL reaction under inert conditions.	Fe_3O_4/IL/W presents an irregular spherical shape on the GO sheet	-	High crystalline	[58]
Bi_2O_3-GO	Microwave	Dust-free synthesis, nanoparticles can be synthesized, high-purity particles, a time-saving process, high yields, no chemicals or water were used, no solvents or water are needed to purify the product after synthesis, sugar used as a source for GO, and chemical-free GO is synthesis.	Energy consumption, expensive technique, maintenance cost, morphology can be damaged under high watts operation, limited penetration depth of the heat radiation into the reaction precursors/medium, and heat-sensitive material cannot be used for surface modification or functionalization.	Sheet morphology	-	High crystalline	[88]

Table 2. *Cont.*

Electrode Materials	Methods	Advantages	Disadvantages	Morphology	Surface Area ($m^2\ g^{-1}$)	Structure	Ref.
NMGO	Ultrasonication	The process is inexpensive, produces highly hybrid particles, high-surface-area particle formation, uses only water as a medium for composite formation, consumes no energy, operates at room temperature, and saves time. It also facilitates the interaction between GO and nanoparticles, enhances composite structural stability, allows for the adsorption of high-density particles on GO surfaces, and does not require the use of solvents or organic media.	Specific morphology cannot be obtained; low-purity particles form on GO due to atmospheric dust contamination; sonication process under long-term operation and high amplitude may develop defects on the particle surface and break GO layers; instrument maintenance; and synthesis of GO, NiO, and MnO_2 to prepare composite structure are expensive and time-consuming.	Nanoparticles on the GO sheet have no specific shape	102.0	Moderate crystalline	[83]
$NCWO_4$/f-MWCNTs/GO	Ultrasonication	Only uses water to produce composites. For ideal composite preparation, first ultrasonicate f-MWCNTs and GO in water. After that, $NCWO_4$ was added to the solution, and the particles had a substantial surface area.	Low-purity particles were observed. It could be an expensive process due to the synthesis of $NCWO_4$ by hydrothermal and then ultrasonication used for composite f-MWCNTs and GO. Additionally, using commercial MWCNTs and their functionalization and commercial GO in this procedure seems to be expensive, and a 2 h continuous ultrasonication process may affect the instrument's functionality.	Urchin-like structure of $NCWO_4$ on the wire-like shape of f-MWCNTs and GO sheet composite	172.2	Low crystalline	[90]
$MnFe_2O_4$/GO	Modified Hummer's method	Carbon-based materials can be synthesized by this technique, a low-temperature process, a time-saving procedure, highly layered GO sheets, high-purity GO sheets, highly dense composite particles, high-surface-area composites, high crystallinity composites, and high structural and morphological stability	The use of the coprecipitation technique to synthesize $MnFe_2O_4$ revealed low purity, moderately expensive procedures, and the need for an ice bath for GO/composite preparation; the precipitation process uses a high molarity ratio for the precursors, commercial graphite is used for the preparation of GO and its composites, vacuum drying is used for the composites, and a high amount of $KMnO_4$ is used for the reaction of GO and its composites.	The $MnFe_2O_4$ has a fine spherical shape on the GO sheet	-	High crystalline	[91]

Table 2. *Cont.*

Electrode Materials	Methods	Advantages	Disadvantages	Morphology	Surface Area ($m^2\ g^{-1}$)	Structure	Ref.
MnO_2-GO/GF	Hydrothermal	The synthesis process is simple, involving only water as the medium for MnO_2-GO formation on GF, an environmentally friendly procedure, a well-structured morphology, the possibility of high-surface-area composites, the use of major precursors for composite preparation, and the absence of any toxic solvents or liquids in the procedure.	The process is costly due to the utilization of commercial GF, a 12 h process, the lack of a detailed explanation for the synthesis procedure, the use of toxic chemicals for pretreating the GF, the production of low-purity particles or composites, the requirement for pretreated GF to dry under vacuum conditions, the absence of a purification procedure after synthesis, the potential expense of using a high-pressure reactor for the synthesis of GO, and potential maintenance issues.	Nanorod shape of MnO_2/GO on GF	-	Low crystalline	[92]
Cu_2O-CoO/GO	Hydrothermal	This process used only an 8 h reaction and may have high-surface-area composites, moderate energy consumption, high-purity composites, highly active sites, and synthesized GO that can directly be added to the precursor for composite preparation using hydrothermal.	The process was moderately expensive, revealed no significant morphology, and revealed low-purity particles. The addition of KOH to the precursor solution for alkaline titration may not be environmentally friendly, the interaction between GO and nanoparticles could be low, the resulting black gel product requires soaking in water for 24 h to purify it, this process is time-consuming, and the final composite particles require calcination in a tubular furnace under an N_2 atmosphere.	No significant morphology; however, GO shows the sheet-layered structure	-	High crystalline	[95]
10%GO/WO_3	Hydrothermal	Using H_2O_2 as a reducing agent in the synthesis could make fine particles, the reaction could directly add GO powder to make the composite, and a methodical process could find the best ratio of GO to WO_3 for the composite formation. Fine-size particles were synthesized, along with high-purity particles, and it has a moderate energy consumption and a high hybrid particle formation.	It is a 12 h process, low-surface-area particles, annealing at 500 °C may damage the morphology, low-activity sites, a high GO ratio may affect the crystallinity of WO_3, using commercial graphite flake for GO synthesis may be costly, and using HCl and H_2O_2 in the synthesis may be an environmentally unfriendly and time-consuming procedure.	Nanorods	33.9	High crystalline	[96]

Table 2. *Cont.*

Electrode Materials	Methods	Advantages	Disadvantages	Morphology	Surface Area (m^2 g^{-1})	Structure	Ref.
GO/CNTs/$CoFe_2O_4$	Simple reaction	The process is low-cost, simple, energy-free, and time-saving. It uses only water as a medium for composite formation under stirring, has rich active sites, is environmentally friendly, and does not use any solvents or organic chemicals.	The process may be costly because it involves synthesizing GO using a modified Hummer's method, which necessitates the use of numerous chemicals, acids, solvents, and an ice bath setup. The solution-combustion method synthesizes $CoFe_2O_4$ but requires calcination at 800 °C. Further, commercial CNTs and graphite flakes are expensive. The preparation of composites in water with simple mixing may not be effective, the particles may not hybridize strongly, and the use of 10% HCl to wash the composites may damage their structure and morphology.	The spherical shape of $CoFe_2O_4$ and the tube shape of CNTs are present on GO sheets	-	High crystalline	[97]
GO-Metal sulfide							
NiS/GO and CoS/GO	Hydrothermal	The process is environmentally friendly and time-saving, as it directly adds GO powder to the precursor solution for composite synthesis, eliminating the need for any additional pH adjustments or reducing agents. This technique can achieve a uniform particle size distribution and deposition. The particle interaction with the GO sheet could be strong, ensuring high stability and purity.	The particles on the GO sheets did not exhibit any notable morphology. They might have a low surface area, undergo a 12 h process, and have few active sites. The process involves the consumption of chemicals for GO, NiS, and CoS synthesis, as well as hydrothermal treatment at high temperatures, which can be somewhat expensive.	No significant shape of nanoparticles was observed on the GO sheet	-	High crystalline	[98]
MnS-La_2S_3/GO	SILAR	Thin film dip coating can be performed at low cost, simple, room temperature, time-saving, large-scale production, no energy consumption, highly active sites, high composite structures, only water used to remove the unbound anion/cations on the film surfaces, an environmentally friendly process, multiple dip coating can be performed, no drying process is required, and no chemical reaction is required to coat the film on the SS substrate.	It takes at least 100 SILAR cycles to obtain a perfect coating. There are concerns about the coating's stability, difficulties regulating the film's thickness, and the possibility of airborne dust particles contaminating thin film surfaces. Handling thin film devices by hand or in a closed pack is challenging due to their susceptibility to scratches or cracks. And dip-coating film on one side of the substrate is challenging.	Thin film/sheet morphology	-	Low crystalline	[100]

Table 2. *Cont.*

Electrode Materials	Methods	Advantages	Disadvantages	Morphology	Surface Area ($m^2 g^{-1}$)	Structure	Ref.
GO-Transition metal chalcogenide							
GO-MnSSe	Simple reaction	Low-cost, simple, environmentally friendly process, no complicated technique or method used, room-temperature process, no energy consumption, mass production, no solvents or reducing agents used for the composite preparation, and low-temperature drying process.	The overnight process used 1 g of GO to prepare Go-MnSSe, low-activity sites, low purity, and potentially low-surface-area particles. The reaction under only water may not form particles on GO, leading to uneven particle formation on GO and no particle formation on GO for the composite GO-MnS. The composite preparation did not calculate a ratio, limiting the possibility of incorporating Mn into GO synthesis using a modified Hummer's method.	The GO sheet contains fine spherical MnSSe nanoparticles	-	Low crystalline	[104]
GO-Layered double hydroxide							
GO/Co_2-Ni_1 LDH	Reflux	The procedure is simple and involves high-surface-area particles, rich active sites, a low-temperature process, high hybrid formation, a sheet type of morphology, highly porous composites, and no energy consumption.	The process is not environmentally friendly, it takes a long time, the particles are not very pure, it uses high molarity ratios, it is moderately expensive because it synthesizes them through reflux in an oil bath, and the clay-like nature of the LDHs may damage the shape of the GO.	Sheets with a wrinkled structure	84.6	Low crystalline	[106]
GO-organic material							
ABQA-GO	Organic reaction	It is simple to synthesize, ABQA and GO are strongly covalently bonded, it has inert atmosphere processes, the process is carried out at low temperatures, there are a lot of active sites and functional groups on the surface of the functionalized GO, and ABQA has a high redox property.	The process is time-consuming, not environmentally friendly, and costly due to the use of expensive organic chemicals and commercial graphite, as well as the use of toxic chemicals and the freeze-drying process. The product's purity may be low, necessitating self-safety precautions before synthesis. The distillation process can be energy-consuming, and only synthesis can take place in fume exhaust hoods. And there are morphology stability issues.	Broken and wrinkled sheet structure	-	Low crystalline	[101]

Table 2. *Cont.*

Electrode Materials	Methods	Advantages	Disadvantages	Morphology	Surface Area ($m^2\ g^{-1}$)	Structure	Ref.
Nd-MOF/GO	Hydrothermal	Easy to synthesize, environmentally friendly, low cost, high-purity particles, well-structured morphology, water and ethanol utilized for reaction, drying particle at less than 90 °C, low-energy consumption, and highly active sites.	The process involves the use of an ice bath before the hydrothermal reaction, treatment at elevated temperatures, and a prolonged duration. High temperatures could not be used to dry the particles; hybrid particles could not form properly, and the morphology was slightly damaged after the composite was mixed with GO.	Hexagonal-rod-shaped Nd-MOFs on the GO sheet	-	High crystalline	[113]
Ni-BTC@GO	Hydrothermal	Cheap, simple to synthesize, drying particles at less than 90 °C, rich active sites, low energy consumption, high-purity particles, high surface area and porous surfaces, and well-structured morphology particles.	It is a time-consuming technique owing to the 48 h hydrothermal process and the 24 h particle drying time. The organic solvents employed for washing the particles may affect their surfaces. Adjusting the dosage of GO in the composite preparation for comparison makes it a bit expensive. Higher doses of GO could affect the crystallinity of MOF, and the procedure is not environmentally friendly because of the usage of organic solvents for the reaction.	Sheet-structured GO on octahedral-shaped Ni-BTC MOF.	-	High crystalline	[102]
SGZ	Simple reaction	The process is cheap, simple, easy to synthesize, conducted at 40 °C, consumes no energy, has highly active sites, may have a high surface area, and uses only methanol for reaction and particle purification.	It is not an environmentally friendly procedure and also not an effective procedure for making composites due to the possibility of contamination, which leads to low purity. Sonication and mechanical stretching destroy the particle shape. Overall, processes such as GO fabrication, S doping on GO, and ZIF-8 MOF to make composite particles are quite expensive. GO affects the crystallinity of MOF.	Irregular shape	-	Low crystalline	[114]

Table 2. *Cont.*

Electrode Materials	Methods	Advantages	Disadvantages	Morphology	Surface Area ($m^2\ g^{-1}$)	Structure	Ref.
GO-MXene							
CGO/PDAAQ/MXene film electrode	Vacuum-assisted filtration	The composite structure exhibits high flexibility and high stability due to the electrostatic interaction between CGO/PDAAQ and MXene. The preparation of the composite film at room temperature and the conducting composite at water are both easy and simple processes.	The procedure is expensive due to the numerous processing stages involved, and the final product requires the use of a freeze-drier. It is not an environmentally friendly procedure for CGO/PDDAQ, and the whole synthesis study seems time-consuming. The pH of the solution should be 6–7 for MXene preparation, ice bath ultrasonication, and vacuum-assisted filtration to obtain the final composite may damage the morphology by the force of the vacuum.	Unsmooth nanotube-blended sheet morphology	-	Low crystalline	[117]
MXene/GO film electrode	Drop-coating and laser-induced graphene technology	Drop-coating is a simple method for achieving film uniformity. GO is used to prevent MXene aggregation. The polyamide substrate provides excellent film deformation-resisting properties, along with good stability and flexibility. The laser-induced graphene (LIG) process was able to demonstrate perfect patterning to obtain a graphene-structured film, a programmable fabrication process, a highly porous structure, and a fibrous structure.	A time-consuming procedure. Laser-induced technology may make it costly. The film drying process should be carried out in inert gas. High-power-watt lasers can damage morphology and contaminate film during drop coating. Self-precaution may be needed to process the patterning electrode. Electrode operation and patterning demand skilled researchers.	Fibrous with stacked structure film	-	-	[118]
WS_2@MXene/GO	Hydrothermal	The procedure is environmentally friendly and easy to synthesize, the composite surface is rich in active sites, and it uses only water for the hydrothermal reaction, resulting in high purity and particle purification through a green process.	The impregnation process for GO with MXene (1 h) and the subsequent addition of sodium tungsten hydrate (2 h) before the reaction result in a moderately time-consuming procedure. The reaction process operates at a high temperature of 200 °C, has a small surface area, produces laboratory-scale samples, and lacks a well-organized morphological structure. This procedure could be moderately expensive due to the use of commercial GO and Ti_2AlC_2 powder.	GO sheets contain grain-shaped WS_2@MXene particles	12.9	High crystalline	[119]

Table 2. *Cont.*

Electrode Materials	Methods	Advantages	Disadvantages	Morphology	Surface Area (m^2 g^{-1})	Structure	Ref.
GO-Bacteria							
B@GO	Simple reaction	The procedure is cheap, simple, environmentally friendly, and time-saving.	Using N_2 gas under calcination may not be effective in dope N on GO. The stirring process for 4 h changes the shape and structure of the GO, resulting in low-quality particles and active sites. Cell culture and incubation of Gram-positive bacteria may be costly and have low-purity particles and active sites. The bacterial solution and the GO were not optimized.	Wrecked sheets	-	Semi-crystalline	[120]

The microwave method is a rapid and efficient method for synthesizing particles with well-structured morphologies, high-purity and crystalline particles, and no reaction medium requirement. The synthesis of composites using this method may result in particles that are well hybridized, and the abundance of active sites may facilitate rapid electrochemical reactions and high performance. Nonetheless, this is a costly method, and prolonged operation at high wattages can damage particles. Many processes for combining GO with other materials include first preparing GO using a modified Hummer's method and then using another approach for combining it. This method may be costly, time-consuming, and energy-intensive. To address these difficulties, GO and its composite with $MnFe_2O_4$ were synthesized utilizing a single procedure, a modified Hummer's approach. This technique revealed good structural stability and highly crystalline particles with no morphological issues. Thus, this type of process may be utilized to construct a GO composite with other materials using a low-cost, easy, and energy-efficient technique and a superior electrode material for electrochemical performance. SILAR technique may be a relatively simple, room temperature process that produces homogeneous thin film electrodes, a variety of coatings on the substrate, and a highly active site and conductivity electrode film. This process is much less expensive than other coating techniques, such as spray coating, spin coating, sputtering, thermal evaporation, pulsed laser deposition, atomic layer deposition, and so on. However, this SILAR may have issues with contamination on the film surfaces, structural stability, the film's low crystallinity and thickness, and the coating process occurring on both sides of the substrate, which is one of its disadvantages. The preparation of thin film electrodes using vacuum-assisted filtration techniques can produce low-cost, easily obtained composites with high particle interaction due to the vacuum force, high flexibility, a highly active site film, and highly porous properties, all of which contribute to high electrochemical performance. However, using a strong vacuum force to filter the particles could damage their morphology and collapse their pores, leading to a poor electrochemical reaction, low ion transportation, and low electrochemical performance. On the other hand, drop-coating is a simple, low-cost method that produces a high-conductivity electrode layer. However, it is not possible to adjust the thickness of the film. The laser-induced graphene technique, which patterns the composite electrode film after drop-coating, is highly effective, fabricating a high fibrous structure and porous morphology that allow contact electrolyte ions to produce quick electrochemical reactions for high electrochemical performance. The reflux/organic chemical approach is a simple, low-temperature way to make particles that have a lot of active sites, form well-hybrid particles, make high-porosity composites, and do not use any energy. However, this approach lacks environmental friendliness; hinders the development of high-purity, crystalline, and surface area particles; and risks damaging particle morphology due to its extended processing time. Therefore, the electrochemical performance of the chemical or organic technique that generates composites may be modest. Therefore, further investigation into the development of technologically advanced composite electrodes may yield cost-effective solutions for implementation in supercapacitors. Such electrodes would possess the following characteristics: a simplified one-step composite preparation process, environmental friendliness, time efficiency, large-scale synthesizing capability, well-structured morphology, substantial surface area, highly hybrid particles, higher crystalline properties, abundant active sites, high structural stability, and high conductivity.

3. GO and Its Composites in Supercapacitor Application

3.1. Fluorine-Functionalized GO

Surface-modified GO might enhance the wettability of the electrodes, so leveraging the pseudocapacitance effect enhances the capacitance of the device. Fluorination is an excellent method to alter the characteristics of graphene since fluorine (F^-) has the greatest electronegativity. Carbon–fluorine bonding in the highest polar states is advantageous for electrochemical reactions because of the interaction between ionic and covalent bonds. Thus, Sim et al. investigated fluorine-functionalized GO (FGO) using a three-electrode configuration in a 1.0 M potassium hydroxide (KOH) solution [77]. Figure 6a shows the CV curves of the FGO and GO electrodes analyzed in the negative potential region at a scan rate of 10 mV s^{-1}. This test shows that FGO had both EDLC and redox activity because F^- functional groups react with OH^- groups. On the other hand, GO only showed EDLC action. The FGO exhibited significantly higher C_s (514 F g^{-1}) in comparison to the GO (231 F g^{-1}). The galvanostatic charge–discharge (GCD) curves of GO and FGO for potential (V) and time (s) are illustrated in Figure 6b. The charge–discharge duration of FGO was longer than that of GO, indicating that FGO has a high charge-storing capacity, which might be attributable to its high C_s. At a current density (CD) of 6 A g^{-1}, the charging and discharging activities of the FGO and GO were 276 and 90 F g^{-1}, respectively. Using electrochemical impedance spectroscopy (EIS), the solution resistance (R_s) and charge-transfer resistance (R_{ct}) of the FGO and GO electrodes were subsequently determined. As shown in Figure 6c, the R_s and R_{ct} values of FGO (0.58 Ω and 0.71 Ω) were observed to be less than GO (1.69 Ω and 2.61 Ω). The diameter of the semi-circular from the EIS plot may be used to compute the R_{ct} values. For ion-diffusion resistance, both electrodes exhibit a distinct vertical line, thereby confirming the better kinetics of the electrodes. Therefore, the FGO electrode stimulates the movement of ions between the electrode and electrolyte, enhancing fast charge-transfer processes. Therefore, FGO exhibits better electron transfer properties than GO. According to the FGO's morphology, the thinner fluorine layer on GO sheets has highly porous (fine pores) structures that facilitate charge storage via EDLC production, while the interconnecting porous channels enhance redox reactions. Furthermore, FGO surfaces may include a large number of fluorine ions, which are responsible for significant redox reactions with OH^- ions, resulting in good electrochemical performance. To facilitate practical implementation, a symmetric hybrid FGO (FGO//FGO) was designed and tested over a potential range of 1.0 to 1.6 V. A sheet of polypropylene serves as the anode in this device. The device also observed EDLC and PC operations. The designed device achieves an energy density (ED) of 25.8 Wh kg^{-1} and power density (PD) of 1280.0 W kg^{-1} at a CD of 1 A g^{-1}. This device demonstrated excellent cycle stability, with no loss of specific capacity after 20,000 cycles, as shown in Figure 6d. The increased stability of the FGO may be due to the improved surface activation resulting from higher levels of F^- sites during electrochemical action, which could be very stable electrochemically. The inset of Figure 6d reveals that the pictures of the hybrid device were programmed to use yellow, blue, green, and white light-emitting diodes (LEDs) at 2, 3, and 4 V, respectively.

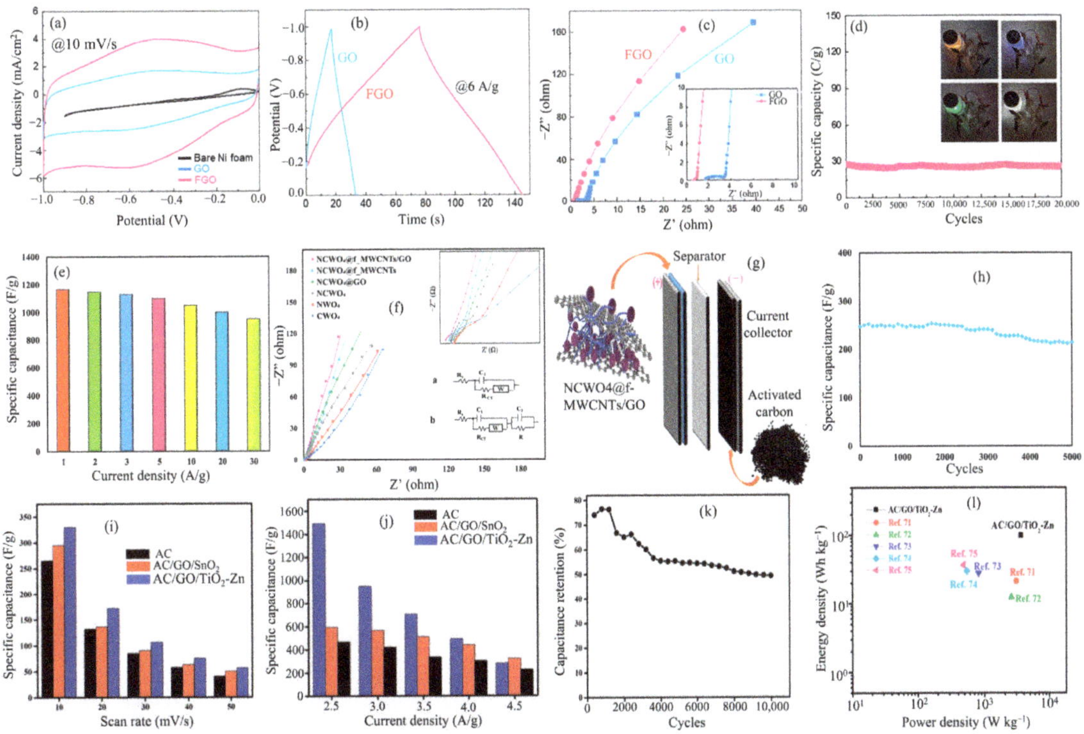

Figure 6. Electrochemical study of electrodes. (**a**) CV curves of GO and others at 10 mV s^{-1}. (**b**) GCD curves of FGO and GO at 6 A g^{-1}. (**c**) EIS spectra. (**d**) Cycle stability of FGO//FGO device with inset picture shows powering up the LED pulps (reproduced with permission from [77]). Copyright 2023, Elsevier. (**e**) Variation of C_s for CD of $NCWO_4$/f-MWCNT/GO. (**f**) EIS spectra of $NCWO_4$/f-MWCNT/GO and others. (**g**,**h**) Schematic diagram and cycle stability of $NCWO_4$@f-MWCNTs/GO//AC device (reproduced with permission from [90]). Copyright 2023, Elsevier. (**i**,**j**) Variation of C_s to scan rate and CD of AC/GO/TiO_2-Zn. (**k**,**l**) Cycle stability and the Ragone plot of AC/GO/TiO_2-Zn//AC/GO/TiO_2-Zn device (reproduced with permission from [79]). Copyright 2023, Elsevier.

3.2. GO-Metal Oxides

Recently, the $Ni_{0.5}Co_{0.5}WO_4$/f-MWCNT/GO ($NCWO_4$/f-MWCNT/GO) composite was used for the three- and two-electrode systems in 6.0 M KOH for supercapacitor application [90]. The enlarged CV curves of $NCWO_4$/CNT/GO, which resulted from the bimetallic synergy, provide insight into the PC mechanism and indicate that the electrode material can operate at high scan rates (100 mV s^{-1}) within the potential ranges of 0.0 to 0.6 V. Figure 6e shows the calculated C_s using the GCD curves under various current densities (CDs). Accordingly, the CD increases from 1 to 30 A g^{-1}, and the C_s decreases slightly from 1166.6 to 950.0 F g^{-1}. As a consequence of this outcome, the composite electrode demonstrated a notable rate capability of 90.3%. It was determined that the electrode exhibited cycle stability with a capacity retention of 95.6% over 5000 cycles. By employing EIS, it was determined that the $NCWO_4$/f-MWCNT/GO electrode possessed excellent conductivity with values of 0.67 Ω and 1.2 Ω, respectively. Accordingly, these values are lower in comparison to $NCWO_4$@f-MWCNTs (0.69 Ω and 1.53 Ω). As illustrated in Figure 6f, a semicircle at high-frequency areas discloses the charge-transfer resistance (R_{ct}) of the electrodes associated with the faradaic processes, while a linear line at lower-frequency regions suggests the EDLC behavior of materials. According to Figure 6f, the NCWO4@f-MWCNTs/GO electrode exhibits a sharper angle than other electrodes, suggesting that the composite is capable of fast ion diffusion. This composite's higher electrochemical properties could be due to the crystalline structure of the electrodes, highly active sites, and the synergistic morphology effect of urchin ($NCWO_4$) and nanotubes (f-MWCNT) on sheets (GO). Further development of an asymmetric device ($NCWO_4$@f-MWCNTs/GO//AC) was implemented within the voltage range of 0.0–1.5 V. Figure 6g depicts a schematic representation of an asymmetric device made from nickel foam (current collector) modified with $NCWO_4$@f-MWCNTs/GO (positive electrode) and AC (negative electrode) and a separator was located between the electrodes to transport ions from the negative electrode to the positive electrode and prevent a short circuit. The device had a sandwich-like circuit design. Additionally, PC behavior was evident due to the presence of redox peaks in CV curves at various scan rates. The device exhibited a high C_s value (266.2 F g^{-1}) at a CD of 1 A g^{-1}, as determined by GCD. The shallow values of R_s (0.22 Ω) and R_{ct} (0.9 Ω) for the device suggest high conductivity. The number of cycles increases to reduce the capacitance retention (%) of the electrode material, as shown in Figure 6h. It was determined that the developed device maintained 87% of its capacity for 5000 cycles. The asymmetric device's $NCWO_4$@f-MWCNTs/GO electrode may be electrochemically stable and have a solid bond to the current collector, resulting in good capacitance retention during the cycle test. The device's maximum PD and ED values were 703.1 W kg^{-1} and 83.3 Wh kg^{-1}, respectively.

Another study examined the three-electrode systems in a 3.0 M KOH electrolyte using AC, AC/GO/SnO_2, and AC/GO/TiO_2-Zn electrodes [79]. The results revealed quasi-triangular CV curves, indicating that all electrodes operated as EDLC-type supercapacitors. Figure 6i plots the C_s of the electrodes using CV curves at various scan rates. Compared to AC (265.3 F g^{-1}) and AC/GO/SnO_2 (295 F g^{-1}), the C_s of the AC/GO/TiO_2-Zn (329.9 F g^{-1}) was higher at 10 mV s^{-1}. The synergistic effect of the electrode materials, along with their highly porous structure and flower-like morphology, could be attributed to the high C_s of the AC/GO/TiO_2-Zn composites, each of which has good crystalline properties. Triangle curves are found by GCD analyses of the electrodes AC and AC/GO/SnO_2, which point to high EDLC performance. On the other hand, because of the faradic charge-transfer reaction, PC behavior was seen in the AC/GO/TiO_2-Zn electrode. The faradaic reaction of the AC/GO/TiO_2-Zn electrode resulted in a longer discharge time than other electrodes, indicating higher C_s. Due to storage processes via the electrochemical diffusion process, the C_s of the electrode materials generally decrease as CD rises. Accordingly, plotting the C_s of the electrodes using GCD curves at different CDs in Figure 6j helps to understand it. The AC, AC/GO/SnO_2, and AC/GO/TiO_2-Zn electrodes had maximum C_s of 467.8 F g^{-1}, 595.2 F g^{-1}, and 1491.6 F g^{-1} at 2.5 A g^{-1}, respectively, at a CD of 2.5 A g^{-1}. Furthermore, compared to AC (11.5 Ω) and AC/GO/SnO_2 (4.4 Ω), the R_s of the AC/GO/TiO_2-Zn (192.2 Ω) electrode was higher. The symmetrical device (AC/GO/TiO_2-Zn//AC/GO/TiO_2-Zn) was built based on the best result obtained from the AC/GO/TiO_2-Zn. A symmetrical device of cycle stability profile is shown in Figure 6k, which demonstrates how capacitance retention steadily declines as the number of cycles increases. It demonstrated low capacitance retention of 61.8% over 10,000 cycles. The device's low capacitance retention rate could be caused by the electrode structure collapsing due to the volume expansion and contraction of the electrode material during the charging and discharging cycles. As a result, while synthesizing ternary compounds, careful consideration should be given to the significant effects on structural suitability and chemical stability during electrochemical reactions. Furthermore, as depicted in Figure 6l, the Ragone plots of the AC/GO/TiO_2-Zn//AC/GO/TiO_2-Zn device's PD vs. ED profile indicate that the device generated a substantial PD of 3572.0 W kg^{-1} despite having an ED of 99.0 Wh kg^{-1} when compared to other reported devices. Due to the Faradaic reaction in 6.0 M KOH, pseudocapacitance (PC) behavior was seen in the Bi_2O_3-GO electrode from the CV analysis in the different scan (5–100 mV s^{-1}) rate [88]. At a CD of 1 A g^{-1}, the electrode's C_s were measured to be 1029.0 F g^{-1}. However, the C_s slightly decreased from 1029 to 808 F g^{-1}, while the CD increased from 1 to 10 A g^{-1}, indicating that as the CD increased, the electrode's C_s remained highly sustained. Though the Bi_2O_3-GO composite did not show significant morphology, they did, however, obtain high C_s due to the highly active sites, high purity, and highly crystalline structure. The R_s values for various wattages, 600 W 120 s, 800 W 90 s, and 800 W 120 s, were 0.91, 0.83, and 1.37 Ω, respectively. The resistance of the electrode exhibited an increase as the wattage was raised. GCD analysis also revealed PC behavior, with the C_s of Bi_2O_3-GO (1029 F g^{-1}) being greater than that of bare Bi_2O_3 (314 F g^{-1}) at a CD of 1 A g^{-1}. Therefore, the electrochemical performance is enhanced by the combination of GO and Bi_2O_3. The electrodes' structural stability contributes to their cycling performance. In this work, the decrease in capacitance retention observed as the number of cycles increases may be caused by an irreversible reaction between Bi_2O_3-GO and KOH electrolyte, which causes structural degradation over a long cycle life. It was found that the electrode maintained approximately 80% of its initial C_s over 3000 cycles. It would be preferable if the cyclic study for the supercapacitor application included at least 10,000 cycles. Consequently, the cycle stability of this particular GO-composite might be compromised, necessitating the optimization of the ratio or functionalization to enhance its structure.

Foroutan et al. investigated the electrochemical performance of $MnFe_2O_4$/GO in 6.0 M KOH and demonstrated the possibility of a redox reaction by obtaining PC behavior as indicated by rectangular CV curves [91]. Semi-rectangular shapes were maintained in the CV curves despite the increase in scan rates (10–200 mV s^{-1}). As determined by GCD analysis, the quasi-rectangular and triangular structures of GO and $MnFe_2O_4$ correspond to EDLC and PC behavior, respectively. In contrast, $MnFe_2O_4$/GO exhibits semi-triangular shapes and demonstrates characteristics of both EDLC and PC. At a CD of 1 A g^{-1}, the C_s of $MnFe_2O_4$/GO (298 F g^{-1}) was found to be greater than that of $MnFe_2O_4$ (285 F g^{-1}) and GO (188 F g^{-1}). Hence, $MnFe_2O_4$ exhibits promise as a material for enhancing the electrochemical characteristics of GO. The high C_s could be the reason for the large number of $MnFe_2O_4$ spherical nanoparticles on GO sheets; their strong crystalline property and highly porous surface may make it easy to connect with the electrolyte ion and promote quicker ion movement to produce high electrochemical performance. Conversely, increasing the CD (0.5–8.0 A g^{-1}) results in a significant reduction in the C_s of the $MnFe_2O_4$/GO (405.0–57.0 F g^{-1}). This could be due to the limitation of ion-diffusion electrolytes to electrode surfaces at higher CDs. Moreover, according to the EIS results, $MnFe_2O_4$/GO (2.16 and 2.45 Ω) may have a higher conductivity than GO (4.7 and 13.8 Ω) and $MnFe_2O_4$ (3.4 and 5.7 Ω), as their R_s and R_{ct} resistance values were lower. In only the 500 cycles, the capacitance retention was found to be 92%. It is possible that strong cycle stability was responsible for the chemical stability of $MnFe_2O_4$/GO in the electrolyte. However, a supercapacitor application that would have performed 10,000 cycles might not find the 500-cycle research relevant or predictable. Regarding the practical applicability, no device preparation or electrochemical investigations were conducted in this study. Similarly, CV analysis showed that the ternary hybrid of AC/TiO_2/GO behaved in both EDLC and PC. In this study, 2.0 M H_2SO_4 was used as the electrolyte solution [80]. However, the authors said that using H_2SO_4 as an electrolyte greatly affects the capacitance rate. The electrochemical studies showed that AC and AC-TiO_2 behaved like EDLC and PC, respectively. At a scan rate of 10 mV s^{-1}, the electrodes' C_s show an AC of 69.7 F g^{-1}, AC-TiO_2 of 159.2 F g^{-1}, and AC-TiO_2-GO of 219.3 F g^{-1}. Additionally, GCD was utilized to examine the C_s of the AC, AC/TiO_2, and AC/TiO_2/GO, which were 130.3, 195.4, and 617.0 F g^{-1}, respectively, at 1 A g^{-1}. Therefore, adding GO to the AC/TiO_2 combination improves its electrochemical properties. The good C_s of the AC/TiO_2/GO was found to possess many nano-scale pores, a large surface area (439.2 m^2 g^{-1}), and good crystallinity, enabling the surface to attract higher amounts of electrolyte ions and achieving high charge storage. This ternary composite's porous structure is highly beneficial for electrical conductivity because it allows for high-ion transport. Figure 7a illustrates the Nyquist plots of the electrode's resistance recorded from 0.01 Hz to 100 kHz at a potential voltage of 10 mV. At the high- and medium-frequency ranges, the plots show semicircles for the different electrodes caused by R_{ct} at the electrode and electrolyte interface. On the other hand, the plots reveal aberrations in the straight-line slope in the low-frequency zone. This demonstrates a decreased diffusion barrier for electrolyte ions. According to Figure 7a, AC-TiO_2-GO has a lower internal resistance than other electrodes. This might be attributed to the faradaic reactions and high conductivity of TiO_2, while the good conductivity of GO contributes to the low charge-transfer resistance. Accordingly, AC/TiO_2/GO showed lower R_s and R_{ct} (1.87 and 2.68 Ω) than AC (5.34 and 7.76 Ω) and AC/TiO_2 (2.67 and 3.40 Ω). Therefore, the addition of GO to AC/TiO_2 improves electrical conductivity. Figure 7b shows the cycle stability profile of the various electrodes, revealing that the electrode's capacitance retention decreased gradually as the number of cycles increased. The ternary composite (AC/TiO_2/GO) structured electrode demonstrated remarkable stability over 1000 cycles of testing. As a result, AC, AC/TiO_2, and AC/TiO_2/GO retained 57.2%, 72.6%, and 80.1% of their capacitance, respectively. A synergistic impact of GO and TiO_2, high conductivity, structural stability, high chemical stability, and extremely low volume expansion and contraction may be responsible for the high cycle stability of ternary composite electrodes. The influence of the structural degradation and non-conductivity of AC causes no synergistic effect on the binary composite electrode; therefore, the moderate cycle stability is achieved exclusively by the effort of TiO_2 owing to structural and chemical stability. The non-conductive sites and the electrochemical processes may be the cause of the poor cycle stability of AC and the collapse of its pore structure. Therefore, only the use of GO improves the cycle stability in this study, which was observed through the cycle profile. This investigation also did not find any devices in preparation for practical applications.

Recently, Lashkenari et al. conducted an electrochemical investigation on GO, $CuFe_2O_4$, N-doped GO (NGO), and N-doped GO/$CuFe_2O_4$ (NGO/$CuFe_2O_4$) using a various electrolyte, such as 1.0 M Na_2SO_4, 1.0 M H_2SO_4, and 1.0 M KOH [85]. NGO/$CuFe_2O_4$ electrode CV profile at 100 mV s^{-1} scan rate in various electrolyte solutions is shown in Figure 7c. Based on the large CV curve and wide working voltage potential, 1.0 M H_2SO_4 outperformed others, and this electrolyte was optimized for the investigations. This study exhibited both EDLC and PC behavior on NGO/$CuFe_2O_4$ due to faradaic reaction, whereas GO, NGO, and $CuFe_2O_4$ demonstrated EDLC, EDLC, and PC behaviors, respectively. Both behaviors on NGO/$CuFe_2O_4$ were observed as a result of the Faradaic process of $CuFe_2O_4$ in the electrolyte (redox reaction) and electrostatic charges on the GO or NGO surface. Because of the synergic impact, NGO/$CuFe_2O_4$'s CV curves were greater than others, suggesting that C_s may be high. Accordingly, NGO/$CuFe_2O_4$ had higher C_s (374.2 F g^{-1}) than GO (138.0 F g^{-1}), NGO (181.2 F g^{-1}), and $CuFe_2O_4$ (214.0 F g^{-1}) at a CD of 1 A g^{-1}. Electrode materials that are a combination of NGO and $CuFe_2O_4$ may exhibit enhanced redox properties, structural stability, conductivity, surface area, active sites, and porousness due to the nanocube shape of $CuFe_2O_4$ on GO sheets. Adding NGO to $CuFe_2O_4$ may also greatly improve the C_s because NGO makes the electrode more electrically conductive and may also increase the specific surface area and porosity. It has a great potential to sustain the C_s from 436 to 201.5 F g^{-1} while increasing the CD from 0.5 to 10 A g^{-1}. The electrodes' Cs decrease with increasing CDs, which could be due to not having enough active material for the redox reaction. The conductivity test results indicated that the R_s of NGO/$CuFe_2O_4$ (6.85 Ω) was marginally lower than that of NGO (7.25 Ω) and $CuFe_2O_4$ (7.1 Ω). Since NGO/$CuFe_2O_4$ has low resistance, it must have a high conductivity. This is because the NGO framework is bonded together, which increases the electrode material's overall conductivity. However, NGO/$CuFe_2O_4$'s R_s value was higher than that of other reported electrode materials [80,95]. In the same way, the R_{ct} values of the electrodes indicate NGO/$CuFe_2O_4$ (1.1 Ω), NGO (2.8 Ω), and $CuFe_2O_4$ (1.3 Ω). Therefore, NGOC may exhibit more capacity for charge transfer than other reactions owing to its low resistance. Figure 7d shows the capacitance retention (%) vs. cycle number profile of $CuFe_2O_4$ and NGO/$CuFe_2O_4$. This study found that increasing the cycle numbers led to a decrease in the capacitance retention of both electrode materials. However, NGO/$CuFe_2O_4$ shows high cycle stability when compared to $CuFe_2O_4$. Interlayer ferrite particles may behave in an unstable form during the charge and discharge processes, which causes $CuFe_2O_4$'s poor cycle stability and high capacitance loss. The NGO-modified $CuFe_2O_4$ particles may help to preserve their rigidity, and $CuFe_2O_4$ can inhibit the restacking of NGO sheets and produce a high-porosity medium channel for the interaction of electrolyte ions. Accordingly, over 2000 cycles, the cycle stability of NGO/$CuFe_2O_4$ (82%) was greater than that of $CuFe_2O_4$ (58.4%). Furthermore, an asymmetric device (NGO/$CuFe_2O_4$//AC) was developed that operated in the voltage range of 0.0–1.7 V. The appearance of quasi-rectangular CV curves caused NGO/$CuFe_2O_4$ and AC to exhibit PC and EDLC characteristics, respectively. The device showed a C_s of 89.1 F g^{-1} at a CD of 1 A g^{-1}. The device achieved a capacitance retention of 78.6% after 500 cycles. Furthermore, this device demonstrated its great potential in high-performance energy storage applications, as it was able to produce ED and PD of 35.7 Wh kg^{-1} and 883.0 W kg^{-1}, respectively, at a CD of 1 A g^{-1}. As seen in Figure 7e, the asymmetric device (NGO/$CuFe_2O_4$//AC) was able to light up the LED, confirming its possible use in practical applications.

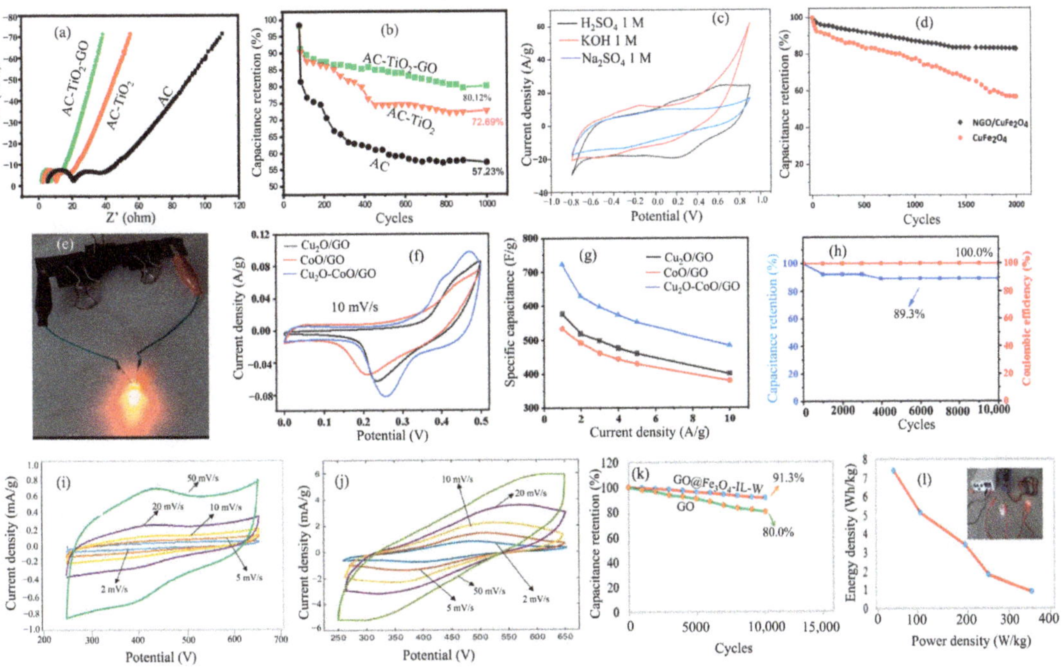

Figure 7. Electrochemical study of (**a**) EIS spectra of AC/TiO_2/GO. (**b**) Cycle stability of AC/TiO_2/GO (reproduced with permission from [80]). Copyright 2023, Elsevier. (**c**) CV curves of NGO/$CuFe_2O_4$ electrode material in different electrolytes. (**d**) Cycle stability of NGO/$CuFe_2O_4$ and others. (**e**) Two NGO/$CuFe_2O_4$//AC in series devices powering up an LED bulb (reproduced with permission from [85]). Copyright 2023, Elsevier. (**f**) CV curve of Cu_2O-CoO/GO and others at 10 mV s^{-1}. (**g**) Variation in C_s with CD of Cu_2O-CoO/GO and others. (**h**) Cycle stability of Cu_2O-CoO/GO//GO device (reproduced with permission from [95]). Copyright 2024, Elsevier. (**i**,**j**) CV curves of GO and GO@Fe_3O_4-IL@W at various scan rates. (**k**) Cycle stability of GO and GO@Fe_3O_4-IL@W. (**l**) Ragone plot and inset picture shows powering up of the LED pulp using GO@Fe_3O_4-IL@W (reproduced with permission from [58]). Copyright 2023, Elsevier.

Similarly, the hydrothermal synthesis of the hierarchical Cu_2O-CoO-modified GO electrode showed PC behavior at 6.0 M KOH due to the formation of redox peaks [95]. Figure 7f shows the electrochemical CV analysis of Cu_2O-CoO/GO with others in the voltage ranges between 0.0 V and 0.5 V. According to the CV curves, Cu_2O/GO and CoO/GO both exhibit a couple of redox peaks, while Cu_2O-CoO/GO has two pairs of redox peaks, indicating that both single-type electrode materials are participating in their pseudocapacitance processes and therefore they could pseudocapacitive materials. The GCD confirms that it exhibits the greatest C_s and excellent reversibility. Therefore, Figure 7g demonstrates the calculation of C_s at various CD plots using GCD curves. The inadequate response is caused by the redox reaction, which only happens on the electrode surface during the charging and discharging process. As the CD rises, the C_s of the electrode materials gradually decrease. However, Cu_2O-CoO/GO still exhibits high C_s when compared to others, which could be the reason for the fast response to the redox reaction. As a result, at a CD of 1 A g^{-1}, Cu_2O-CoO/GO (723 F g^{-1}) was achieved with higher C_s than Cu_2O/GO (576 F g^{-1}) and CoO/GO (533 F g^{-1}). Cu_2O-CoO/GO demonstrated good retention of capacitance, increasing the CD from 1 to 10 A g^{-1} while decreasing the CD gradually from 723 to 484 F g^{-1}. The high C_s of Cu_2O-CoO/GO could be due to its highly active sites, high conductivity, and high specific surface area, which could be attributed to the spherical kind of Cu_2O-CoO nanoparticles on GO sheets. Cu_2O-CoO/GO (0.1 Ω) has a lower resistance (R_s) than Cu_2O/GO (0.5 Ω) and CoO/GO (0.31 Ω), potentially making it a high-conductivity electrode. These resistance values have shown that Cu_2O-CoO/GO had lower resistance than other electrode materials, such as Cu_2O and CoO. This suggests that the TMO composite on GO can greatly improve the material's ability to conduct electricity while also increasing the pseudocapacitive response, which may show high electrochemical performance. Furthermore, an asymmetric device (Cu_2O-CoO/GO//GO) was developed and tested in the voltage range of 0.0–1.6 V. It showed a C_s of 125.0 F g^{-1} at a CD of 1 A g^{-1}. This was able to make a middling ED of 44.1 Wh kg^{-1} and a low PD of 794.0 W kg^{-1}. Figure 7h illustrates the cycle stability profile of an asymmetric device, demonstrating low capacitance reduction and high coulombic efficiency over 10,000 cycles, suggesting high structural stability in the electrochemical reaction. Accordingly, it kept its capacitance well (89.3% after 10,000 cycles) at a CD of 10 A g^{-1}. Another study found that the best GO/CNT/COF weight ratio (2.4: 2.1: 1.0) had a higher C_s of 544.91 F g^{-1} at 1 mV s^{-1} and 175.09 F g^{-1} at 1 A g^{-1} [97]. Due to their porous structure, high electrolyte ion adsorption could be attributed to the high C_s achievement, which may account for the CNT and GO materials' high specific surface area, while the COF contributes to their high redox activity. In general, GO and CNTs are high-conductivity materials. However, COF in the composite structure could be the reason that affects the overall conductivity of GO/CNT/COF, which is attributable to R_{ct} (11.34 Ω) and R_s (2.25 Ω), which were found to be somewhat greater than the reported literature, suggesting moderate conductivity. Because of its high structural stability in the electrolyte solution, the electrode demonstrated high cycle stability (84.76% over 8000 cycles), and there may be no volume expansion or contraction of the GO, CNT, and COF combinations. Furthermore, the developed symmetric supercapacitor (GO/CNT/COF//GO/CNT/COF) produced 24.3 Wh kg^{-1} of mild ED and 248.95 W kg^{-1} of low PD; therefore, this means that this kind of device could be used to store low amounts of energy.

Another ternary composite was investigated in 1.0 M H_2SO_4 and designated GO@Fe_3O_4-ionic liquid@Na_2WO_4 (GO@Fe_3O_4-IL@W) [58]. Figure 7i,j show the CV curves for GO and GO@Fe_3O_4-IL-W at different scan rates. According to the CV curves, when the scan rate rises, the current steadily increases, confirming the electrode materials' good capacitive performance and both electrodes' pseudocapacitive behaviors. GO@Fe_3O_4-IL@W showed higher stability and reversibility in comparison to GO, and larger CV curves were caused by the faradaic reaction. At a CD of 0.2 A g^{-1}, the high C_s were identified as GO@Fe_3O_4-IL@W (332.4 F g^{-1}). In contrast, GO showed only 122 F g^{-1}. The composite has high C_s when compared to GO because IL and tungstate (W) have qualities such as high electrical conductivity and high redox properties, whereas Fe_3O_4 nanoparticles are structurally stable but have low conductivity. Therefore, the combination of Fe_3O_4, IL, and W could enhance the electrochemical performance of GO. In addition, high Fe_3O_4-IL and W nanoparticles on the GO sheet could allow highly active sites, a high synergistic effect, and a high surface area for a high electrochemical reaction. Even with a significant fall in the electors' C_s due to increased CD, the ternary composite retained a higher retention rate than GO. It was found that the ternary composite's resistance (R_s and R_{ct}) was less than GO's due to the high conductivity of IL and W. Thus, GO@ Fe_3O_4-IL@W (5.47 and 9.35×10^{-5} Ω) and GO (7.60 and 0.002 Ω) will also be. In Figure 7k, the cycle stability profile of GO and its composite shows that the capacitance retention (%) of both electrode materials decreased with increasing cycles. However, the GO@Fe_3O_4-IL@W exhibits higher cycle performance than the GO. Accordingly, it was shown that GO@Fe_3O_4-IL@W exhibited notable cycle stability (91.3% capacitance retention over 10,000 cycles). Moreover, the capacitance retention of GO is only 80%. The high number of nanoparticles on the GO sheet, which are associated with its high chemical stability and high reversibility, may be the cause of the high cycle stability of GO@Fe_3O_4-IL@W. The GO also did well in cycles but not as well as the composite because it has less conductivity and only has a sheet-like structure, which means it may undergo some moderate structural deformation due to volume expansion and contraction. Figure 7l shows the Ragone plot, which measured the ED and PD of the GO@Fe_3O_4-IL@W. It shows that as PD increased, the electrode materials' ED decreased. The inset picture reveals that the composite electrode can light up a red LED using the cell setup. The electrode's ED and PD were determined to be very low using the three-electrode method, measuring 7.38 Wh kg^{-1} and 40.0 W kg^{-1}, respectively. These low values indicate that the electrode material may have low energy storage, but it may provide high values when it is in device form with a negative electrode (symmetric or asymmetric configuration). Similarly, the ternary alloy NiO/MnO_2/GO (NMGO) was tested in 1.0 M KOH and demonstrated PC behavior when compared to GO and MnO_2 (EDLC behavior) [83]. This means that NiO, not GO and MnO_2, can start the faradaic reaction in the electrolytic process. NMGO had a high C_s (402 F g^{-1}) compared to MGO (297.0 F g^{-1}), NiO/GO (283.0 F g^{-1}), and GO (82.0 F g^{-1}) at a CD of 1 A g^{-1}. Thus, MnO_2 and NiO have the potential to improve the electrochemical efficacy of GO. The high C_s of NMGO might also be a synergistic impact of (NiO and MnO_2) nanoparticles densely accomplished on GO surfaces covered evenly, which may increase conductivity, highly active sites, and high surface area (102 m^2 g^{-1}) when compared to merely MnO_2 and NiO_2 nanoparticles on GO. The low C_s of GO may be related to its sheet-structured shape, which is characterized by weak conductivity, low activity sites, and low surface area (94 m^2 g^{-1}). Additionally, it was observed that NMGO (34 Ω) had a lower R_{ct} value than MGO (68 Ω), NGO (87 Ω), and GO (138 Ω). It was determined that the low conductivity of GO may have contributed to the reaction's restricted charge transfer and, consequently, low electrochemical permeances. Moreover, it was observed that the ternary composite (NMGO) exhibited superior capacitance retention (93%) and cycle stability (over 14,000 cycles). The cycle stability of NMGO was good because it had dense nanoparticles on GO sheet shapes that made it more conductive, had a lot of surface area, was chemically and structurally stable, and did not expand or contract structurally during the electrochemical reaction. Additionally, in 1.0 M Na_2SO_4, an asymmetric device (NMGO//MWCNT) was fabricated and investigated. At a CD of 1 A g^{-1}, it exhibits C_s of 90.0 F g^{-1}. On the other hand, the capacitance retention of the device is low as a result of the increase in CD from 1 to 4 A g^{-1}, which gradually reduces C_s from 90.0 to 42.0 F g^{-1}. Because of its good cycle stability (retaining 88% of capacitance after 6000 cycles), the developed asymmetric device could have high chemical and structural stability in the electrolyte solution. Therefore, the high structural combination of NMGO and MWCNT devices could be very effective in electrochemical performance. The asymmetric device may have moderate energy storage because it yielded values of ED and PD, measuring 28.0 Wh kg^{-1} and 750 W kg^{-1}, respectively. Consequently, the proposed ternary composite may be utilized as a material for supercapacitors instead of the raw materials.

Redox activity (pseudocapacitive behavior) owing to Faradaic reactions occurring at the electrode/electrolyte interface was detected on the electrodes of 5%GO/WO_3,

10%GO/WO_3, and 15%GO/WO_3 composite in an electrolyte of 2.0 M KOH [96]. CV study revealed that the C_s values for 5%, 10%, and 15% GO/WO_3 were 508, 668, and 475 F g^{-1} at 5 mV s^{-1}, respectively. Increasing the Wt.% of GO increases the C_s; however, after 10wt.%, the C_s drops, which might be due to a limiting of charge transport from an electrolyte to the electrode surface. As the wt.% of GO on WO_3 nanorods increases from 5 to 10, C_s may rise. This might be because the surfaces have a large number of active sites and pores, shorten the distance for electrolyte ions diffusion, and increase the electrolyte ion transportation, resulting in rapid redox reactions. The decreasing C_s for the composites with 15% GO/WO_3 was due to a high wt.% of GO sheets restacking with WO_3, which limited the ion diffusion and conductivity, resulting in low performance. The GCD analysis confirmed that 10%GO/WO_3 (738.0 F g^{-1}) had a higher C_s than 5%GO/WO_3 (590 F g^{-1}) and 15%GO/WO_3 (341 F g^{-1}). All prepared electrodes exhibited solution resistance (R_s) values ranging from 1.05 to 2.35 Ω; consequently, the 10% GO/WO_3 electrode showed a low R_s value of 1.05 Ω, suggesting that its high electrical conductivity may contribute to its high electrochemical performances. Moreover, over 7000 cycles, the 10%GO-WO_3 maintained 88% of its charge storage, demonstrating its exceptional cycle stability. The electrode's high cycle stability may be due to the decorated WO_3 nanorod morphology with GO, which may have active sites for high interfacial connections for the electrolyte ions and chemically strong structures. Moreover, the asymmetrical device (GO/WO_3//GO-NF) was developed, exhibiting a C_s of 213 F g^{-1} at a rate of 5 mV s^{-1}. It was found that 25 Wh kg^{-1} of mild ED and 1000 W kg^{-1} of high PD. Over 3000 cycles, the device's cycle stability showed that 87% of its capacitance was still in it. Due to their structural and chemical stability, GO/WO_3 electrode materials are also capable of producing high cycle stability in device applications.

3.3. GO-Metal Sulfide

For supercapacitor applications, metal sulfides like MnS and La_2S_3 perform poorly because of their sluggish kinetics, low ED, low conductivity, and poor cyclic performance. Metal sulfides could be composited with GO to enhance electrochemical performance. Thus, Mane et al. investigated the electrochemical performance of MnS, MnS-La_2S_3, and MnS-La_2S_3/GO in a 1.0 M Na_2SO_4 [100]. In Figure 8a, the CV curve for MnS-La_2S_3/GO was wider than the others, and it looked like a quasi-rectangular shape. At 100 mV s^{-1}, the PC behavior was found for all electrodes. Through GCD analysis, it was discovered that, at a CD of 2 A g^{-1}, the C_s for MnS-La_2S_3/GO (890 F g^{-1}) was greater than that of MnS-La_2S_3 (694 F g^{-1}) and MnS (666 F g^{-1}). The sheet-like MnS-La_2S_3 particles on GO morphology could have a high porosity, high surface area, highly active sites, high conductivity, and high synergistic effect, which may be ascribed to shorter ion path and ease of electrolyte ion adsorption and make high quicker to boost the electrochemical performance and thus they obtained high C_s than other electrode materials. Regarding MnS-La_2S_3, their morphology resembles cotton candy, and they may have a good surface area, porosity, active sites, conductivity, synergistic effect, and moderate ion route that contributes to moderate electrochemical performance and C_s. Further, no significant morphology was observed on MnS. Its low surface area, low porosity, low-activity sites, low conductivity, and lack of a synergistic effect may be attributed to its higher ion path, which results in slower ion transportation. This, in turn, leads to a low electrochemical performance and low C_s. Moreover, as the CD was increased from 2 to 5 A g^{-1}, the C_s of the electrodes MnS-La_2S_3/GO (890 to 545 F g^{-1}), MnS-La_2S_3 (694 to 415 F g^{-1}), and MnS (666 to 320 F g^{-1}) exhibited a gradual decrease. This can be attributed to the fact that at low CD, ions can penetrate extensively into the electrode surface, given sufficient time to do so. In contrast, migration of ions into the electrode surface is restricted at higher CD values due to time constraints. Figure 8b shows Nyquist plots of various electrode materials to study their resistances, and the plots show electrodes with semi-circular curves in the higher frequency region. The straight line in a low-frequency zone is caused by diffusive resistance, which acknowledges short ionic routes. This is advantageous for rapid ion transportation and great electrochemical performance. The electrodes' diameter of the semi-circle equals the actual axis of the Nyquist plot and the R_{ct} of the electrodes caused by the reversible redox reaction on their surfaces. Subsequently, the resistance values (R_s and R_{ct}) of MnS-La_2S_3/GO (0.6 and 18.2 Ω cm^{-2}) were found to be poorer than those of MnS-La_2S_3 (0.9 and 39.7 Ω cm^{-2}) and MnS (1.1 and 80.4 Ω cm^{-2}). As a result, a charge-transfer mechanism that is more rapid than others may be feasible on the MnS-La_2S_3/GO. The MnS-La_2S_3/GO material exhibited remarkable cycle stability, retaining 89% of its capacitance over 4000 cycles. The high cycle stability could be due to the electrodes' composite nanosheet structure, which works well together and maintains shape during electrochemical reactions and cycles. It is also possible that the electrodes had good conductivity, had no binder on their surface, had a shorter ion path, and were highly porous with active sites. Moreover, an investigation was conducted on the symmetrical device (MnS-La_2S_3/GO//MnS-La_2S_3/GO) using a PVA-Na_2SO_4 electrolyte with an operational voltage range of 0.0–1.8 V. A high C_s of 151 F g^{-1} was produced by the device at a CD of 7 A g^{-1}. A device was discovered to exhibit high cycle stability, with capacitance retention of 92.5% over 10,000 cycles. The developed electrode materials exhibit high cycle stability in three and two electrode setups, potentially due to their high chemical stability and lack of structural degradation during the cycle test. It also has been determined that the improved interface between the MnS-La_2S_3/GO and PVA-Na_2SO_4 gel electrolyte contributes to excellent cyclic stability. Figure 8c depicts the Ragone plot of the developed symmetric device, which represents that the device's ED decreased gradually with an increase in PD. This device achieved moderate ED and PD values of 54.2 Wh kg^{-1} and 1300 W kg^{-1}, respectively. To implement a real-time application, two symmetric devices were linked in a series of LEDs. Figure 8d shows a picture of the LED setup panel at two different discharge times (30 s and 150 s), along with two symmetric devices (MnS-La_2S_3/GO//MnS-La_2S_3/GO) that were connected in series. Each device was charged at 3.2 V for 30 s and discharged for 150 s via the LEDs' illumination. As a result, the fabricated symmetrical device can be utilized in industrial and real-time settings.

In another study, NiS/CoS/GO was investigated for supercapacitors in 3.0 M KOH and demonstrated PC behavior [98]. At a CD of 5 A g^{-1}, high C_s (1114 F g^{-1}) were observed. This high accomplishment by NiS/CoS-modified GO might be highly pours morphology, homogeneous particle distribution on GO sheets, structurally stable, highly active sites, high conductivity, and a very shorter ion route that may enable very faster ion transmission for high C_s. Resistance (R_{ct}) was found to be only 1.99 Ω cm^2, and therefore, it may have had a high charge-transfer process in an electrolyte. High cycle stability also revealed that 80% of capacitance was retained after 8000 cycles. Decorating CoS and NiS particles around GO might make the structure more stable, increase the surface area, and create more active sites. This could make it easier for high-ion movement during the cycle test's electrochemical reaction. Therefore, they could maintain stability for a long time. However, single compositions such as CoS/GO, NiS/GO, and GO show lower cycle stability when compared to CoS/NiS/GO. There may be structural volume changes during the charge–discharge test that contribute to low chemical stability. When all materials combine to enhance both structural and chemical stability, this could potentially lead to high cycle stability. The higher the ED, the higher the device's PD, which may be a high-energy storage material. Therefore, this CoS/NiS/GO electrode demonstrated high ED and PD values of 674.12 Wh kg^{-1} and 8089.44 W kg^{-1}, respectively. However, the device preparation for practical application was not found in this study, and the actual ED and PD can be measured from either symmetrical or asymmetrical devices.

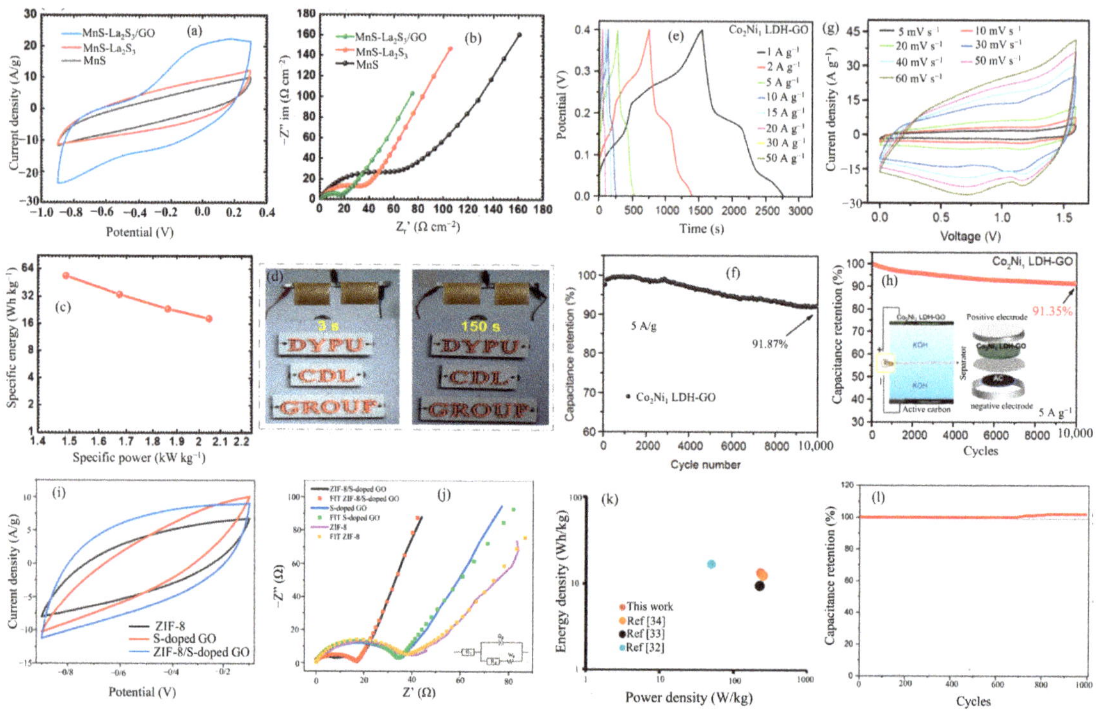

Figure 8. Electrochemical analysis of (**a**) CV curves of MnS-La_2S_3/GO at 100 mV s^{-1}. (**b**) EIS spectra of MnS-La_2S_3/GO. (**c**) Ragone plot of MnS-La_2S_3/GO//MnS-La_2S_3/GO device. (**d**) 211 LEDs were turned on using MnS-La_2S_3/GO//MnS-La_2S_3/GO device that had been charged for 30 s (photo image of after 3 s) and discharged (photo image of after 150 s) (reproduced from [100]). Copyright 2023, Elsevier. (**e**) GCD plots of Co_2Ni_1-GO. (**f**) Cyclic stability of Co_2Ni_1-GO. (**g**) CV curves of Co_2Ni_1-GO//AC device. (**h**) Cycle stability of device at 5 A g^{-1} (reproduced from [106]). Copyright 2023, Springer Nature. (**i**) CV curve of ZIF-8/SGO and others. (**j**) EIS spectra of ZIF-8/SGO and others. (**k**) Ragone plot of ZIF-8/SGO electrode and (**l**) cycle stability of ZIF-8/SGO (reproduced from [114]). Copyright 2023, John Wiley & Sons, Inc.

3.4. GO-Transition Metal Chalcogenides

Transition metal chalcogenides (TMDs) have a 2-D sheet-like layered structure, a large surface area, good redox properties, and low electronegativity, among other characteristics. However, the poor electrical conductivity and high-volume variations make it unsuitable for commercial uses. Thus, it might be used with high electrical conductivity-based materials to boost electrochemical performance. In addition, adding sulfur (S) and selenide (Se) compounds to the metals improved charge transfer between the metals and carbon-based materials. Yasoda et al. have investigated a GO composite with MnSSe (GO-MnSSe) in a 1.0 M KCl solution for supercapacitor applications [104]. As a result, EDLC and PC behavior were identified for the GO-MnSSe, while GO only demonstrated EDLC behavior. As a result of MnSSe's redox properties, a faradaic reaction occurred when GO-MnSSe was used. The C_s of the GO-MnSSe (300 F g^{-1}) was found to be high at 10 mV s^{-1}, and it showed excellent reversibility since the C_s were highly sustained even at a high scan rate of 100 mV s^{-1}. The GCD study revealed EDLC and PC in GO-MnSSe, with a C_s of 603 F g^{-1} at 0.1 A g^{-1}. The cycle stability was modest, with 67% capacitance retention after 9000 cycles. The insufficient cycle stability of GO may account for its poor chemical stability and structural deterioration, as the decorated MnSSe nanoparticles did not succeed in improving the functionality of GO on account of their uniform and limited layering on GO sheets. Consequently, the potential benefits of adjusting the ratio of MnSSe to GO include better electrochemical performance and cycle stability. This research did not include an investigation of the two-electrode setup. The research found that using a three-electrode setup for supercapacitor applications yielded superior results due to high C_s and cycle stability.

3.5. GO-Layered Double Hydroxides

LDH has a layered structure with a large surface area, allowing for numerous active sites and strong redox characteristics. LDHs' layered structure with large layer spacing promotes ion insertion and deintercalation, which aids in C_s improvement. However, LDHs have poor electrical conductivity and sluggish ion diffusion, which may compromise cycle stability [121]. It may be feasible to improve the conductivity and other characteristics by combining carbon-based materials. Therefore, Zhao et al. developed an optimal ratio of Co_2Ni_1 LDH composite with GO (Co_2Ni_1-GO) for the electrochemical study in 3.0 M KOH [106]. The electrode exhibited PC behavior due to the redox reaction caused by LDH in the electrolytes. The GCD curves of Co_2Ni1 LDH-GO's charging and discharging concerning potential vs. time are shown in Figure 8e. These curves show that the electrode's discharging time drastically decreased as the CD increased. However, it retains an acceptable symmetry curve, strong charge–discharge properties, and reversibility while increasing the CD. At 0.1 A g^{-1}, the electrode Co_2Ni_1-GO had a very high C_s (3317.5 F g^{-1}), whereas bare Co_2Ni_1 exhibited only 793 F g^{-1}. LDH nanosheets on GO layers might improve the shorter ion route, resulting in much quicker ion transport and the ability to store more charges. The significant storage charges may be due to the LDHs-GO combination having a high surface area, porosity, active sites, conductivity, structural stability, and chemical stability when compared to bare LDHs. GO might be a useful adjunct electrode material for LDH to enhance its C_s for supercapacitor applications. Furthermore, the produced composite electrode may have great conductivity owing to its very low resistance. The Co_2Ni_1-GO electrode has R_s and R_{ct} values of 0.88 and 0.029 Ω, respectively. In contrast, bare Co_2Ni_1 LDHs exhibited a higher resistance than Co_2Ni_1-GO, which may imply poor conductivity. Figure 8f depicts Co_2Ni_1-GO's cycle stability profile, which plots capacitance retention (%) against cycle number at a CD of 5 A g^{-1}. It shows that the capacitance retention (%) decreases with increasing cycle number. However, Co_2Ni_1-GO has very good cycle stability, with 91.8% capacitance retention after 10,000 cycles. Both sheets (Co_2Ni_1-GO LDH and GO) in composite form may have excellent structural stability, highly active sites, mechanical strength, and chemical stability, and, therefore, an excellent response to the charging–discharging cycle test. The asymmetric device (Co_2Ni_1-GO//AC) was developed for practical use and functioned throughout a voltage range of 0–1.6 V. When the scan rate (5–50 mV s^{-1}) is increased, the CV curves exhibit high symmetry, indicating that the device can transport electrons effectively as shown in Figure 8g. The device achieved a C_s of 328.7 F g^{-1} at a CD of 1 A^{-1} and maintained a high C_s (114.2 F g^{-1}) at a CD of 50 A g^{-1}. Figure 8h shows the cycle stability profile of an asymmetric device by plotting capacitance (%) against cycle numbers. As the cycle numbers increase from 0 to 10,000, the capacitance (%) slowly decreases. The device demonstrated 91.3% cycle stability over 10,000 cycles at a CD of 5 A g^{-1}. Furthermore, the device's ED and PD were found to be high, at 94.7 W h kg^{-1} and 750.0 W kg^{-1}, respectively. According to its high C_s, high cycle stability, and high ED, Co_2Ni_1-GO proved to be a potential candidate for energy storage.

3.6. GO-Organic Materials

MOFs have a large surface area and active surface; however, their poor conductivity limits their electrochemical effectiveness [122]. Additionally, sulfur doping into GO (SGO) impacts graphene's electroneutrality and develops charged ion sites, resulting in increased electrical conductivity. Recently, ZIF-8 MOFs were composited with SGZ (ZIF-8/SGO) for supercapacitor applications in 6.0 M KOH [114]. Figure 8i represents the CV analysis of ZIF-8, SGO, and ZIF-8/SGO (SGZ) electrodes at a scan rate of 100 mV s^{-1}. According to the CV plots, SGZ showed PC behavior due to sulfur-doped GO contributing to the redox reaction, and its CV rectangular curve was steep. ZIF-8 and SGO, on the other hand, demonstrated EDLC and PC tendencies, respectively, as shown by small CV rectangular and quasi-rectangular curves. However, the synergistic effect of SGZ showed a high area CV curve when compared to ZIF-8 and SGO, indicating that it may have high electrochemical properties that enhance the electrochemical reaction for high energy storage. At the CD of 1 A g^{-1}, the SGZ (261 F g^{-1}) had greater C_s than ZIF-8 (150 F g^{-1}) and SGO (246 F g^{-1}). However, the C_s of the SGZ (298–80 A g^{-1}) rapidly reduced when the CD increased from 0.5–8.0 A g^{-1}. This electrode material of moderate C_s and high reduction of C_s at elevated CD could be due to morphological and purity issues with SGZ. Because no significant morphological structure was found, it may have low porosity, low-activity sites, and moderate conductivity, which could contribute to moderate electrochemical performance. Furthermore, the R_{ct}-related semicircle in the high-frequency area assigned to the redox reactions and the straight line in the low-frequency region represented by Warburg impedance are attributed to the electrolyte diffusion within the SGZ electrode, as shown in Figure 8j. SGZ exhibited lower resistance (R_{ct}) (14.7 Ω) than SGO (29.6 Ω) and ZIF8 (35.4 Ω). Thus, SGZ may have higher conductivity than SGO and ZIF-8. However, these resistance values are relatively high compared to the other known electrode materials [96,102,106,113]. The poor conductivity of the ZIF-8 MOF could reduce the total composite electrode material's electrical conductivity, affecting electrochemical performance. Figure 8k depicts Ragone plots of SGZ vs. other reported electrode materials, demonstrating that SGZ has more ED than the others. However, the electrode may be a low-energy-storage electrode material since its high resistance caused it to exhibit poor PD (231 W kg^{-1}) and ED (13.4 Wh kg^{-1}) values. As shown in Figure 8l, there was no loss in capacitance retention (%) up to 700 cycles; However, after 1000 cycles, the capacitance (%) rose to 102%, owing to the SGZ sites' self-activation over time. The good cycle stability of SGZ was shown, which might be attributed to chemical stability. Due to its high resistance, moderate C_s, and cycle stability, SGZ may not be ideal for high-energy-storage applications, and it may be necessary to optimize the ratio of ZIF-8 and SGO for the efficient composite to enhance the electrochemical characteristics. The two-electrode setup has not been found in this investigation.

Similarly, neodymium-MOF-modified GO (Nd-MOFs/GO) was investigated for supercapacitor applications in 3.0 M KOH [113]. In this study, Nd-MOFs and Nd-MOFs/GO exhibited EDLC behavior; however, the rectangular CV curve was larger for Nd-MOFs/GO; therefore, the composite may have high conductivity and electrochemical performance. Accordingly, Nd-MOFs/GO may have greater C_s than Nd-MOFs themselves. Modified MOFs (677.6 F g^{-1}) demonstrated greater C_s than Nd-MOFs (11.1 F g^{-1}) at a scan rate of 10 mV s^{-1}. The GCD also revealed that modified MOFs had larger C_s (633.5 F g^{-1}) than Nd-MOFs (11.3 F g^{-1}). Despite having highly active sites, the nanorod morphology of Nd-MOFs may contribute to their low conductivity, resulting in a very low C_s value. Conversely, composites containing GO can achieve high C_s because of their highly active sites, high porosity, high conductivity, and synergistic effect, which accelerates the electrochemical reaction through faster ion transportation. However, when exploring the morphology of Nd-MOFs/GO, the nanorod structure of MOFs was found to be loosely associated with GO, and the composition of GO was found to be higher than that of Nd-MOFs. As a result, GO is responsible for the majority of the outstanding electrochemical performance. As a result, GO might be used in tandem with Nd-MOFs to store large amounts of electrochemical energy. The modified MOFs demonstrated high cycle stability (88.7% capacitance retention over 4000 cycles), and the resistance (R_s) of modified MOFs (0.8 Ω) was lower than Nd-MOFs (0.92 Ω). As a result, modified MOFs may have very high conductivity, allowing for improved electrochemical performance. This finding indicates that Nd-MOFs/GO is a long-lasting electrode material in the electrolyte solution that can successfully serve as a super-capacitive electrode that has high cycle stability. The electrode setup resulted in high ED (31.4 Wh kg^{-1}) and PD (1500 W kg^{-1}), and these values indicate that this composite electrode material may be useful in energy storage.

Another study examined the optimized Ni-BTC MOFs/GO 2 composite in 3.0 M KOH [102]. In this case, the Faraday reaction of the nickel (Ni^{2+}) ion in the electrolyte caused the GO composite with MOFs to exhibit PC behavior. Using GCD, Ni-BTC@GO 2 demonstrated a high C_s of 1199.0 F g^{-1}, whereas Ni-BTC only showed 625.6 F g^{-1} at a CD of 1 A g^{-1}. Therefore, incorporating GO into MOFs may be a viable way to improve the electrochemical reactivity of MOFs. It was found that the electrochemical performance of MOFs was not dependent on morphology but rather on particle size, conductivity, and crystallinity. As a result, MOFs composited with GO had high C_s, but bare MOFs exhibited poor C_s despite having an octahedral shape and micron-sized particles. It is also possible that the MOF's low performance was due to their bulk structure, which contributed to their poor electrochemical performance. As a result of their high conductivity and crystalline structure, the thin layer of GO sheets over micro-sized octahedral MOFs had the greatest contribution to electrochemical performance. However, when a large number of GO aggregates on MOF surfaces, the conductivity may be reduced. The Ni-BTC MOFs/GO 2 were able to retain high capacitance (1199 to 676 F g^{-1}) when the CD was increased from 1 to 20 A g^{-1}. Because the resistance of R_s and R_{ct} was discovered to be 0.79 and 7.0 Ω, the modified MOFs may have a moderate conductivity, whereas the naked Ni-BTC MOFs may have a poor conductivity material because their resistance was 0.66 and 7.64 Ω. As a result, the higher conductivity of Ni-BTC/GO may serve as a better electrode material for energy storage than bare MOFs. The Ni-BTC MOFs/GO 2 showed high cycle stability, with 84.47% capacitance retention even after 5000 cycles. The high cycle stability was due to the GO sheets improving the structural properties of Ni-BTC, which increased the contact region between the electrode and the electrolyte, reducing the ion diffusion path and retaining the activated sites of Ni-BTC MOFs for energy storage during the electrochemical reaction. With the assistance of GO functionalization on MOFs, the electrochemical reaction between electrolyte ions reduces the dimension change of Ni-BTC MOFs, which improves the cycle stability of composite electrode materials. Additionally, during operation in the 0–1.6 V range, the asymmetric device (Ni-BTC MOFs/GO 2//AC) demonstrated a moderate ED of 42.89 Wh kg^{-1} at a PD of 800 W kg^{-1}. Over 10,000 cycles, this device was able to achieve 70% cycle stability in capacitance retention. The Ni-BTC MOFs/GO 2 have shown promising properties as an electrode material for a device application, including good conductivity, high-electrolyte contact surfaces, and high stability. When compared to other MOF/GO composites, the findings suggest that Ni-BTC MOFs/GO 2 might be an exceptional electrode.

In the electrolyte of 1.0 M H_2SO_4, the bioinspired organic molecule ABQA-modified GO/carbon paper (ABQA-GO/CP) showed both PC and EDLC behavior [101]. On the other hand, ABQA and GO/carbon paper, respectively, were the cause of the PC and EDLC behaviors. The ABQA-GO/CP electrode performs well because of the synergistically high charge transfer that results from the covalent binding of GO and ABQA. Accordingly, by using CV measurement, the maximum C_s of the ABQA-GO/CP was identified to be 203.2 F g^{-1} at 5 mV s^{-1}, while GCD revealed 235.5 F g^{-1} at 0.5 A g^{-1}. The rising scan rates and CDs were 203.2 to 7.8 F g^{-1} and 235.5 to 205.5 F g^{-1}, respectively, from 5 to 500 mV s^{-1} and 0.5 to 10 A g^{-1}. The findings reveal that raising the scan rate did not sustain the electrode's C_s, while increasing the CD accomplished it. The cause might be insufficient time for ion diffusion from the electrolyte to the electrode surface at a higher scan rate or CD. In contrast to other composites consisting of GO-organic materials, the electrochemical performance of this organic molecule functionalized GO/CP was notably poor, as evidenced by its low capacitance (C_s), increased scan rate, and CDs that were incapable of sustaining its initial capacitance. The potential cause for the collapsed or deformed morphology of GO following ABQA functionalization is the presence of low-activity sites, fewer pores, reduced ion diffusion, and the potential for delayed ion transportation, all of which contribute to diminished electrochemical performance. The designed symmetric device (ABQA-GO/CP//ABQA-GO/CP) was evaluated in the range of −0.2 to 1.2 V and indicated lower C_s than the three-electrode system. Accordingly, the GCD investigation revealed 134.1 F g^{-1} at 0.5 A g^{-1} but with good capacitance retention (113.4 F g^{-1}) even at 10 A g^{-1}. On the other hand, the device's organic molecule-functionalized GO/CP exhibits a lower resistance, potentially enhancing its electrochemical performance. Accordingly, the device's low resistances (R_s and R_{ct}) of 0.63 and 0.79 Ω cm^2 suggest the potential for excellent conductivity. The device achieved high ED and PD of 32.8 Wh kg^{-1} and 1256 W kg^{-1}, respectively, at 0.5 A g^{-1}. This device demonstrated good cycle stability, retaining 106% of capacitance over 5000 at 5 A g^{-1}. The ABQA-GO/CP electrodes may have good chemical stability, in addition to the combined contribution of the faradaic redox reaction by ABQA and EDLC by GO, which may account for the overall symmetric device performance. As a result, the organic material composite containing GO demonstrated excellent electrochemical performance, making it potentially valuable in industrial applications. Overall, the MOF performance.

3.7. GO-MXene

Laser-induced graphene (LIG) has been consistently investigated recently to analyze graphene's scalability in energy storage applications. LIG has found applications in various domains, including energy storage devices and wearable sensors, owing to its elementary fabrication process. Electrodes based on LIG have a more porous structure and superior conductivity. Despite this, supercapacitors continue to struggle with high capacitance issues as a result of the porous structure of laser-induced graphene electrode materials, which restricts conductivity and is inadequate for ion transport. Therefore, to improve electrochemical performance, high-conductivity composites should be used to increase the conductivity of laser-induced graphene. As a result, Fu et al. designed the self-assembled MXene-GO composite on polyamide film-enhanced LIG (LIG-C) to investigate electrochemical energy storage in the H_3PO_4-PVA electrolyte [118]. They also studied LIG on bare polyamide (LIG-PI) surfaces for comparison. Accordingly, the electrodes based on LIG-PI and LIG-C exhibited high capacitance retention of 98.75% and 102.4%, respectively, over 1000 cycles, indicating excellent stability. They also tested that the LIG-C film electrode was capable of showing a high capacitance retention of 98.9% even after 6 months. The MXene-GO uniformly drop-coated polyamide film could be the reason for its chemical stability due to its high resistance to chemicals and high structural stability. They also use LIG technology to shape the film electrode, which has a graphene structure that is made up of very tightly stacked layers with porous structures that can effectively interact with many electrolyte ions. During the cycle test, the film structure may not be involved in volume expansion and contraction, thus achieving its high cycle stability. In contrast, the resistance values of these electrode materials are significantly higher than those of other GO composites. As a result, LIG and LIG-C have respective R_s values of 242 and 182 Ω. These high resistivity values may be due to the polyamide film's insulating properties, which could make MXene and GO less conductive. Consequently, the overall conductivity of the film is very low. However, LIG-C film has a lower resistivity than LIG-PI film. This might be because laser-induced graphene technology creates unique atomic structures of graphene on MXene-GO-polyamide film, which makes it possible for it to have low resistivity. Despite the use of LIG, there are still challenges to obtaining a low resistivity for the films, which may affect their electrochemical performance. Additionally, the LED lights were illuminated by the series-connected optimized, integrated device operating within the voltage range of 0.8–2.4 V. Similarly, utilizing CO_2 laser-induced patterning, a self-assembled MXene-GO electrode was fabricated and investigated in H_3PO_4-PVA electrolyte [103]. In contrast to MXene, the MXene-GO electrode exhibited elevated integral CV curves, which resulted in correspondingly lower C_s for MXene. The capacitance retention of this composite electrode remains constant at 92.6% for 1000 cycles. The elevated capacitance retention observed in the MXene-GO can be attributed to the efficient ion transport facilitated by the GO composite. In addition, it may be because the MXene-

GO electrode has a highly uniform film coating, is structurally and chemically stable, has conductivity, has many active sites, and has channels for moving high ions. Moreover, an MXene-GO (planar 3) supercapacitor device was devised, demonstrating exceptional capacitance retention (94.8%) throughout 1000 cycles. The prepared MXene-GO film proved capable of maintaining high cycle stability in device applications, which could be the reason for uniform film morphology, structural stability, and chemical stability. The results of this investigation demonstrated that an MXene-GO composite featuring optimized mass ratios exhibited superior ion transportation and higher C_s compared to bare MXene. The LIG or other laser-induced method for etching the film electrodes for supercapacitors may be used in flexible or other electronic applications. Better electrochemical performances, however, need to be optimized in terms of the material ratio, film morphology, and film structure to increase porosity and conductivity.

In 1.0 M H_2SO_4, PC behavior was observed on CTAB-functionalized GO@polyaminoanthraquinone-MXene (CGO/PDAAQ-MXene) film [117]. According to Figure 9a, the CV curves of MXene and CGO/PDAAQ films were substantially identical, indicating that they exhibited excellent capacitance properties. However, CGO/PDAAQ produced a slightly larger curve, indicating that it may function better than MXene. On the other hand, the strong synergistic impact of CGO/PDAAQ-MXene might be a greater electrochemical performance with quite a larger CV integral area than MXene and CGO/PDAAQ-MXene. Consequently, the composite electrode's C_s could be higher than that of MXene and CGO/PDAAQ. The authors claimed that the elevated capacitance was a result of the substantial electron transfer facilitated by benzoquinone in PDAAQ, followed by the MXene contribution. At 1 A g^{-1}, the C_s of CGO/PDAAQ-MXene was determined to be 346 F g^{-1}, while CGO/PDAAQ exhibited 183 F g^{-1}. The high C_s are also due to the morphology of the composite film, which shows a high number of nanotubes (PDAAQ) with sheets (CGO) on MXene film. Rough surfaces may have highly active sites and high porosity, allowing for high electrolyte ion contact and shorter ion paths for faster ion transportation for high energy storage. The C_s may be contingent on the electrodes' conductivity. Nevertheless, the resistance (R_{ct}) of the CGO/PDAAQ-MXene (1.15 Ω) was higher than that of MXene (0.89 Ω) in this study. The poor conductivity of the composite film might be due to a difference in the electronic structure of the materials that were mixed to produce a composite form, resulting in conductivity variations. Figure 9b depicts the cycle stability profile of a composite film electrode, illustrating capacitance retention (%) and coulombic efficiency vs. cycle numbers. It shows that increasing the cycle numbers from 0 to 5000 steadily decreases capacitance retention while maintaining coulombic efficiency. Accordingly, CGO/PDAAQ-MXene demonstrated exceptional cycle stability, retaining 83% of its capacitance and 100% of its coulombic efficiency after completing 5000 cycles. The chemical and structural stability of CTAB-modified GO functionalized with conducting polymer (PDAAQ) decorated on MXene structures may be attributed to its high cycle stability. An additional asymmetric device (CGO/PDAAQ-MXene//rGO) was utilized for electrochemical performance at 1.8 V. At 0.5 A g^{-1}, the device's C_s was measured at 114 F g^{-1}, and it maintained capacitance retention of 70.3% over 10,000 cycles. The Ragone plots of ED and PD values were calculated from the GCD curves for this asymmetric device and are shown in Figure 9c. The CGO/PDAAQ-MXene/rGO comparison with reported devices shows these plots. As the device's ED decreased, there was an increase in the PD. Generally, supercapacitors have such high PD that they can release energy rapidly, but due to their low ED, they store only low energy. Therefore, the research could focus on improving the ED of the supercapacitor by making an effective electrode design and structure for more energy storage to use fast charging and discharging. The device observed a low PD of 404 W kg^{-1} and a high ED of 41 Wh kg^{-1}, and this device shows higher values compared to other reported devices. Furthermore, Figure 9d depicts the LED function achieved by connecting two devices in series, which may light up a series of LED bulbs. Therefore, based on the C_s, cycle stability, and device application, this CGO/PDAAQ-MXene film may be a suitable candidate for high-energy-storage supercapacitor applications.

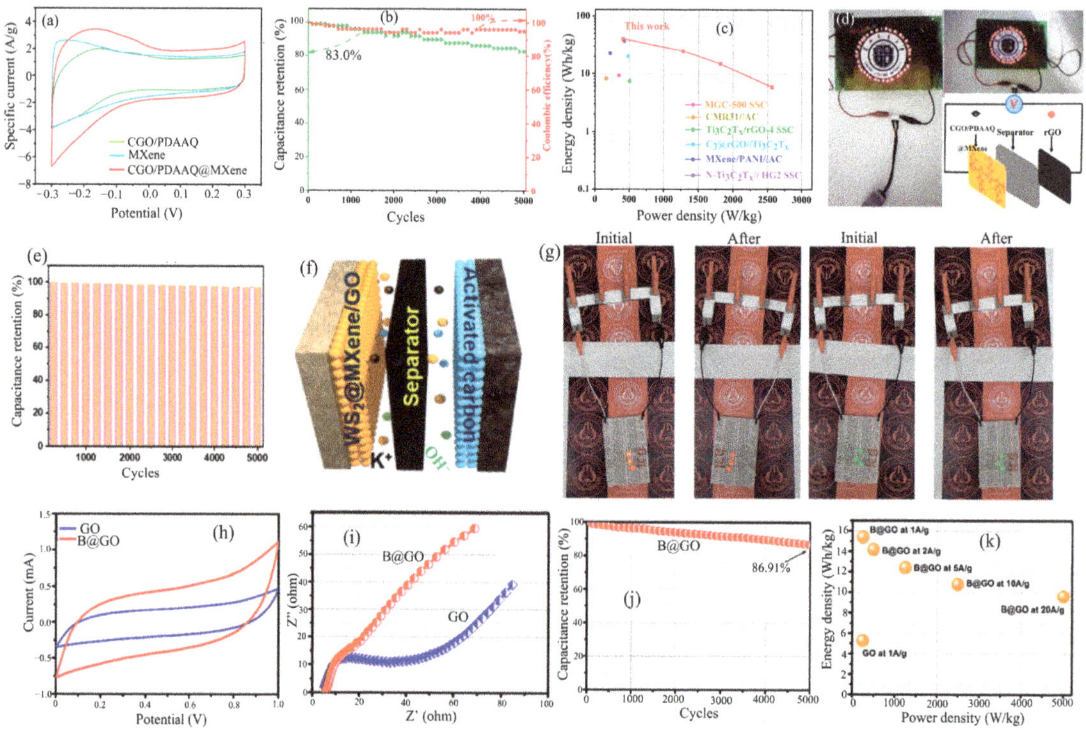

Figure 9. (**a**) CV curves of CGO/PDAAQ@MXene and other electrodes at 10 mV s^{-1}. (**b**) Cyclic stability of CGO/PDAAQ@MXene. (**c**) Ragone plot of the CGO/PDAAQ@MXene//rGO device. (**d**) Powering up of the LED light by CGO/PDAAQ@MXene//rGO device (reproduced with permission from [117]). Copyright 2023, Elsevier. (**e**) Cycle stability of WS_2@MXene/GO electrode at 10 A g^{-1}. (**f**) Schematic diagram of the developed WS_2@MXene/GO//AC device. (**g**) Initial and after 20 min lighting of red and green LEDs by the WS_2@MXene/GO//AC device (reproduced with permission from [119]). Copyright 2023, Elsevier. (**h**) CV curve of GO and BGO. (**i**) EIS spectra of GO and BGO. (**j**) Cycle stability of BGO. (**k**) The Ragone plot of BGO (reproduced with permission from [120]). Copyright 2023, Elsevier.

The electrochemical use of MXene has been severely hampered by several issues, including restacking due to weak hydrogen bonding and van der Waals forces, as well as oxidation weaknesses in colloidal solutions [119,123,124]. Furthermore, tungsten sulfide (WS_2) exhibits inadequate ion transport, low electrical conductivity, and deficient C_s. Carbon-based materials, characterized by high electrical conductivity, excellent active sites, and high structural stability, ought to be incorporated into MXene and WS_2 composites to mitigate these concerns and attain desirable electrochemical properties. Therefore, Hussain et al. synthesized a WS_2-MXene hybrid composite (WS_2@MXene/GO) containing GO for an electrochemical investigation in 1.0 M KOH [119]. This composite material exhibited battery-like activity, as determined by CV analysis. The C_s of the WS_2@MXene/GO electrode (1111 F g^{-1}) was found to be higher than that of MXene (270 F g^{-1}), GO (138 F g^{-1}), WS_2 (229 F g^{-1}), and WS_2@MXene (852 F g^{-1}) at a CD of 2 A g^{-1}. Comparatively, WS_2@MXene/GO exhibited the highest electrochemical performance because GO functioned exceptionally well with the WS_2@MXene composite. Furthermore, the composite material exhibits a high conductivity due to sufficient exposure to electroactive sites, elevated ion diffusion, and high porosity. Although the composite material composition (WS_2@MXene/GO) lacked distinct morphology, exhibited low crystallinity, and had a small surface area (12.9 m^2 g^{-1}), it achieved a high C_s, which could potentially account for its low resistance, highly active sites, and structural stability. Even though the CD increased from 2 to 20 A g^{-1}, the C_s remained high at 601 F g^{-1}. Additionally, the electrode's conductivity may be quite high as a result of the comparatively low resistance values (R_s and R_{ct}, 2.3 and 0.41 Ω, respectively). The cycle stability profile of WS_2@MXene/GO is illustrated in Figure 9e. From 0 to 5000 cycles, the capacitance retention (%) decreased very slightly. Accordingly, the WS_2@MXene/GO composite maintained 97.15% of its capacitance over 5000 cycles. The material composition of the composite contributed to its long-cycling stability and facilitated rapid ion transport in the electrolyte. Figure 9f depicts a schematic representation of the structural arrangement of an asymmetric device with anode and cathode electrodes of WS_2@MXene/GO and activated carbon (AC), respectively, sandwiched by a KOH-dipped Whatman filter paper as a separator. An asymmetric device (WS_2@MXene/GO//AC) was built and examined in 1.0 M KOH, showing a C_s of 320 F g^{-1} at a CD of 2 A g^{-1}. Furthermore, this asymmetric device achieved high ED and PD values of 95 Wh kg^{-1} and 1000.4 W kg^{-1}, respectively. The device also demonstrated high cycle stability, with a capacitance retention of 93.1% after 15,000 cycles. For practical application, Figure 9g depicts the initial and after 20 min of red and green LED illumination by five serially connected WS_2@MXene/GO//AC devices. This serially connected asymmetric device was tested to light up the red and green 4 V LEDs for 20 min after charging for 40 s. Finally, the GO composite with MXene-based electrode films or materials demonstrated very high cycle stability owing to their composition and structural stability during charging–discharging cycles, as well as high ED. Because the film electrode is coated over insulated materials like glass, polyamide, and other low-conductivity polymer films, its conductivity is lower than that of powder-formed electrodes. As a result, these substrates may influence the conductivity of active materials on them. Thus, the powder form of electrode materials exhibits strong conductivity without the assistance of base substrates. Thus, the powder electrode material form of GO-MXene may provide higher performance than film-type GO-MXene-based electrodes.

3.8. GO-Bacteria

The robust cell walls of bacteria enable them to endure a wide variety of hostile environments. They are renewable resources that are inexpensive, abundantly available, and environmentally friendly. Consequently, microorganism-based materials have gained prominence as appealing biological substances for supercapacitor applications in recent times. Also, a bacterial cell culture derived from Bacillus subtilis was employed as the N_2 source for the electrode material [125]. In light of the merits associated with this biological material, Verma et al. investigated the electrochemical properties of a Bacillus subtilis@GO composite (BGO) [120]. Figure 9h depicts the CV curves of GO and BGO, indicating that the rectangular form area of B@GO was greater than GO. As a result, BGO might have a stronger electrochemical performance than GO. Thus, the C_s of GO were observed to be improved after its composite with Bacillus subtilis. As a result, B@GO (111 F g^{-1}) exhibited higher C_s than GO (38.1 F g^{-1}). High capacitance values of bacteria composites with GO may be attributable to the presence of bacterial heteroatoms (N), which contribute to surface area and sufficient active sites. However, the weakened sheet shape of BGO continues to offer a highly active site for electrochemical reactions. Furthermore, Figure 9i depicts Nyquist plots of GO and B@GO, with straight and sloping lines in the high- and low-frequency ranges. With a small slope and vertical line, the BGO electrode may have capacitive performance and low charge-transfer resistance. With a steeper slope and a less vertical line, GO electrodes demonstrate poor capacitive performance and high charge-transfer resistance. Accordingly, it was observed that the charge-transfer resistance of B@GO (29.2 Ω) was comparatively lower than that of GO (47.6 Ω), indicating that B@GO potentially possesses a good conductivity. As a result, the bacterial solution may have strong conductivity properties that may boost the conductivity of GO. Figure 9j displays the cycle stability profile of B@GO, showing that capacitance retention (%) steadily declines as the number of cycles increases from 0 to 5000. According to the cycle stability profile, B@GO demonstrated significant capacitance retention (86.9%) over 5000 cycles at a CD of 1 A g^{-1}. ED and PD were found to be lower than expected, at 15.4 Wh kg^{-1} and 249 W kg^{-1}, respectively, at a CD of 1 A g^{-1}, through this electrode. Figure 9k shows the Ragone plots of the B@GO electrode at different CDs for plotting the ED against PD. The ED of B@GO gradually decreased with increasing CDs (1 to 20 A g^{-1}) at the corresponding PD, whereas GO shows very low ED compared to B@GO. Conversely, GO only showed an ED of 5.2 Wh kg^{-1} at a PD of 248 W kg^{-1}. On the other hand, B@GO revealed 15.42 Wh kg^{-1} at a PD and CD of 249 W kg^{-1} and 1 A g^{-1}, respectively. For practical applications, the B@GO has not been utilized to construct any devices for electrochemical investigations. As a result, the B@GO electrodes performed better than the GO electrodes due to their typical compositional properties, surface affinity by bacteria functionalization, improved conductivity, structural stability, and highly active sites. Also, the heteroatom doping by bacteria on GO surfaces helps with electrochemical kinetics.

3.9. Discussion

The electrochemical results of GO and its different composites were studied using three- and two-electrode systems, and the results are shown in Table 3. Using three electrodes, it is evident that GO/CNT/COF, 10%GO/WO_3, and Nd-MOFs/GO provide better CV results when compared to other composites, with high C_s of 544.9, 688.0, and 677.0 F g^{-1}, respectively. The reason for this might be that certain electrode material compositions have a highly active site, porosity, capacitive and redox activity, and electrolyte ion permeability. The AC/TiO_2/GO and ABQA-GO/CP had lower C_s than the others because they may have slow-moving charge-transfer ions, fewer electrochemical sites, a lower redox property, etc. In addition, GCD analysis showed that some other electrodes worked better on high C_s. These included $NCWO_4$/f-MWCNT/GO (1166.6 F g^{-1}), AC/GO/TiO_2-Zn (1491.0 F g^{-1}), Bi_2O_3/GO (1029.0 F g^{-1}), Co_2Ni_1-GO (3317.5 F g^{-1}), Ni-BTC MOF/GO 2 (1199.0 F g^{-1}), and WS_2@MXene/GO (1111.0 F g^{-1}). High-conductivity electrodes, high charge-transfer ions, high electrochemical properties, and highly active sites that lead to high conductivity and allow the high electrochemical performances and quick ion transport from electrolyte to electrode surfaces might be the cause of the high C_s. Generally, C_s values of electrode materials vary with scan rate during CV. As a result, at high scan rates, the sluggish migration of electrolyte ions toward the electrode's active sites induces a drop in C_s. Low scan rates, on the other hand, provide for easy passage through electro-active regions as well as rapid diffusion of electrolyte ions, resulting in high C_s. The same may be observed for C_s at high and low current densities during GCD. Low C_s at higher current density might be due to inadequate time for electrolyte ions to reach electro-active sites, resulting in low C_s, while high C_s at low-current density indicate enough time for electrolyte ions to acquire electro-active sites and high C_s. Because studies use different scan rates and current densities during CV and GCD analysis, respectively, the C_s values compared to electrode materials seem quite strange. However, the electrode materials that may exhibit high C_s at fast scan rates and high-current density may have significant potential for energy storage.

Table 3. Results of GO and its composites in three and two-electrode systems.

	Three-Electrode System				Two-Electrode System						
	CV	GCD	EIS	Cycles							
Electrodes	C_s		R_s and R_{ct}	CS and CR	Device	C_s	ED	PD	CS and CR	EL	Ref.
GO-Fluorine											
FGO	514.0	276.0	0.58; 0.71	-	FGO//FGO (Symmetry)	-	25.8	1280.0	20,000; 99.0%	1.0 M KOH	[77]
GO-metal oxides											
$NCWO_4$/f-MWCNT/GO	-	1166.6	0.67; 1.2	5000; 95.6%	$NCWO_4$/f-MWCNT/GO//AC (Asymmetry)	266.2	83.3	703.1	5000; 87.0%	6.0 M KOH	[90]
AC/GO/TiO_2-Zn	329.9	1491.6	192.2; -	-	AC/GO/TiO_2-Zn//AC/GO/TiO_2-Zn (Symmetry)	-	99.0	3572.0	10,000; 61.8%	3.0 M KOH	[79]
Bi_2O_3/GO	-	1029.0	0.91; -	3000; 80%	-	-	-	-	-	6.0 M KOH	[88]
$MnFe_2O_4$/GO	-	298.0	2.16; 2.45	500; 92.0%	-	-	-	-	-	6.0 M KOH	[91]
AC/TiO_2/GO	219.3	617.0	1.87; 2.68	1000; 80.1%	-	-	-	-	-	2.0 M H_2SO_4	[80]
NGO/$CuFe_2O_4$	-	374.2	6.85; 1.1	2000; 82%	NGO/$CuFe_2O_4$//AC (Asymmetry)	89.1	35.7	883.0	500; 78.6%	1.0 M H_2SO_4	[85]
Cu_2O-CoO/GO	-	723.0	0.1; -	-	Cu_2O-CoO/GO//GO (Asymmetry)	125.0	44.1	794.0	10,000; 89.3%	6.0 M KOH	[95]
GO/CNT/COF	544.9	175.0	2.25; 11.34	8000; 84.7%	GO/CNT/COF//GO/CNT/COF (Symmetry)	-	24.3	248.9	-	1.0 M H_2SO_4	[97]
GO@ Fe_3O_4-IL@W	-	332.4	5.47; 9.35×10^{-5}	10,000;91.3%	-	-	7.3	40.0	10,000; 91.3%	1.0 M H_2SO_4	[58]
NMGO	-	402.0	-; 34.0	14,000; 93%	NMGO//MWCNT (Asymmetry)	90.0	28.0	750.0	6000; 88%	1.0 M KOH or Na_2SO_4	[83]
10%GO/WO_3	668.0	738.0	1.05; -	7000; 88.0%	GO/WO_3//GO (Asymmetry)	213.0	25.0	1000.0	3000; 87%	2.0 M KOH	[96]
GO-metal sulfides											
MnS-La_2S_3/GO	-	890.0	-	4000; 89.0%	MnS-La_2S_3/GO//MnS-La_2S_3/GO (Symmetry)	151.0	54.2	1300.0	10,000; 92.5%	1.0 M Na_2SO_4 or PVA-Na_2SO_4	[100]
NiS/CoS/GO	-	1114.0	-	8000; 80.0%	-	-	674.1	8089.4	-	3.0 M KOH	[98]
GO-Transition metal chalcogenides											
GO-MnSSe	300.0	603.0	-	9000; 67.0%	-	-	-	-	-	1.0 M KOH	[104]
GO-Layered double hydroxides											
Co_2Ni_1-GO	-	3317.5	0.88; 0.029	10,000; 91.8%	Co_2Ni_1-GO//AC (Asymmetry)	328.7	94.7	750.0	10,000; 91.3%	3.0 M KOH	[106]
GO-organic materials											
SGZ	-	261.0	-; 14.7	1000; 102%	-	-	-	-	-	6.0 M KOH	[114]
Nd-MOFs/GO	677.6	633.5	0.8; -	4000; 88.7%	-	-	-	-	-	3.0 M KOH	[113]
Ni-BTC MOF/GO 2	-	1199.0	0.79; 7.0	5000; 84.4%	Ni-BTC MOF/GO 2//AC (Asymmetry)	-	42.8	800.0	10,000; 70%	3.0 M KOH	[102]
ABQA-GO/CP	203.2	235.5	-	-	ABQA-GO/CP//ABQA-GO/CP (Symmetry)	134.1	32.8	1256.0	5000; 106%	1.0 M H_2SO_4	[101]

Table 3. *Cont.*

	Three-Electrode System				Two-Electrode System						
	CV	GCD	EIS	Cycles							
Electrodes		C_s	R_s and R_{ct}	CS and CR	Device	C_s	ED	PD	CS and CR	EL	Ref.
GO-MXene											
LIG-C film	-	-	182; -	1000; 102.4%	LIG-C-based supercapacitor	-	-	-	1000; 103%	H_3PO_4-PVA	[118]
MXene-GO film	-	-	8.5; -	1000; 92.6%	Planar supercapacitor	-	-	-	1000; 94.8%	H_3PO_4-PVA	[103]
CGO/PDAAQ-MXene film	-	346.0	-; 1.15	5000; 83%	CGO/PDAAQ-MXene//rGO (Asymmetry)	114.0	41.0	404.0	10,000; 70.3%	1.0 M H_2SO_4	[117]
WS_2@MXene/GO	-	1111.0	2.3; 0.41	5000; 97.1%	WS_2@MXene/GO//AC (Asymmetry)	320.0	95.0	1000.4	15,000; 93.1%	1.0 M KOH	[119]
GO-Bacteria											
BGO	-	111.0	-; 29.2	5000; 86.9%	-	-	-	-	-	0.5 M Na_2SO_4	[120]

C_s = specific capacitance (F g^{-1}); CV = cyclic voltammetry; GCD =galvanostatic charge–discharge; R_s = solution resistance (Ω); R_{ct} = charge-transfer resistance (Ω); ED = energy density (Wh kg^{-1}); PD = power density (W kg^{-1}); CS&CR = cycle stability (cycles) and capacitance retention (%); EL = electrolyte.

Fast charge/electron transfer made possible by the conductivity of the electrode materials is a key characteristic linked to high energy storage. The electrical structure of the electrode materials affects the high and low conductivities. In general, the zero-bandgap electronic structure of carbon-based materials results in excellent conductivity. The conductivity is also affected by the electro materials' impurities. Thus, the material used to synthesize or fabricate an electrode should be of high purity. To obtain better electrochemical performance, it is preferable to develop electrode material composites with high conductivity that include carbon-based materials. Furthermore, film-type electrodes (CGO/PDAAQ-MXene film) exhibit higher resistance in comparison to powder-based electrode materials (FGO, $NCWO_4$/f-MWCNT/GO, 10%GO/WO_3, Bi_2O_3/GO, Cu_2O-CoO/GO, Co_2Ni1-GO, Nd-MOFs/GO and WS_2@MXene/GO). This is attributed to the formation of coatings on insulating substrate materials, including glass, polyamide, and other low-conductivity polymer materials. Consequently, it may be advantageous to enhance the total conductivity through the use of the coated substrate's conductivity characteristic. On the other hand, film-type electrode materials are more expensive to fabricate than powder-based electrode materials. However, the film-type electrode might be practical for flexible electronic applications. So, films with high conductivity and low cost should be the core of research.

The cycle test is useful for studying the electrode materials to determine how much initial capacitance they maintain with increasing cycle numbers at a particular current density. The capacitance (%) of the electrode materials is almost equal to 100%, while increasing the cycle numbers might mean those electrode materials are highly structurally stable during charging–discharging, and the energy storage is accomplished over the physical adsorption and desorption of ions on the electrode and electrolyte interface, which can be observed in electrode materials such as $NCWO_4$/f-MWCNT/GO, $MnFe_2O_4$/GO, GO@ Fe_3O_4-IL@W, NMGO, Co_2Ni_1-GO, WS_2@MXene/GO, and MXene-GO film. In another case, the capacitance retention (%) increases by more than 100% as the cycle numbers increase due to self-activation; however, after the self-activation step, the capacitance retention rate gets to the initial capacitance value or is slightly less than 100%, as seen in SGZ, ABQA-GO/CP, and LIG-C films. On the other hand, the capacitance (%) of electrode materials (Bi_2O_3/GO, NGO/$CuFe_2O_4$, NiS/CoS/GO, and GO-MnSSe) progressively decreases as the cycle number increases owing to modest structural change but no deterioration; these materials are relatively stable when charging and discharging. It might also be the cause of charge transfer during chemical reactions and phase transformations, as well as volume expansion and contraction of electrode materials throughout the cycling process. Therefore, the electrode materials synthesized should have high structural stability and chemical resistance. Although the structure and physio-chemical properties of electrodes have an impact on cyclic stability, EDLC-based electrodes exhibit greater cyclic stability than PC or battery behavior-type electrodes, and test conditions also play a role [126]. In this review, the performance of GO-metal oxide, GO-MXene, and GO-Layered double hydroxides electrode materials in the three-electrode system was found effective.

The PD and ED results of the device might be determined by its overall electrochemical performance. However, in general, the PD of the supercapacitor is high; consequently, it may be more prudent to concentrate on improving the device's ED, given that its energy storage is determined by its ED values. Higher and lower energy storage on the device corresponds to higher and lower ED, respectively. Consequently, devices that have higher ED can be found, including AC/GO/TiO_2-Zn//AC/GO/TiO_2-Zn, Co_2Ni_1-GO//AC, WS_2@MXene/GO//AC, $NCWO_4$/f-MWCNT/GO//AC, and MnS-La_2S_3/GO//MnS-La_2S_3/GO. The electrolyte is an essential part of supercapacitors that transfers and balances energies between the two electrodes. The choice of electrolyte is critical to the safety and performance of supercapacitor devices since most electrode materials are stable in base electrolytes rather than acids. Aqueous electrolytes have higher conductivity than acids and ionic electrolytes, which helps to reduce resistance and improve device performance. One potential explanation for the solution's elevated conductivity, leading to a greater abundance of electrolyte ions, is the potential for increased energy and power density. Because of this, good conductivity and a lot of electrolyte solutions could make the devices work better overall. The ternary composite structure-based devices were also found to be more stable over a wide range of cycles than other structured composites. The additive effect of the ternary-type compound may be what makes the electrolyte solution so good at conducting electricity and moving charges around. The overall performance of the three-electrode system was better than the two-electrode system. This is because the three-electrode system uses a single working electrode, which allows for high conductivity and fast charge transfer. However, the precise measurement of energy storage may be considered for the two-electrode system's use. High cycle stability of the devices for the composition of GO with fluorine, metal sulfides, layered double hydroxides, and MXene was shown to be good due to high structural stability and faster ion transportation. According to device ED findings, GO-metal oxides, GO-layered double hydroxides, and GO-MXene performed better than other types of devices in the two-electrode setup due to their good C_s and high cycle performance.

4. Simulation Studies

The electrode materials' electrical structure, which can influence the energy storage performance, can be obtained by simulation study. The design of materials for supercapacitors that have higher ED and PD may now be assisted by density functional theory (DFT) simulation studies. DFT simulations, often used to study the physics and chemistry of solid states, provide a computationally efficient method for calculating structural, electrical, and thermodynamic aspects from quantum mechanics. The basis of this simulation is the solution of the system's body's Schrodinger equation using different wave functions. As can be shown in the previous reports, there are two DFT theorems for studying the system's energy states and the electron density using the Kohn–Sham equation [127,128]. Additionally, the widely used and accessible density functional computation to examine the electron density location is the local density approximation (LDA) [129]. Later, the accuracy of LDA was enhanced by the use of the generalized gradient approximation (GGA) functional theory. In solids computing, the most often used GGA functional is the Perdew–Burke–Ernzerhof (PBE) functional [130]. The self-interaction error that occurs when electrons are randomly transferred between many positions is not eliminated by GGA or LDA, which may lead to multiple calculation mistakes in the redox energy. The Hubbard-U correction approach was used to resolve this problem. The exchange (J) and Coulomb energies (U) may be used to describe the on-site Coulomb interactions. Furthermore, in comparison to the Hubbard-U correction technique, the hybrid DFT methodology is employed to explore the independence of structure and more precise predictions for electrode attributes. The Hartree–Fock (HF) exchange technique was introduced to the hybrid DFT to correct the electron self-interaction mistakes and avoid the problem of excessive delocalized states in the local approximation. The Heyd–Scuseria–Ernzerhof (HSE06) functional is the only one that combines fractional PBE–GGA exchange and exact partial exchange to complete PBE correlation. It is also used to forecast precise poor gap values. As a result, DFT simulations provide fundamental properties of materials, including charge transport, lattice stability, and electronic structure.

To predict the physical properties of the electrode material, such as crystal structure, electronic interaction, ion diffusivity, and phase transition, as well as the atomic-level structures of the electrode material, DFT calculations are used to study the interactions between nuclei atoms and electrons and the bonding between atoms. Software like Quantum Espresso (version 6.3), VASP (version 6.4.1), DMol^3 (version 7.0), Gaussian 16 (Revision C.01), CASTEP (version 23.1.1), WIEN2K (version 23.2), Turbomole (version 7.7), etc., are used for DFT calculations. An essential factor in controlling an electrode material's electrical characteristics is its electronic density of states (DOS). Accurate band structure and electronic states were studied using quantum mechanical techniques, and the DOS investigations immediately provide insight into a material's conductivity, magnetic properties, bonding mechanism, and doping impact [127]. Partial density of states, or PDOS, is a study of orbital hybridization and bonding process. It describes the contribution of a particular atom's orbital to the overall DOS of the corresponding material. Moreover, the increase or decrease in the number of states at the Fermi level of a particular orbital determines the transfer of charge.

A thorough understanding of the electrochemical performance of materials may be gained by theoretical research on the electrode–electrolyte kinetic behavior by estimating various parameters using DFT. Consequently, the ions' mobility, which contributes to improving the charge transfer in the supercapacitor, may be developed by a diffusion energy barrier of ions that has little impact on the electrode surface. Therefore, a better way to understand the mobility of electrolytic ions and maybe value their charge storage ability is to forecast their diffusion energy using simulation tools. Additionally, Luryi introduced the concept of quantum capacitance (QC) for two-dimensional electron gas in 1988 [131]. The differential in electric charge that relates to the electrochemical potential is also known as chemical capacitance. QC predominates in 2D systems-based materials (carbon) used as supercapacitor electrode materials [132]. Measuring the energy storage capability of materials requires an understanding of the relationship between electrode voltage and quantum capacitance. It includes a kinetic constant by the DOS as well as an electrostatic component that analyses the interactions between electrons and their connections. While the molarity of the solution affects the electrochemical capacitance, the inner quantum structure of the electrode determines the QC [119]. The equation for the quantum capacitance of electrode materials may be found in the previous study [133,134]. Furthermore, the charge storage capability of a material may be theoretically assessed by determining the voltage induced by ions [135].

Karmakar et al. recently conducted a study on the structure and electrical characteristics of $NiGa_2O_4$/rGO using VASP combined with a PAW-GGA exchange-correlation functional code [132]. The study revealed that the weak van der Waals forces were responsible for the interactions between the (311) surface of $NiGa_2O_4$ and rGO. Figure 10a depicts the DFT-optimized structure of (311) surface rGO-supported $NiGa_2O_4$. According to this figure, the distance between $NiGa_2O_4$ and rGO was measured to be 0.33 nm. The electronic state of $NiGa_2O_4$/rGO was determined to be greater than that of $NiGa_2O_4$ around the Fermi level in the DOS analysis. Composite with rGO may enhance the conductivity of $NiGa_2O_4$ by improving the electron states around the Fermi level. Consequently, increased conductivity results in larger C_s of the material. In the partial density of states (PDOS), the electronic states associated with the carbon 2p orbital are found to be more pronounced around the Fermi level in the hybrid structure compared to the unmodified rGO. This enhancement is attributed to the substantial transfer of electrons from $NiGa_2O_4$ to rGO. The charge transfer occurred due to the orbital interactions involving the 3d orbital of Ni, as well as the 4s and 4p orbitals of Ga, with the 2p orbital of C. In addition, Bader charge analysis revealed that a charge of 0.072e was transferred to rGO. The quantum capacitance (QC) is higher for the low-dimensional system with a hybrid structure compared to the bare electrode materials when considering a broad range of potential, as shown in Figure 10b. Hence, the orbital interactions and charge transfer from $NiGa_2O_4$ to rGO significantly contribute to enhancing the electrochemical performances of this hybrid material.

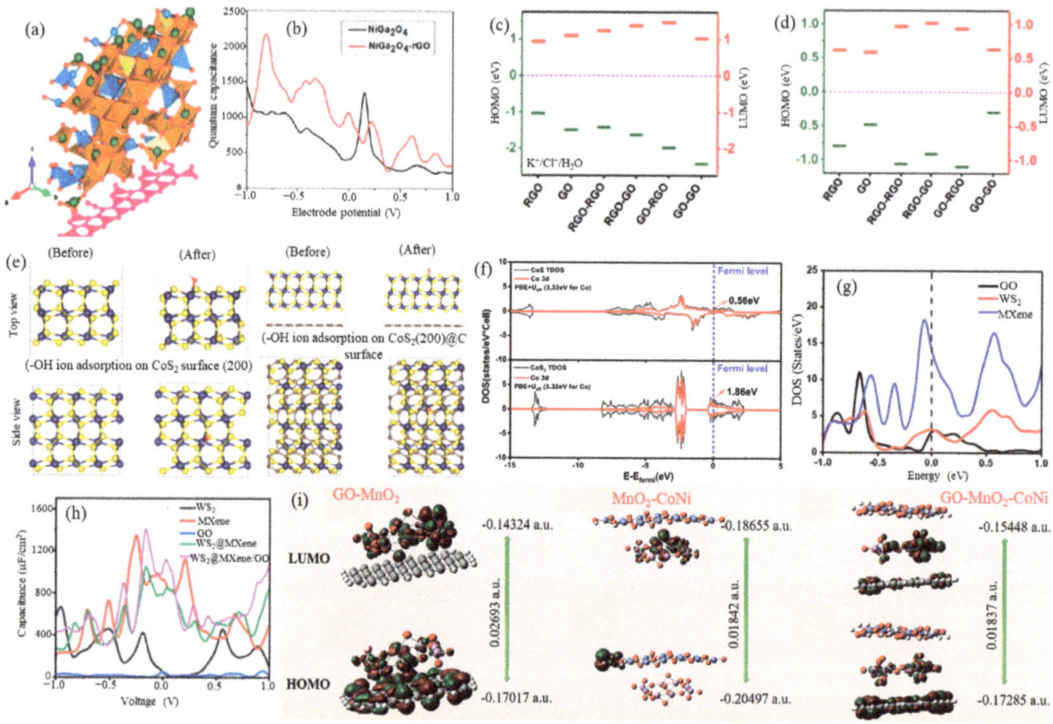

Figure 10. DFT studies of (**a**) geometry-optimized composite structure of (311) surface of $NiGa_2O_4$/rGO. (**b**) Variation of QC with the applied potential for the surface of $NiGa_2O_4$/rGO (reproduced with permission from [132]). Copyright 2021, The American Chemical Society. (**c**) HOMO–LUMO molecular orbitals for the optimized structures of various electrodes using DFT. (**d**) HOMO–LUMO molecular orbitals for the optimized structures of various electrodes with K^+ and Cl^- ions using the COSMO solvation system (reproduced from [136]). Copyright 2023, The American Chemical Society. (**e**) CoS_2(2 0 0) and CoS_2(2 0 0)@C surfaces of before and after OH^- adsorption onto Co atom. (**f**) TDOS-optimized configuration of CoS and CoS_2 (reproduced with permission from [137]). Copyright 2023, Elsevier. (**g**) PDOS of WS_2@MXene/GO, (**h**) QC of GO, WS_2, MXene, WS_2@MXene, and WS_2@MXene/GO obtained using DFT (reproduced with permission from [119]). Copyright 2023, Elsevier. (**i**) Electrostatic potential mapping of the surface model of GO-MnO_2, MnO_2-CoNi, and GO-MnO_2-CoNi (reproduced with permission from [138]). Copyright 2024, Elsevier.

Another work utilized SIESTA version 4 to analyze the DOS and QC of graphene treated with epoxide–hydroxyl and N-doped graphene [139]. The electrical conductivity of both functionalized graphene may be higher than that of bare graphene because electronic states are very near to the Fermi level, which intensely modifies its DOS owing to the presence of functional groups on the graphene's surface. Furthermore, QC was computed from the DOS, and greater QC values were seen for both functionalized graphene and naked graphene. The simulation studies determined that N-doped graphene (CQ = 0) and epoxide–hydroxyl-treated graphene (CQ = 34.6 $\mu F\ cm^{-2}$) could potentially serve as positive and negative electrodes, respectively, in an asymmetric supercapacitor, based on their respective CQ values. Islam et al. recently employed DFT simulation to examine the electronic structure of single-layer (GO/rGO) and double-layer (rGO-GO, rGO-rGO, GO-GO, and GO-rGO) screen-printed electrodes by analyzing their highest occupied molecular orbital (HOMO) and highest occupied molecular orbital (LUMO) energy gaps [136]. The energy gap of HOMO and LUMO was discovered to be greater in a double-layer screen-primed electrode than in a single layer, as shown in Figure 10c. Additionally, the energy gap of HOMO and LUMO was observed to be greater in hydrated electrodes (containing K^+ and Cl^- ions) compared to non-hydrated electrodes, as shown in Figure 10d. This suggests that hydrated electrodes may possess a more stable electronic structure.

DFT simulations were recently used to investigate the performance and calculation of OH ios adsorption energy and its adsorption on CoS_2@C sites for the CoS_2@gC/rGO electrode material [137]. Before and after OH^- ion adsorption on the surfaces of CoS_2(200) and CoS_2(200)@C are depicted in Figure 10e. The adsorption energy of -OH ions on the Co site of the CoS_2(200) surface was observed to be higher in comparison to the CoS_2(200)@C surface. Moreover, according to the Bader effective charge values, CoS_2(200) (−0.079 eV) acquired negative charges as a result of a greater number of electrons being adsorbed, whereas CoS_2(200)@C exhibited positive Bader effective charges (0.029 eV) as a consequence of electrons being neglected on Co atoms. Thus, OH^- adsorption towards CoS_2(200)@C may be higher than that of the CoS_2(200) slab surface. Additionally, the material's total density of states (TDOS) was investigated using PBE + U_{eff}. Accordingly, Figure 10f depicts the total TDOS of each unit cell and Co 3d contributions in CoS_2 and CoS by PBE + U_{eff} functional. The TDOS of CoS_2 was observed to be higher than that of CoS, possibly because the energy peaks near the Fermi level are higher in CoS_2. Consequently, the electronic conductivity of CoS_2 may be higher. High electronic conductivity and high electron transfer in the electrode material will, therefore, permit high electrochemical performance in supercapacitor applications.

On the plane of (001), a DFT first-principles calculation was performed on the electrode materials WS_2-embedded MXene/GO [119]. The bond lengths of GO (C–C (1.42 Å) and C–O (1.46 Å)) were observed to be comparatively shorter than MXene ((Ti–C (2.03 Å)) and WS_2 (W–S (2.42 Å)). This discrepancy in bond length was attributed to the covalent bond between WS_2 and MXene. Additionally, Figure 10g represents the PDOS results, which demonstrate that the high conductivity of WS_2@MXene/GO is due to the high overlap between the valence and conduction bands near the Fermi level, which confirms the metallic character of the electrodes. Furthermore, Figure 10h shows quantum capacitance (QC) values for the WS_2@MXene/GO structure (1024.3 µF cm^{-2}) were significantly higher than those for WS_2@MXene (931.6 µF cm^{-2}), owing to the dominance of Ti *d*-states at the Fermi level. However, it produces an exceptionally high QC value of 1404.4 µF cm^{-2} while operating at a negative potential voltage of −0.16 V. WS_2@MXene/GO may be a promising electrode material to boost the electrochemical performance in supercapacitors because of its high energy levels and good QC. Similarly, the van der Waals interaction between $CoSe_2$ and rGO resulted in zero band gap values for $CoSe_2$ and $CoSe_2$/rGO was observed by DOS calculation, and the DOS peak intensity was higher for $CoSe_2$-rGO near the Fermi level than for $CoSe_2$, suggesting that both are metallic and may have high conductivity [140]. The difference in charge density indicates that both positive and negative charges are collected near the $CoSe_2$ and rGO surfaces. Consequently, as compared to rGO, a higher amount of OH^- ion could be adsorbed in the electrolyte using $CoSe_2$ surfaces. Therefore, compared to $CoSe_2$-rGO (−3.0 eV), the adsorption energy of the OH^- ion on $CoSe_2$ (−2.6 eV) was lower. Moreover, for $CoSe_2$-rGO and $CoSe_2$, the distance between the OH-ion and the Co site was 1.999 Å and 2.035 Å, respectively. After wrapping $CoSe_2$ with rGO, these DFT analyses demonstrate that $OH^-$is strongly adsorbed on the $CoSe_2$ surface. Because of this beneficial effect, $CoSe_2$-rGO may be a suitable electrode material to improve electrochemical performances. This is because of its metallic nature, higher adsorption energy, and short distances to adsorb OH^- ions.

Yang et al. investigated chrysin-modified rGO (CHY-rGO) for use in supercapacitors [141]. To do this, they used DFT modeling to study the adsorption of chrysin on GO. They found that the π–π parallel approach is very stable when chrysin is combined with rGO and that 1.18 eV is the maximum adsorption energy. It was discovered that there was a significant amount of charge (electrons) accumulating on the CHY-rGO surfaces as a result of the oxygen atom on the chrysin surfaces, which readily causes electrostatic interaction between rGO and chrysin. The CHY-rGO electrode's strong charge buildup and high adsorption energy may thus contribute to its high C_s, ED, PD, and cycle stability. Additionally, CHY-rGO's DOS peak intensity was greater than that of bare rGO, suggesting that CHY-rGO may be a high capacitance electrode. Furthermore, the DOS and PDOS of carbon (C) on CHY-rGO overlap (π-π stacking interactions) close to the Fermi level, suggesting that there exist forces between the CHY and rGO's C-2p orbitals in the CHY-rGO. It may be an electrode material with high conductivity due to the overlap around the Fermi level. DFT was used to examine the combinations of GO-MnO_2-CoNi for the supercapacitor application [138]. As a result, the positive charges of CoNi and the negative charges of GO and MnO_2 electrostatically attracted to one another. Figure 10i illustrates the study of the surface model of electrodes using electrostatic potential mapping. Accordingly, compared to GO-MnO_2 (0.02693 a.u.) and MnO_2-CoNi-LDH (0.01842 a.u.), the energy level difference between HOMO and LUMO of GO-MnO_2-CoNi (0.01837 a.u.) was less, indicating that the combined three electrodes had high conductivity. According to DFT research, GO, MnO_2, and CoNi-LDH are linked by electrostatic interactions, proving that this heterostructure may be constructed and used in supercapacitors.

A supercapacitor was recently fabricated using highly oxidized GO, and the DFT analysis was conducted to determine how this GO was formed from a graphite sheet [142]. Accordingly, a bilayer graphite sheet revealed that the surface of the material had negative charges and that the C-C, C-H, and interlayer bond lengths were, respectively, 1.588, 1.105, and 1.60 Å. Furthermore, there was a repulsion between negatively charged graphite and negatively charged acids as well as $KMnO_4$ due to the high electronegativities of H_2SO_4, H_3PO_4, and $KMnO_4$. On the other hand, graphite molecules could adsorb via an O (H_2SO_4), OH (H_3PO_4), or O&K ($KMnO_4$). The electronic energy of bilayer graphite is around 7708.251 Hartree, but bilayer graphite takes on an additional −72.061 Hartree upon adsorption with H_2SO_4, H_3PO_4, and $KMnO_4$. Upon adsorption of H_2SO_4 with graphite, the S–O bond lengths may increase from 1.567 to 1.628 Å, as confirmed by the HOMO pattern. The energy required for the formation of GO with a single O atom is −7830.079 Hartree. The presence of O and OH on the graphite leads to the formation of GO through the disruption of interlayer bonds.

Similarly, for supercapacitor use, guanidine-functionalized GO (G-GO) was studied through an amide process [143]. Also, the ease of designing electrodes (functionalization), their electronic states, the distance between layers, and the quality were all confirmed using the Quantum-ESPRESSO by DFT. At first, the carbon atoms of GO and dicyclohexylcarbodiimide react with each other, creating a double bond. Then, 4-dimethylaminopyridine was used to get rid of the dicyclohexylurea molecule by attacking the carbonyl group. Finally, it was predicted that guanidine would functionalize GO by attacking the carbonyl and substituting 4-dimethylaminopyridine in the structure. Because functionalization increases interlayer distance and prevents GO from restacking, it may be a useful strategy for GO synthesis. It was found that the interlayer gap between guanidine and GO was 2.17 Å. It was found that the QC value rose along with the DOS value increasing at the Fermi level and becoming positively charged electrodes as a result of electron loss after guanidine functionalization. Consequently, G-GO's QC value (3572 F g^{-1}) was higher than G's (2290 F g^{-1}). Owing to the superior QC of G-GO electrode material, supercapacitor applications may be possible.

Supercapacitor electrode design is still beset with difficulties, such as finding stable, suitable architectures and persistent features. Therefore, efforts may be made to develop better designs for high-performance electrodes with unique features with the use of DFT modeling. In light of this, simulated studies based on DFT predictions were examined alongside recent reports on GO-based electrode materials for supercapacitors. To comprehend electrode design and electronic structure, electrode interaction, adsorption energies, charge density difference, interlayer distance, and quantum capacitance, the DFT simulation is very helpful. To identify innovative materials, DFT-assisted simulation may be used. Finding viable candidates with the capacity to have high energy storage, high electrical conductivity, and high stability is made easier with the use of a simulation-based computational analysis before the synthesis of electrode materials.

5. Conclusions and Future Perspectives

This review represents a recent study on synthesizing graphene oxide (GO) and its composites with diverse materials to develop electrode materials for supercapacitor applications. As a sign of recent progress in the field of GO material-based electrode material for supercapacitor uses, the need for energy storage devices has been brought up for discussion. The experimental conditions, processing time, and cost associated with the synthesis of a variety of materials, including fluorine, metal oxides, metal sulfide, metal chalcogenides, LDHs, organic material, MXene, biological material, and its composite with GO, were evaluated. It was found that metal oxides, metal sulfides, metal chalcogenides, and LDHs with GO composites are inexpensive and straightforward to produce. The present review outlines the diverse composites of GO electrode materials and electrolytes employed in electrochemical investigations for supercapacitors. To this degree, the scan rates, charging–discharging rates, resistances, and cyclic stability of a three-electrode setup are provided. To achieve Faradaic processes, GO composites containing additional 2D materials exhibit either pseudocapacitive or battery-like characteristics. The primary determinant of electrochemical performance is the choice of electrode materials featuring nanostructured surfaces; this should result in a reduction of the supercapacitor's capital expenditure. Based on the current analysis, it is evident that GO incorporating electrode materials, including metal oxides, LDHs, MOFs, and MXene, results in exceptionally high supercapacitor performance. Compared to undoped GO materials, GO functioning with other substances exhibit high C_s. Thus, GO could be a viable material for improving the conductivity, C_s, energy and power density, cycle performances, and conductivity of other low-cost materials to enhance their electrochemical performance. Furthermore, it could serve as a strategy for fabricating inexpensive and high-energy storage devices.

Supercapacitor devices based on GO composites with metal oxides, sulfides, carbides, polymers, MOFs, MXene, biological materials, etc., should prioritize easy and cost-effective preparation procedures in light of prospects. Furthermore, GO nanosheets encounter extra difficulties, such as agglomerations and self-restacking, which hinder performance by decreasing surface area, redox sites, and ion transport. Composites of GO and other nanomaterials may improve the materials' surface area and function as separators between the GO sheets, mitigating these problems and making them ideal for use as supercapacitors. Establishing pores through the introduction of organic or heteroatom dopants can serve as a viable strategy to promote accelerated ion diffusion in GO and other composite materials, thereby augmenting the overall activity. Also, more research should be done on the design features, and new ways of making GO and its hybrid materials work more effectively combined to improve total electrochemical performance. Concerning the electrochemical performance of a variety of GO-based composite electrode materials, effective electrodes for energy storage devices of the next generation should be developed.

To optimize the design of electrodes with particular capacitance, ion diffusion, and stability, DFT might be utilized to precisely tailor the electrode materials at the nanoscale level. It may be accomplished by investigating how supercapacitor performances are impacted by size, structure, shape, and functionalization. DFT may assist researchers studying energy storage in identifying effective electrode materials with high energy storage capacities. With DFT modeling, novel electrolytes, surface functionalization, or electrode hybrids may be investigated to enhance the electrochemical performance of electrode materials.

Author Contributions: Conceptualization, G.S. and M.A.; methodology, G.S. and S.D.; software, S.D.; validation, K.D. and V.R.S.; formal analysis, M.D.K. and S.D.; investigation, G.S.; resources, M.A. and K.D.; data curation, V.R.S.; writing—original draft preparation, G.S.; writing—review and editing, M.A. and V.R.S.; visualization, K.D. and M.D.K.; supervision, T.H.O.; project administration, T.H.O.; funding acquisition, T.H.O. All authors have read and agreed to the published version of the manuscript.

Funding: This research was funded by the National Research Foundation of Korea (NRF) by the Korean Government (MSIT), grant number 2022R1A2C100428.

Data Availability Statement: Not applicable.

Conflicts of Interest: The authors declare no conflicts of interest.

References

1. Wang, J.; Guo, Z.; Li, F.; Yang, H.; Che, H.; Zhang, Z.; Wang, Y.; Zhang, X.; Li, S.; Mu, J. Rational design of ternary $NiCo_2Al_x$-LDH coupled with PANI coated Nitrogen-doped carbon capsule for High-Performance asymmetric supercapacitors. *J. Electroanal. Chem.* **2023**, *931*, 117189. [CrossRef]
2. Nasrin, K.; Arunkumar, M.; Koushik Kumar, N.; Sudharshan, V.; Rajasekar, S.; Mukhilan, D.; Arshad, M.; Sathish, M. A rationally designed hetero-assembly of 2D/2D Nitrogen-doped MXene/Graphene via supercritical fluid processing for high energy durable supercapacitors. *Chem. Eng. J.* **2023**, *474*, 145505. [CrossRef]
3. Li, M.; Zhou, S.; Cheng, L.; Mo, F.; Chen, L.; Yu, S.; Wei, J. 3D Printed Supercapacitor: Techniques, Materials, Designs, and Applications. *Adv. Funct. Mater.* **2023**, *33*, 2208034. [CrossRef]
4. Park, H.W.; Roh, K.C. Recent advances in and perspectives on pseudocapacitive materials for Supercapacitors—A review. *J. Power Sources* **2023**, *557*, 232558. [CrossRef]
5. Balasankar, A.; Arthiya, S.E.; Ramasundaram, S.; Sumathi, P.; Arokiyaraj, S.; Oh, T.; Aruchamy, K.; Sriram, G.; Kurkuri, M.D. Recent Advances in the Preparation and Performance of Porous Titanium-Based Anode Materials for Sodium-Ion Batteries. *Energies* **2022**, *15*, 9495. [CrossRef]
6. Mouraliraman, D.; Thiagarajan, A.; Deepa, S.; Sriram, G.; Aruchamy, K.; Oh, T.H.; Shin, D. 3D printed lithium-ion batteries: An in-depth examination of the advancements in flexibility and stand-alone capability. *J. Energy Storage* **2024**, *81*, 110395. [CrossRef]
7. Nishchith, B.S.; Kalegowda, Y.; Ashoka, S.; Shreenivasa, L.; Sriram, G.; Kurkuri, M.D.; Ajeya, K.V.; Jung, H.-Y. Electrochemical kinetic study and performance evaluation of surface-modified mesoporous sodium carbonophosphates nanostructures for pseudocapacitor applications. *J. Alloys Compd.* **2023**, *939*, 168711. [CrossRef]
8. Anwar, M.I.; Asad, M.; Ma, L.; Zhang, W.; Abbas, A.; Khan, M.Y.; Zeeshan, M.; Khatoon, A.; Gao, R.; Manzoor, S.; et al. Nitrogenous MOFs and their composites as high-performance electrode material for supercapacitors: Recent advances and perspectives. *Coord. Chem. Rev.* **2023**, *478*, 214967. [CrossRef]
9. Conway, B.E. Transition from "Supercapacitor" to "Battery" Behavior in Electrochemical Energy Storage. *J. Electrochem. Soc.* **1991**, *138*, 1539. [CrossRef]
10. Jiang, Y.; Liu, J. Definitions of Pseudocapacitive Materials: A Brief Review. *Energy Environ. Mater.* **2019**, *2*, 30–37. [CrossRef]
11. Bhat, V.S.; Kanagavalli, P.; Sriram, G.; John, N.S.; Veerapandian, M.; Kurkuri, M.; Hegde, G. Low cost, catalyst free, high performance supercapacitors based on porous nano carbon derived from agriculture waste. *J. Energy Storage* **2020**, *32*, 101829. [CrossRef]
12. Shaheen Shah, S.; Abu Nayem, S.M.; Sultana, N.; Saleh Ahammad, A.J.; Abdul Aziz, M. Preparation of Sulfur-doped Carbon for Supercapacitor Applications: A Review. *ChemSusChem* **2022**, *15*, e202101282. [CrossRef] [PubMed]
13. Tong, Y.; Yang, J.; Li, J.; Cong, Z.; Wei, L.; Liu, M.; Zhai, S.; Wang, K.; An, Q. Lignin-derived electrode materials for supercapacitor applications: Progress and perspectives. *J. Mater. Chem. A* **2023**, *11*, 1061–1082. [CrossRef]
14. Zhao, X.; Li, G.; Feng, M.; Wang, Y. Semi-continuous electrochemical extraction of lithium from brine using CF-NMMO/AC asymmetric hybrid capacitors. *Electrochim. Acta* **2020**, *331*, 135285. [CrossRef]
15. Tamang, S.; Rai, S.; Bhujel, R.; Bhattacharyya, N.K.; Swain, B.P.; Biswas, J. A concise review on GO, rGO and metal oxide/rGO composites: Fabrication and their supercapacitor and catalytic applications. *J. Alloys Compd.* **2023**, *947*, 169588. [CrossRef]
16. Iqbal, M.Z.; Zakar, S.; Haider, S.S. Role of aqueous electrolytes on the performance of electrochemical energy storage device. *J. Electrochem. Soc.* **2020**, *858*, 113793. [CrossRef]
17. Zhong, C.; Deng, Y.; Hu, W.; Qiao, J.; Zhang, L.; Zhang, J. A review of electrolyte materials and compositions for electrochemical supercapacitors. *Chem. Soc. Rev.* **2015**, *44*, 7484–7539. [CrossRef] [PubMed]
18. Cheng, B.; Li, X.; Li, H.; Zhang, W.; Han, R.; Zhang, Y.; Xiang, J. Morphology-controllable synthesis of bimetallic oxide $Bi_2Fe_4O_9$ as novel electrode materials for high-performance supercapacitors. *J. Energy Storage* **2024**, *79*, 110137. [CrossRef]
19. Zhang, Y.; Jing, X.; Yan, X.-H.; Gao, H.-L.; Gao, K.-Z.; Cao, Y.; Hu, S.; Zhang, Y.-Y. Rational design of NiMn-based electrode materials for high-performance supercapacitors. *Coord. Chem. Rev.* **2024**, *499*, 215494. [CrossRef]
20. Zeng, J.; Wei, L.; Guo, X. Bio-inspired high-performance solid-state supercapacitors with the electrolyte, separator, binder and electrodes entirely from kelp. *J. Mater. Chem. A* **2017**, *5*, 25282–25292. [CrossRef]

21. Xiong, C.; Zhang, Y.; Ni, Y. Recent progress on development of electrolyte and aerogel electrodes applied in supercapacitors. *J. Power Sources* **2023**, *560*, 232698. [CrossRef]
22. Zhang, L.; Hu, X.; Wang, Z.; Sun, F.; Dorrell, D.G. A review of supercapacitor modeling, estimation, and applications: A control/management perspective. *Renew. Sustain. Energy Rev.* **2018**, *81*, 1868–1878. [CrossRef]
23. Şahin, M.E.; Blaabjerg, F.; Sangwongwanich, A. A Comprehensive Review on Supercapacitor Applications and Developments. *Energies* **2022**, *15*, 674. [CrossRef]
24. Poonam; Sharma, K.; Arora, A.; Tripathi, S.K. Review of supercapacitors: Materials and devices. *J. Energy Storage* **2019**, *21*, 801–825. [CrossRef]
25. Noori, A.; El-Kady, M.F.; Rahmanifar, M.S.; Kaner, R.B.; Mousavi, M.F. Towards establishing standard performance metrics for batteries, supercapacitors and beyond. *Chem. Soc. Rev.* **2019**, *48*, 1272–1341. [CrossRef] [PubMed]
26. Boulanger, N.; Talyzin, A.V.; Xiong, S.; Hultberg, M.; Grimm, A. High surface area activated carbon prepared from wood-based spent mushroom substrate for supercapacitors and water treatment. *Colloids Surf. A* **2024**, *680*, 132684. [CrossRef]
27. Gao, R.; Guo, W.; Zhang, Y.; Zhang, Q.; Li, Y.; Wang, J. Enhancement of gelatinization on electrochemical performance of corn starch-based porous carbon as electrode material in supercapacitors. *Diam. Relat. Mater.* **2024**, *141*, 110598. [CrossRef]
28. Yang, K.; Fan, Q.; Zhang, Y.; Ren, G.; Huang, X.; Fu, P. Hierarchical porous carbon aerogels as a versatile electrode material for high-stability supercapacitors. *RSC Adv.* **2024**, *14*, 1123–1133. [CrossRef] [PubMed]
29. Zhang, R.; Zhang, Z.; Xu, P.; Xu, J.; Gao, Y.; Gao, G. Cellulose nanofiber hydrogel with high conductivity electrolytes for high voltage flexible supercapacitors. *Carbohydr. Polym.* **2024**, *326*, 121654. [CrossRef] [PubMed]
30. Yang, S.; Wang, Z.; Xie, Y.; Gao, X.; Yao, F.; Bai, H.; Zhang, X.; Zhang, H.; Yue, H. Polyaniline nanowire arrays on biomass-derived carbon nanotubes with typha longbracteata for high-performance symmetric supercapacitors. *Diam. Relat. Mater.* **2024**, *141*, 110608. [CrossRef]
31. Liu, J.; Yang, G.; Lu, K.; Ye, C.; Ma, Y.; Ye, J. One-step process to obtain manganese-assisted laser-induced multi-layer graphene-like carbon supercapacitor. *Electrochim. Acta* **2024**, *474*, 143505. [CrossRef]
32. Varghese, S.M.; Mohan, V.V.; Suresh, S.; Bhoje Gowd, E.; Rakhi, R.B. Synergistically modified $Ti_3C_2T_x$ MXene conducting polymer nanocomposites as efficient electrode materials for supercapacitors. *J. Alloys Compd.* **2024**, *973*, 172923. [CrossRef]
33. Wang, W.; Yang, C.; Han, D.; Yu, S.; Qi, W.; Ling, R.; Liu, G. Ni_3S_2/Ni_2O_3 heterojunction anchored on N-doped carbon nanosheet aerogels for dual-ion hybrid supercapacitors. *J. Colloid Interface Sci.* **2024**, *654*, 709–718. [CrossRef] [PubMed]
34. Zheng, W.; Fan, L.; Meng, Z.; Zhou, J.; Ye, D.; Xu, W.; Xu, J. Flexible quasi-solid-state supercapacitors for anti-freezing power sources based on polypyrrole@cation-grafted bacterial cellulose. *Carbohydr. Polym.* **2024**, *324*, 121502. [CrossRef] [PubMed]
35. Wan, L.; Chen, J.; Zhang, Y.; Du, C.; Xie, M.; Hu, S. High-mass-loading cobalt iron phosphide@nickel vanadium layered double hydroxide heterogeneous nanosheet arrays for hybrid supercapacitors. *J. Colloid Interface Sci.* **2024**, *654*, 539–549. [CrossRef]
36. Wu, M.-S.; Hong, L.-R. Cyclic voltammetric formation of hollow porous γ-MnO_2 microspheres as stable electrodes for high-performance supercapacitors. *J. Energy Storage* **2024**, *75*, 109721. [CrossRef]
37. Cao, X.-M.; Liu, D.; Sun, Z.-J.; Zhang, Q. In situ construction of core–shell structured cobalt oxide@nickel–cobalt-layered double hydroxide nanorods with abundant oxygen vacancies towards boosting electrochemical energy storage. *Inorg. Chem. Front.* **2024**, *11*, 789–798. [CrossRef]
38. Sriram, G.; Kurkuri, M.; Oh, T.H. Recent Trends in Highly Porous Structured Carbon Electrodes for Supercapacitor Applications: A Review. *Energies* **2023**, *16*, 4641. [CrossRef]
39. Anwer, A.H.; Zubair, M.M.; Mashkoor, F.; Benamor, A.; Hasan, I.; Shoeb, M.; Jeong, C. Enhancing the electrochemical performance of hybrid supercapacitors with in-situ grown ultrasound-mediated heterostructure bi-metallic and dual-linker MOF nanoarchitecture by harnessing charge storage mechanisms. *J. Alloys Compd.* **2024**, *970*, 172512. [CrossRef]
40. Xu, M.; Zhu, X.; Lai, Y.; Xia, A.; Huang, Y.; Zhu, X.; Liao, Q. Production of hierarchical porous bio-carbon based on deep eutectic solvent fractionated lignin nanoparticles for high-performance supercapacitor. *Appl. Energy* **2024**, *353*, 122095. [CrossRef]
41. Navaneeth, P.; Jayakanthan, S.; Kumar, M.S.; Kumar, A.; Pradeep, A.; Suneesh, P.V.; Babu, T.S. High surface area cobalt aluminium layered double hydroxide printed electrodes for flexible supercapacitor and on-chip electrochemical bacterial lysing. *Electrochim. Acta* **2024**, *476*, 143744. [CrossRef]
42. Shinde, S.K.; Karade, S.S.; Truong, N.T.N.; Veer, S.S.; Shaikh, S.F.; Al-Enizi, A.M.; Jagadale, A.D.; Yadav, H.M.; Tamboli, M.S.; Park, C.; et al. Highly stable and thin layer of conducting PANI decorative on porous 3D nanoflakes like-$NiCo_2O_4$/CF nanomaterials for efficient Hybrid Supercapacitors application. *J. Energy Storage* **2024**, *78*, 109960. [CrossRef]
43. Li, X.; Ren, J.; Sridhar, D.; Xu, B.B.; Algadi, H.; El-Bahy, Z.M.; Ma, Y.; Li, T.; Guo, Z. Progress of layered double hydroxide-based materials for supercapacitors. *Mater. Chem. Front.* **2023**, *7*, 1520–1561. [CrossRef]
44. Liu, Y.; Chen, H.; Li, L. Applications and challenges of porous carbon with different dimensions in supercapacitors-a mini review. *Front. Energy Res.* **2022**, *10*, 951701. [CrossRef]
45. Singal, S.; Yadav, A.; Sharma, R.K. A high-performance asymmetric supercapacitor based on morphologically tuned CoB anode and heterostructured graphitic carbon nitride/$MoSe_2$ cathode. *J. Energy Storage* **2024**, *80*, 110300. [CrossRef]
46. He, M.; Wu, L.; Yu, A.; Li, X.; Guan, S.; Han, Q.; Wang, H.; Zhou, B.; Melvin, G.J.H.; Wang, M.; et al. Surface functionalization of vertical graphene significantly enhances the energy storage capability for symmetric supercapacitors. *Carbon* **2024**, *216*, 118511. [CrossRef]

47. Huang, Z.; Zhang, Z.; Qi, X.; Ren, X.; Xu, G.; Wan, P.; Sun, X.; Zhang, H. Wall-like hierarchical metal oxide nanosheet arrays grown on carbon cloth for excellent supercapacitor electrodes. *Nanoscale* **2016**, *8*, 13273–13279. [CrossRef] [PubMed]
48. Cai, D.; Wang, D.; Liu, B.; Wang, Y.; Liu, Y.; Wang, L.; Li, H.; Huang, H.; Li, Q.; Wang, T. Comparison of the Electrochemical Performance of $NiMoO_4$ Nanorods and Hierarchical Nanospheres for Supercapacitor Applications. *ACS Appl. Mater. Interfaces* **2013**, *5*, 12905–12910. [CrossRef] [PubMed]
49. Korkmaz, S.; Kariper, İ.A. Graphene and graphene oxide based aerogels: Synthesis, characteristics and supercapacitor applications. *J. Energy Storage* **2020**, *27*, 101038. [CrossRef]
50. Ghosh, R.; Siddiqui, M.S.; Kalita, H. Fabrication of highly conductive graphene paper for supercapacitors with a one-step hydrothermal method. *Carbon Trends* **2024**, *14*, 100317. [CrossRef]
51. Pandey, M.; Deshmukh, K.; Raman, A.; Asok, A.; Appukuttan, S.; Suman, G.R. Prospects of MXene and graphene for energy storage and conversion. *Renew. Sustain. Energy Rev.* **2024**, *189*, 114030. [CrossRef]
52. Nwankwo, M.C.; Ezealigo, B.; Nwanya, A.C.; Nkele, A.C.; Agbogu, A.; Chime, U.; Asogwa, P.U.; Ezekoye, B.A.; Ekwealor, A.B.C.; Osuji, R.U.; et al. Syntheses and characterizations of GO/Mn_3O_4 nanocomposite film electrode materials for supercapacitor applications. *Inorg. Chem. Commun.* **2020**, *119*, 107983. [CrossRef]
53. Chen, J.; Li, Y.; Huang, L.; Jia, N.; Li, C.; Shi, G. Size Fractionation of Graphene Oxide Sheets via Filtration through Track-Etched Membranes. *Adv. Mater.* **2015**, *27*, 3654–3660. [CrossRef]
54. Gund, G.S.; Dubal, D.P.; Patil, B.H.; Shinde, S.S.; Lokhande, C.D. Enhanced activity of chemically synthesized hybrid graphene oxide/Mn_3O_4 composite for high performance supercapacitors. *Electrochim. Acta* **2013**, *92*, 205–215. [CrossRef]
55. Du, W.; Lv, Y.; Lu, H.; Chen, Z.; Wright, D.S.; Zhang, C. Surface modification by graphene oxide: An efficient strategy to improve the performance of activated carbon based supercapacitors. *Chin. Chem. Lett.* **2017**, *28*, 2285–2289. [CrossRef]
56. Iqbal, M.F.; Yousef, A.K.M.; Hassan, A.; Hussain, S.; Ashiq, M.N.; Ul Mahmood, H.; Razaq, A. Significantly improved electrochemical characteristics of nickel sulfide nanoplates using graphene oxide thin film for supercapacitor applications. *J. Energy Storage* **2021**, *33*, 102091. [CrossRef]
57. Chen, X.; Paul, R.; Dai, L. Carbon-based supercapacitors for efficient energy storage. *Natl. Sci. Rev.* **2017**, *4*, 453–489. [CrossRef]
58. Dalvand, S.; Khoushab, Z.; Mousavi-Khoshdel, S.M.; Ghafuri, H.; Esmaili Zand, H.R.; Omidvar, M. Graphene oxide@Fe3o4@Tungstate modified ionic liquid as a novel electrode material for high-performance supercapacitor. *Int. J. Hydrogen Energy* **2023**, *48*, 10098–10107. [CrossRef]
59. Kuila, T.; Mishra, A.K.; Khanra, P.; Kim, N.H.; Lee, J.H. Recent advances in the efficient reduction of graphene oxide and its application as energy storage electrode materials. *Nanoscale* **2013**, *5*, 52–71. [CrossRef]
60. Naser, S.A.E.; Badmus, K.O.; Khotseng, L. Synthesis, Properties, and Applications of Metal Organic Frameworks Supported on Graphene Oxide. *Coatings* **2023**, *13*, 1456. [CrossRef]
61. Abdillah, O.B.; Rus, Y.B.; Ulfa, M.; Dedi; Iskandar, F. Recent progress on reduced graphene oxide and polypyrrole composites for high performance supercapacitors: A review. *J. Energy Storage* **2023**, *74*, 109300. [CrossRef]
62. Mousavi, S.M.; Hashemi, S.A.; Kalashgrani, M.Y.; Gholami, A.; Binazadeh, M.; Chiang, W.-H.; Rahman, M.M. Recent advances in energy storage with graphene oxide for supercapacitor technology. *Sustain. Energy Fuels* **2023**, *7*, 5176–5197. [CrossRef]
63. Sahoo, P.K.; Kumar, N.; Jena, A.; Mishra, S.; Lee, C.-P.; Lee, S.-Y.; Park, S.-J. Recent progress in graphene and its derived hybrid materials for high-performance supercapacitor electrode applications. *RSC Adv.* **2024**, *14*, 1284–1303. [CrossRef]
64. Luo, Z.; Quan, J.; Ding, T.; Xu, B.; Li, W.; Mao, Q.; Ma, W.; Li, M.; Xiang, H.; Zhu, M. Recent advances in iron oxide/graphene composites for flexible supercapacitors. *J. Alloys Compd.* **2024**, *980*, 173614. [CrossRef]
65. Nargatti, K.I.; Ahankari, S.S.; Dizon, J.R.C.; Subramaniam, R.T. Environmentally Friendly Water-Based Reduced Graphene Oxide/Cellulose Nanofiber Ink for Supercapacitor Electrode Applications. *ACS Omega* **2024**, *9*, 11730–11737. [CrossRef]
66. Ndagijimana, P.; Rong, H.; Ndokoye, P.; Mwizerwa, J.P.; Nkinahamira, F.; Luo, S.; Guo, D.; Cui, B. A review on activated carbon/graphene composite-based materials: Synthesis and applications. *J. Clean. Prod.* **2023**, *417*, 138006. [CrossRef]
67. Shruti, M.S.; Khilari, S.; Samuel, E.J.J.; Han, H.; Nayak, A.K. Recent trends in graphene assisted vanadium based nanocomposites for supercapacitor applications. *J. Energy Storage* **2023**, *63*, 107006. [CrossRef]
68. Philip, A.; Ruban Kumar, A. Recent advancements and developments employing 2D-materials in enhancing the performance of electrochemical supercapacitors: A review. *Renew. Sustain. Energy Rev.* **2023**, *182*, 113423. [CrossRef]
69. Zhang, Y.; Zhou, C.-G.; Yan, X.-H.; Cao, Y.; Gao, H.-L.; Luo, H.-W.; Gao, K.-Z.; Xue, S.-C.; Jing, X. Recent advances and perspectives on graphene-based gels for superior flexible all-solid-state supercapacitors. *J. Power Sources* **2023**, *565*, 232916. [CrossRef]
70. Sahoo, B.B.; Pandey, V.S.; Dogonchi, A.S.; Thatoi, D.N.; Nayak, N.; Nayak, M.K. Synthesis, characterization and electrochemical aspects of graphene based advanced supercapacitor electrodes. *Fuel* **2023**, *345*, 128174. [CrossRef]
71. Yang, H.; Zheng, H.; Duan, Y.; Xu, T.; Xie, H.; Du, H.; Si, C. Nanocellulose-graphene composites: Preparation and applications in flexible electronics. *Int. J. Biol. Macromol.* **2023**, *253*, 126903. [CrossRef] [PubMed]
72. Ray, P.K.; Mohanty, R.; Parida, K. Recent advancements of NiCo LDH and graphene based nanohybrids for supercapacitor application. *J. Energy Storage* **2023**, *72*, 108335. [CrossRef]
73. Kumar, R.; Sahoo, S.; Joanni, E.; Singh, R.K.; Yadav, R.M. Graphene-metal oxide hybrid materials with 2D and 3D morphologies for advanced supercapacitor electrodes: Status, challenges and prospects. *Mater. Today Nano* **2023**, *24*, 100399. [CrossRef]
74. Ahmad, F.; Zahid, M.; Jamil, H.; Khan, M.A.; Atiq, S.; Bibi, M.; Shahbaz, K.; Adnan, M.; Danish, M.; Rasheed, F.; et al. Advances in graphene-based electrode materials for high-performance supercapacitors: A review. *J. Energy Storage* **2023**, *72*, 108731. [CrossRef]

75. Chaudhuri, H.; Yun, Y.-S. A critical review on the properties and energy storage applications of graphene oxide/layered double hydroxides and graphene oxide/MXenes. *J. Power Sources* **2023**, *564*, 232870. [CrossRef]
76. Na, W.; Jun, J.; Park, J.W.; Lee, G.; Jang, J. Highly porous carbon nanofibers co-doped with fluorine and nitrogen for outstanding supercapacitor performance. *J. Mater. Chem. A* **2017**, *5*, 17379–17387. [CrossRef]
77. Sim, Y.; Surendran, S.; Cha, H.; Choi, H.; Je, M.; Yoo, S.; Chan Seok, D.; Ho Jung, Y.; Jeon, C.; Jin Kim, D.; et al. Fluorine-doped graphene oxide prepared by direct plasma treatment for supercapacitor application. *Chem. Eng. J.* **2022**, *428*, 132086. [CrossRef]
78. El-Shafai, N.M.; Ramadan, M.S.; Alkhamis, K.M.; Aljohani, M.M.; El-Metwaly, N.M.; El-Mehasseb, I.M. A unique engineering building of nanoelectrodes based on titanium and metal oxides nanoparticles captured on graphene oxide surface for supercapacitors and energy storage. *J. Alloys Compd.* **2023**, *939*, 168685. [CrossRef]
79. Veeresh, S.; Ganesha, H.; Nagaraju, Y.S.; Vijeth, H.; Devendrappa, H. Activated carbon incorporated graphene oxide with SnO_2 and TiO_2-Zn nanocomposite for supercapacitor application. *J. Alloys Compd.* **2023**, *952*, 169907. [CrossRef]
80. Abdallah, F.; Arthur, E.K.; Mensah-Darkwa, K.; Gikunoo, E.; Baffour, S.A.; Agamah, B.A.; Nartey, M.A.; Agyemang, F.O. Electrochemical performance of corncob-derived activated carbon-graphene oxide and TiO_2 ternary composite electrode for supercapacitor applications. *J. Energy Storage* **2023**, *68*, 107776. [CrossRef]
81. Wang, Q.; Zhang, Y.; Xiao, J.; Jiang, H.; Hu, T.; Meng, C. Copper oxide/cuprous oxide/hierarchical porous biomass-derived carbon hybrid composites for high-performance supercapacitor electrode. *J. Alloys Compd.* **2019**, *782*, 1103–1113. [CrossRef]
82. El-Shafai, N.M.; Alamrani, N.A.; Al-bonayan, A.M.; Abu-Melha, S.; El-Metwaly, N.M.; El-Mehasseb, I. Building electrons clouds of redesigned copper oxide nanorods captured on the graphene oxide surface for supercapacitors and energy storage. *Surf. Interfaces* **2023**, *38*, 102757. [CrossRef]
83. Arshad, N.; Usman, M.; Adnan, M.; Ahsan, M.T.; Rehman, M.R.; Javed, S.; Ali, Z.; Akram, M.A.; Demopoulos, G.P.; Mahmood, A. Nanoengineering of NiO/MnO_2/GO Ternary Composite for Use in High-Energy Storage Asymmetric Supercapacitor and Oxygen Evolution Reaction (OER). *Nanomaterials* **2023**, *13*, 99. [CrossRef] [PubMed]
84. Wang, J.; Deng, Q.; Li, M.; Jiang, K.; Zhang, J.; Hu, Z.; Chu, J. Copper ferrites@reduced graphene oxide anode materials for advanced lithium storage applications. *Sci. Rep.* **2017**, *7*, 8903. [CrossRef] [PubMed]
85. Lashkenari, M.S.; Ghasemi, A.K.; Khalid, M.; Shahgaldi, S. Facile synthesis of N-doped graphene oxide decorated with copper ferrite as an electrode material for supercapacitor with enhanced capacitance. *Electrochim. Acta* **2023**, *465*, 142959. [CrossRef]
86. Nithya, V.D.; Sabari Arul, N. Progress and development of Fe_3O_4 electrodes for supercapacitors. *J. Mater. Chem. A* **2016**, *4*, 10767–10778. [CrossRef]
87. Sumboja, A.; Liu, J.; Zheng, W.G.; Zong, Y.; Zhang, H.; Liu, Z. Electrochemical energy storage devices for wearable technology: A rationale for materials selection and cell design. *Chem. Soc. Rev.* **2018**, *47*, 5919–5945. [CrossRef] [PubMed]
88. Emmanuel Hounkanrin, S.R.J.; Guo, Z.; Luo, J. Microwave-synthesized Bismuth oxide-graphene oxide composite as an electrode for supercapacitors. *Int. J. Electrochem. Sci.* **2023**, *18*, 100086. [CrossRef]
89. Dhanabalan, K.; Bhosale, M.; Murugan, N.; Aruchamy, K.; Sriram, G.; Sadhasivam, T.; Oh, T.H. A versatile heterostructure junction of NiO–C-MoO_x (X= 2 & 3) composite electrocatalysts for hydrogen evolution reaction. *J. Solid State Chem.* **2024**, *335*, 124702.
90. Abdolahi, B.; Gholivand, M.B.; Shamsipur, M.; Taherpour, A.; Mirzaei, M.; Daramroudi, A.H. $Ni_{0.5}Co_{0.5}WO_4$@f-MWCNTs/GO nanocomposite with improved supercapattery activity as the electrode for asymmetric supercapacitor device. *J. Energy Storage* **2023**, *73*, 109181. [CrossRef]
91. Foroutan, N.; Lashkenari, M.S.; Alizadeh, E.; Sedighi, M. Synthesis of manganese ferrite/graphene oxide nanocomposite and investigation of its supercapacitor behaviors. *Int. J. Hydrogen Energy* **2023**, *48*, 25859–25868. [CrossRef]
92. Pang, N.; Wang, M.; Wang, X.; Xiong, D.; Xu, S.; Lu, X.; Wang, L.; Jiang, L.; Chu, P.K. Graphene-oxide-modified MnO_2 composite electrode for high-performance flexible quasi-solid-state zinc-ion batteries. *Mater. Sci. Eng. B* **2024**, *299*, 116981. [CrossRef]
93. Mao, X.; Zou, Y.; Xu, F.; Sun, L.; Chu, H.; Zhang, H.; Zhang, J.; Xiang, C. Three-Dimensional Self-Supporting Ti_3C_2 with MoS_2 and Cu_2O Nanocrystals for High-Performance Flexible Supercapacitors. *ACS Appl. Mater. Interfaces* **2021**, *13*, 22664–22675. [CrossRef] [PubMed]
94. Basu, P.; Mahesh, R.; Harish, S.; Joseph, S.; Sagayaraj, P. One-pot hydrothermal preparation of Cu_2O-CuO/rGO nanocomposites with enhanced electrochemical performance for supercapacitor applications. *Appl. Surf. Sci.* **2018**, *449*, 474–484.
95. Yu, Z.; Wang, S.; Xiao, Z.; Xu, F.; Xiang, C.; Sun, L.; Zou, Y. Graphene-supported Cu_2O-CoO heterojunctions for high performance supercapacitors. *J. Energy Storage* **2024**, *77*, 110009. [CrossRef]
96. Shembade, U.V.; Gurav, S.R.; Gaikwad, M.A.; Wategaonkar, S.B.; Ghatage, S.R.; Sonkawade, R.G.; Kim, J.H.; Moholkar, A.V. Hydrothermal synthesis of graphene oxide interspersed in non-uniform tungsten oxide nanorod and its performance towards highly efficient hybrid supercapacitor. *Ceram. Int.* **2024**, *50*, 340–350. [CrossRef]
97. Verma, S.; Verma, B. Synergistic optimization of nanostructured graphene oxide based ternary composite for boosting the performance of supercapacitor electrode material via response surface methodology. *Colloids Surf. A* **2024**, *682*, 132893. [CrossRef]
98. Chandraraj, S.S.; Xavier, J.R. Facile synthesis of graphene based mixed metal sulphide nanocomposite for energy storage applications. *Surf. Interfaces* **2023**, *36*, 102515. [CrossRef]
99. Rao, S.S.; Kanaka Durga, I.; Naresh, B.; Jin-Soo, B.; Krishna, T.N.V.; In-Ho, C.; Ahn, J.-W.; Kim, H.-J. One-Pot Hydrothermal Synthesis of Novel Cu-MnS with PVP Cabbage-Like Nanostructures for High-Performance Supercapacitors. *Energies* **2018**, *11*, 1590. [CrossRef]

100. Mane, V.J.; Lokhande, A.C.; Nikam, R.P.; Padalkar, N.S.; Lokhande, V.C.; Dhawale, D.S.; Lokhande, C.D. MnS-La_2S_3/GO composite electrodes for high-performance flexible symmetric supercapacitor. *Appl. Surf. Sci. Adv.* **2023**, *15*, 100399. [CrossRef]
101. Biradar, M.R.; Rao, C.R.K.; Bhosale, S.V.; Bhosale, S.V. Bio-inspired adenine-benzoquinone-adenine pillar grafted graphene oxide materials with excellent cycle stability for high energy and power density supercapacitor applications. *J. Energy Storage* **2023**, *58*, 106399. [CrossRef]
102. Chen, T.; Shen, T.; Wang, Y.; Yu, Z.; Zhang, W.; Zhang, Y.; Ouyang, Z.; Cai, Q.; Ji, Y.; Wang, S. In Situ Synthesis of Ni-BTC Metal–Organic Framework@Graphene Oxide Composites for High-Performance Supercapacitor Electrodes. *ACS Omega* **2023**, *8*, 10888–10898. [CrossRef] [PubMed]
103. Fu, X.-Y.; Ma, C.-J.; Shu, R.-Y.; Zhang, Y.-Y.; Jiang, H.-B. Electrostatic self-assembled MXene–graphene oxide composite electrodes for planar supercapacitors. *Appl. Phys. Lett.* **2023**, *122*, 113901. [CrossRef]
104. Yasoda, K.Y.; Baji, D.S.; Kumar, M.S.; Santhanagopalan, D.; Batabyal, S.K. Sustainable development of manganese sulfoselenide nanoparticles anchored graphene oxide nanocomposite for high-performance supercapacitor and lithium-ion battery applications. *J. Alloys Compd.* **2023**, *930*, 167282. [CrossRef]
105. Sriram, G.; Thangarasu, S.; Selvakumar, K.; Kurkuri, M.; Dhineshbabu, N.R.; Oh, T.H. Effective removal of Rose Bengal using Ni-Co-Zn layered triple hydroxide: Studies on batch adsorption, mechanism, selectivity, co-ions, and reusability. *Colloids Surf. A* **2024**, *685*, 133199. [CrossRef]
106. Zhao, W.; Liu, T.; Wu, N.; Zhou, B.; Yan, Y.; Ye, Y.; Gong, J.; Yang, S. Bimetallic electron-induced phase transformation of CoNi LDH-GO for high oxygen evolution and supercapacitor performance. *Sci. China Mater.* **2023**, *66*, 577–586. [CrossRef]
107. Shi, J.; Du, N.; Zheng, W.; Li, X.; Dai, Y.; He, G. Ultrathin Ni-Co double hydroxide nanosheets with conformal graphene coating for highly active oxygen evolution reaction and lithium ion battery anode materials. *Chem. Eng. J.* **2017**, *327*, 9–17. [CrossRef]
108. Sriram, G.; Uthappa, U.T.; Losic, D.; Kigga, M.; Jung, H.-Y.; Kurkuri, M.D. Mg–Al-Layered Double Hydroxide (LDH) Modified Diatoms for Highly Efficient Removal of Congo Red from Aqueous Solution. *Appl. Sci.* **2020**, *10*, 2285. [CrossRef]
109. Sriram, G.; Bendre, A.; Altalhi, T.; Jung, H.-Y.; Hegde, G.; Kurkuri, M. Surface engineering of silica based materials with Ni–Fe layered double hydroxide for the efficient removal of methyl orange: Isotherms, kinetics, mechanism and high selectivity studies. *Chemosphere* **2022**, *287*, 131976. [CrossRef] [PubMed]
110. Bao, E.; Ren, X.; Wang, Y.; Zhang, Z.; Luo, C.; Liu, X.; Xu, C.; Chen, H. Advanced hybrid supercapacitors assembled with CoNi LDH nanoflowers and nanosheets as high-performance cathode materials. *J. Energy Storage* **2024**, *82*, 110535. [CrossRef]
111. Hummers, W.S., Jr.; Offeman, R.E. Preparation of Graphitic Oxide. *J. Am. Chem. Soc.* **1958**, *80*, 1339. [CrossRef]
112. Nanaji, K.; Sarada, B.V.; Varadaraju, U.V.; Rao, T.N.; Anandan, S. A novel approach to synthesize porous graphene sheets by exploring KOH as pore inducing agent as well as a catalyst for supercapacitors with ultra-fast rate capability. *Renew. Energy* **2021**, *172*, 502–513. [CrossRef]
113. Siddiqui, R.; Rani, M.; Ahmad Shah, A.; Razaq, A.; Iqbal, R.; Neffati, R.; Arshad, M. Fabrication of tricarboxylate-neodymium metal organic frameworks and its nanocomposite with graphene oxide by hydrothermal synthesis for a symmetric supercapacitor electrode material. *Mater. Sci. Eng. B* **2023**, *295*, 116530. [CrossRef]
114. Khakpour, F.; Ghafouri Taleghani, H.; Soleimani Lashkenari, M. Metal–organic framework/sulfur-doped graphene oxide nanocomposite for high efficiency electrochemical double-layer capacitors. *Appl. Organomet. Chem.* **2023**, *37*, e6969. [CrossRef]
115. Zhang, H.; Wei, Z.; Wu, J.; Cheng, F.; Ma, Y.; Liu, W.; Cheng, Y.; Lin, Y.; Liu, N.; Gao, Y.; et al. Interlayer-spacing-regulated MXene/rGO Foam for Multi-functional Zinc-ion Microcapacitors. *Energy Storage Mater.* **2022**, *50*, 444–453. [CrossRef]
116. Sun, P.; Liu, J.; Liu, Q.; Yu, J.; Chen, R.; Zhu, J.; Sun, G.; Li, Y.; Liu, P.; Wang, J. An inkjet-printing ink based on porous NiS/N-MXene for high-performance asymmetric micro-supercapacitors and self-powered microelectronics. *Chem. Eng. J.* **2023**, *474*, 145466. [CrossRef]
117. An, N.; Li, W.; Shao, Z.; Zhou, L.; He, Y.; Sun, D.; Dong, X.; Hu, Z. Graphene oxide coated polyaminoanthraquinone@MXene based flexible film electrode for high-performance supercapacitor. *J. Energy Storage* **2023**, *57*, 106180. [CrossRef]
118. Fu, X.-Y.; Shu, R.-Y.; Ma, C.-J.; Zhang, Y.-Y.; Jiang, H.-B.; Yao, M.-N. Self-assembled MXene-graphene oxide composite enhanced laser-induced graphene based electrodes towards conformal supercapacitor applications. *Appl. Surf. Sci.* **2023**, *631*, 157549. [CrossRef]
119. Hussain, S.; Vikraman, D.; Ali Sheikh, Z.; Taqi Mehran, M.; Shahzad, F.; Mujasam Batoo, K.; Kim, H.-S.; Kim, D.-K.; Ali, M.; Jung, J. WS_2-embedded MXene/GO hybrid nanosheets as electrodes for asymmetric supercapacitors and hydrogen evolution reactions. *Chem. Eng. J.* **2023**, *452*, 139523. [CrossRef]
120. Verma, S.; Verma, B. Synergistic interaction of bacteria with graphene oxide for high performance supercapacitor. *Bioresour. Technol. Rep.* **2023**, *21*, 101354. [CrossRef]
121. Li, X.; Wu, H.; Guan, C.; Elshahawy, A.M.; Dong, Y.; Pennycook, S.J.; Wang, J. (Ni,Co)Se_2/NiCo-LDH Core/Shell Structural Electrode with the Cactus-Like (Ni,Co)Se_2 Core for Asymmetric Supercapacitors. *Small* **2019**, *15*, 1803895. [CrossRef] [PubMed]
122. Sriram, G.; Bendre, A.; Mariappan, E.; Altalhi, T.; Kigga, M.; Ching, Y.C.; Jung, H.-Y.; Bhaduri, B.; Kurkuri, M. Recent trends in the application of metal-organic frameworks (MOFs) for the removal of toxic dyes and their removal mechanism—A review. *Sustain. Mater. Technol.* **2022**, *31*, e00378. [CrossRef]
123. Feng, S.; Wang, X.; Wang, M.; Bai, C.; Cao, S.; Kong, D. Crumpled MXene Electrodes for Ultrastretchable and High-Area-Capacitance Supercapacitors. *Nano Lett.* **2021**, *21*, 7561–7568. [CrossRef] [PubMed]

124. Wu, J.; Li, Q.; Shuck, C.E.; Maleski, K.; Alshareef, H.N.; Zhou, J.; Gogotsi, Y.; Huang, L. An aqueous 2.1 V pseudocapacitor with MXene and V-MnO_2 electrodes. *Nano Res.* **2022**, *15*, 535–541. [CrossRef]
125. Hashem, A.; Tabassum, B.; Fathi Abd_Allah, E. Bacillus subtilis: A plant-growth promoting rhizobacterium that also impacts biotic stress. *Saudi J. Biol. Sci.* **2019**, *26*, 1291–1297. [CrossRef] [PubMed]
126. Wu, Q.; He, T.; Zhang, Y.; Zhang, J.; Wang, Z.; Liu, Y.; Zhao, L.; Wu, Y.; Ran, F. Cyclic stability of supercapacitors: Materials, energy storage mechanism, test methods, and device. *J. Mater. Chem. A* **2021**, *9*, 24094–24147. [CrossRef]
127. Patra, A.; Namsheer, K.; Jose, J.R.; Sahoo, S.; Chakraborty, B.; Rout, C.S. Understanding the charge storage mechanism of supercapacitors: In situ/operando spectroscopic approaches and theoretical investigations. *J. Mater. Chem. A* **2021**, *9*, 25852–25891. [CrossRef]
128. Ali, S.; Ahmad, T.; Tahir, M.Y.; Usman, M.; Chhattal, M.; Hussain, I.; Khan, S.; Hassan, A.M.; Assiri, M.A.; Rosaiah, P.; et al. The emergence of density functional theory for supercapacitors: Recent progress and advances. *J. Energy Storage* **2023**, *73*, 109100. [CrossRef]
129. Becke, A.D. Perspective: Fifty years of density-functional theory in chemical physics. *J. Chem. Phys.* **2014**, *140*, 18A301. [CrossRef]
130. Perdew, J.P.; Burke, K.; Ernzerhof, M. Generalized Gradient Approximation Made Simple. *Phys. Rev. Lett.* **1996**, *77*, 3865–3868. [CrossRef]
131. Luryi, S. Quantum capacitance devices. *Appl. Phys. Lett.* **1988**, *52*, 501–503. [CrossRef]
132. Karmakar, S.; Mistari, C.D.; Shajahan, A.S.; More, M.A.; Chakraborty, B.; Behera, D. Enhancement of Pseudocapacitive Behavior, Cyclic Performance, and Field Emission Characteristics of Reduced Graphene Oxide Reinforced $NiGa_2O_4$ Nanostructured Electrode: A First Principles Calculation to Correlate with Experimental Observation. *J. Phys. Chem. C* **2021**, *125*, 7898–7912. [CrossRef]
133. Kolavada, H.; Gajjar, P.N.; Gupta, S.K. Unraveling quantum capacitance in supercapacitors: Energy storage applications. *J. Energy Storage* **2024**, *81*, 110354. [CrossRef]
134. Ghosh, S.; Behera, S.K.; Mishra, A.; Casari, C.S.; Ostrikov, K.K. Quantum Capacitance of Two-Dimensional-Material-Based Supercapacitor Electrodes. *Energy Fuels* **2023**, *37*, 17836–17862. [CrossRef]
135. Pathak, M.; Jose, J.R.; Chakraborty, B.; Rout, C.S. High performance supercapacitor electrodes based on spinel $NiCo_2O_4$@MWCNT composite with insights from density functional theory simulations. *J. Chem. Phys.* **2020**, *152*, 064706. [CrossRef] [PubMed]
136. Islam, T.; Hasan, M.M.; Sarker, S.; Ahammad, A.J.S. Intrinsic Properties of GO/RGO Bilayer Electrodes Dictate Their Inter-/Intralayer Intractability to Modulate Their Capacitance Performance. *ACS Omega* **2023**, *8*, 14013–14024. [CrossRef] [PubMed]
137. Reddy, P.A.K.; Han, H.; Kim, K.C.; Bae, S. Synthesis of ZIF-67-derived CoS_2@graphitic carbon/reduced graphene oxide for supercapacitor application. *Chem. Eng. J.* **2023**, *471*, 144608. [CrossRef]
138. Li, X.; Lin, Z.; Wang, C.; Wang, H.; Feng, S.; Li, T.; Ma, Y. Hierarchical graphene oxide/manganese dioxide/cobalt-nickel layered double hydroxide electrodes for high energy density asymmetric flexible supercapacitor. *Chem. Eng. J.* **2024**, *484*, 149430. [CrossRef]
139. da Silva, D.A.C.; Paulista Neto, A.J.; Pascon, A.M.; Fileti, E.E.; Fonseca, L.R.C.; Zanin, H.G. Combined Density Functional Theory and Molecular Dynamics Simulations To Investigate the Effects of Quantum and Double-Layer Capacitances in Functionalized Graphene as the Electrode Material of Aqueous-Based Supercapacitors. *J. Phys. Chem. C* **2021**, *125*, 5518–5524. [CrossRef]
140. Li, Z.; Hu, B.; Yu, R.; Tian, T.; Guo, Z.; Mu, J.; Zhang, X.; Wang, G.; Liu, X.; Liu, A. Hollow cobalt selenide nanospheres wrapped with reduced graphene oxide nanosheets as electrodes for hybrid supercapacitor. *Appl. Surf. Sci.* **2023**, *608*, 155237. [CrossRef]
141. Yang, Y.; Qian, D.; Yang, J.; Xiong, Y.; Chen, Y.; He, Y.; Hu, Z. Graphene modification with chrysin molecules as a high performance electrode material for supercapacitor. *Appl. Surf. Sci.* **2023**, *639*, 158267. [CrossRef]
142. Kumar, N.; Verma, S.; Kumar, P.; Khan, A.A.; Park, J.; Srivastava, V.C. Efficient one-pot two-step electrochemical synthesis of highly oxidized graphene oxide for enhanced energy and environmental applications: Structure elucidation through DFT simulations. *Carbon* **2024**, *218*, 118722. [CrossRef]
143. Mohammadi, S.; Mahdi Homayounfard, A.; Morteza Mousavi-Khoshdel, S. Experimental and theoretical investigation of high-performance supercapacitor based on guanidine functionalized graphene oxide. *Electrochem. Commun.* **2024**, *163*, 107703. [CrossRef]

inorganics

MDPI

Article

Ni and Co Catalysts on Interactive Oxide Support for Anion Exchange Membrane Electrolysis Cell (AEMEC)

Katerina Maksimova-Dimitrova, Borislava Mladenova, Galin Borisov and Evelina Slavcheva *

Acad. Evgeni Budevski Institute of Electrochemistry and Energy Systems, Bulgarian Academy of Sciences, Acad. G. Bonchev bl. 10, 1113 Sofia, Bulgaria; katerina.maksimova@iees.bas.bg (K.M.-D.); borislava.mladenova@iees.bas.bg (B.M.); gal.rusev@iees.bas.bg (G.B.)

* Correspondence: eslavcheva@iees.bas.bg

Abstract: The work presents novel composite catalytic materials—Ni and Co deposited on Magneli phase titania—and describes their complex characterization and integration into membrane electrode assemblies to produce hydrogen by electrochemical water splitting in cells with anion exchange membranes (AEMEC). Chemical composition, surface structure, and morphology were characterized by XRD and SEM analysis. The activity in the evolution of the partial electrode reactions of hydrogen (HER) and oxygen (OER) was assessed in an aqueous alkaline electrolyte (25 wt.% KOH) using linear sweep voltammetry. The interactive role of the support was investigated and discussed. Among the tested samples, the sample with 30 wt.% Co (Co30/MPT) demonstrated superior performance in the OER. The reaction started at 1.65 V, and at 1.8 V, the current density reached 75 mA cm^{-2}. The HER is most efficient on the sample containing 40 wt.% Ni (Ni40/MPT), where the current density reaches 95 mA at a potential of −0.5 V. The change in catalytic efficiency compared to that of the unsupported Ni and Co is due to synergism resulting from electronic interactions between the transition metal having a hyper-d-electron character and hypo-d-electron support. The pre-selected catalysts were integrated in membrane electrode assembly (MEA) using commercial and laboratory-prepared anion-conductive membranes and tested in a custom-made AEMEC. The performance was compared to that of MEA with a commercial carbon-supported Pt catalyst. It was found that the MEA with newly prepared catalysts demonstrated better performance in long-term operation (50 mA cm^{-2} at 1.8 V in a 60 h durability test), which, combined with the higher cost efficiency, gave credence to considering this combination of materials as promising for AEMEC applications.

Keywords: non-precious catalysts; oxide support; hydrogen generation; AEM water electrolysis

1. Introduction

Hydrogen has been considered the energy vector of the future for more than two decades. Recently, it became clear that the large-scale production of green hydrogen is the key success factor for the transformation to a zero-emission economy and that the electrolysis of water powered by renewable energy sources is the only mature technology for this purpose. A new trend in the development of this technology is electrolysers working with polymer anion-exchange membranes (AEMs). The AEM provides high OH-conductivity, separates the H_2 and O_2 gases produced on the cathode and anode of the cell, and enables both compact system design and high operating pressure. The AEM electrolysis cells with a polymer membrane are expected to combine the advantages of the classical alkaline electrolysers working with aqueous electrolyte and the electrolysis cells using a proton exchange membrane electrolysis cell (PEMEC)—namely, by using non-precious metal catalysts to achieve high operating current densities and a dynamic response to the varying power supplied from renewable sources.

In classical alkaline electrolysis, the electrodes are made from stainless steel or foam nickel. In AEM technology, similarly to PEM water electrolysis, the efficiency of the process can be enhanced by the use of nanosized catalysts dispersed on catalyst carriers with highly developed surfaces [1–4]. This type of catalyst shows higher activity than the non-supported catalysts. The effect is attributed to the redistribution of the electronic density on the active surface, which optimizes the adsorption strength of the intermediates and facilitates the kinetics of the partial electrode reactions. As a result, the catalytic loading of the electrodes can be reduced essentially to less than 0.5 mg cm^{-2} while at the same time preserving the efficiency of the electrolysis [5–7].

The most commonly used catalysts in AEM water electrolysers for the hydrogen evolution reaction (HER) and the oxygen evolution reaction (OER) are transition metals like Ni and Co, as well as their alloys. One of the most favorable metal oxide materials for OER in alkaline media is $NiCo_2O_4$. This material is known for its high electrical conductivity and high number of catalytically active sites. It has a spinel structure and is widely studied as a non-platinum catalyst due to its ability to exhibit a catalytic efficiency comparable to that of platinum-based catalysts [8–10]. Another interesting alloy for these partial reactions is NiFeCoP, which has been explored in our previous research. These materials were electrodeposited and investigated in alkaline media, including in AEM water electrolysis and zero-gap configuration with hybrid electrical connections [11,12].

The electrical efficiency and long-term durability of AEM electrolysers are highly dependent on the type of catalytic carrier used. The most commonly used catalytic supports are carbon materials, such as carbon blacks, nanotubes, fibers, etc., known for their excellent electric conductivity and high surface area. The drawback of carbon materials is their inclination to corrosion at the high anode potentials where the oxygen evolution takes place, leading to gradual degradation of the electrodes and consequent failure in the MEA performance. Therefore, in the last decade, intensive research has been carried out on the application of non-carbon-supported catalysts. A variety of materials combining high electrical conductivity and resistance to corrosion, such as metal oxides (SnO_2, Ta_2O_5, TiO_2, Nb_2O_5, WOx, etc.), carbides, or nitrides, have been investigated, and some very promising results have been reported [13,14]. The goal is to develop new generation advanced corrosion-resistant catalysts with zero PGM content, capable of ensuring the superior cost efficiency of the AEMEC and to justify this type of electrolysis as an economically attractive technology for the industrial production of green hydrogen.

This goal is also addressed in this study through the synthesis of Co and Ni nanoparticles deposited on Magneli phase titania (MPT), their complex characterization, and the assessment of their applicability in AEMEC with a laboratory-prepared OH-conducting polymer membrane electrolyte. The aim is to develop composite catalysts with improved efficiency and stability by combining supporting materials with excellent electrical conductivity and high corrosion resistance and transition metals with proven catalytic activity and chemical stability in alkaline media.

According to the literature, the structure of the Magneli phase titania Ti_4O_7 is a combination of rutile and corundum and has a deficiency of oxygen (Figure 1). It consists of four-unit rutile-like chains bounded by corundum-like planes [15].

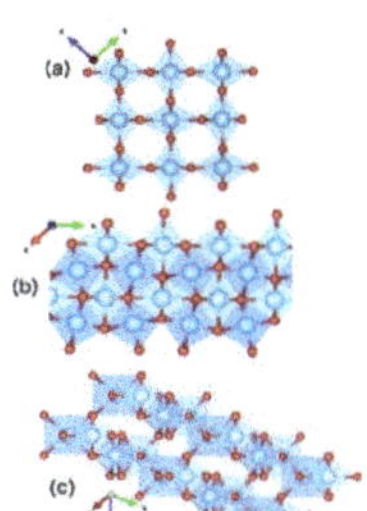

Figure 1. Schematic structure of (**a**) rutile, (**b**) corundum and (**c**) MPT, which is a combination of both (**a**) and (**b**). Ti atoms are the blue spheres enclosed by blue octahedra. The oxygen atoms are the red spheres on the edges of the octahedra.

2. Results

The structure and phase composition of the synthesized materials were examined by X-ray diffraction (XRD). In Figure 2a are presented the results from XRD analysis of the selected catalyst support (MPT). They show that MPT is a mixture of two well recognized crystalline phases—Magneli phase titania Ti_4O_7 and hollandite $K_{1.28}Ti_8O_{16}$ ($1 < x < 1.2$). The peaks at 2θ 26.38, 20.78, and 31.76 correspond to the main Ti_4O_7 Magneli phase (ref. code: 00-072-1722], while the presence of $K_{1.28}Ti_8O_{16}$ is substantiated by the appearance of peaks at 2θ 27.69, 36.11, and 40.35 [ref code: 01-084-2058]. The SEM images in Figure 2b show homogeneously distributed MPT particles with a size in the range of 100–150 nm.

The composition, surface structure and morphology of Ni and Co catalysts supported on MPT were also examined by means of scanning electron microscopy (SEM) and X-ray diffraction (XRD). The SEM results showed that the metallic nanoparticles were homogeneously distributed on the support, forming cauliflower aggregates that are illustrated by the microscopic images of selected samples presented in Figure 3. The increase in the active metal of more than 40% wt. leads to agglomeration of the particles and formation of clusters, which in turn, decreases the active surface area, and thus the catalyst efficiency.

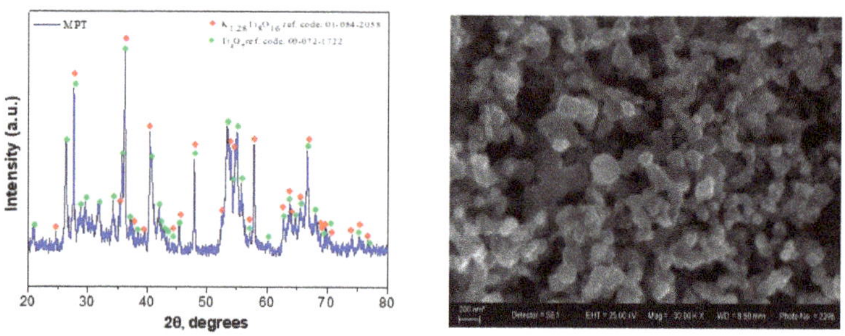

(a) (b)

Figure 2. Structural data of the MPT under study: (**a**) X-ray diffractograms, (**b**) SEM images.

Figure 3. SEM images of (**a**) Ni40/MPT and (**b**) Co30/MPT at different magnifications.

XRD patterns of the synthesized nickel and cobalt catalysts are presented in Figure 4. Figure 4a shows the peak with the greatest intensity at 44.38 2θ, as well as peaks at 51.67 and 76.1 corresponding to the (111), (200), and (220) planes of symmetry of Ni cubic structure and space Fm-3m group (ref. code: 00-065-0380). The results in Figure 4b demonstrate the presence of metallic cobalt in two crystallographic states—hexagonal (ref. code: 00-001-1278), corresponding to the peak with greatest intensity at 47.39 (101), and cubic (111), substantiated with the peak at 44.43 (ref. code: 00-001-1259). Both figures show that the phases of MPT are detected in all samples under study.

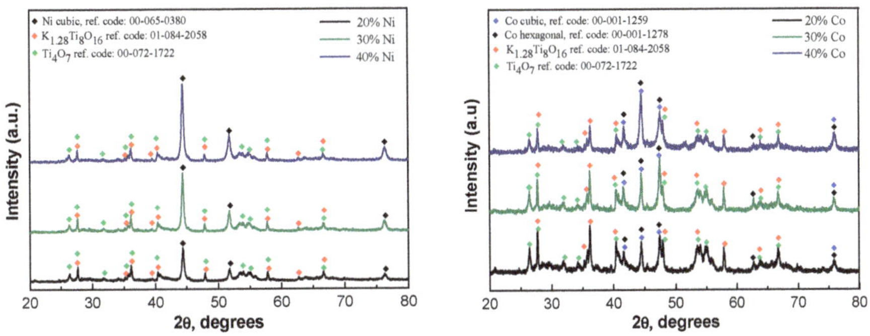

(a) (b)

Figure 4. X-ray diffractograms of the synthesized MPT-supported (**a**) Ni and (**b**) Co catalysts.

The size of the catalyst crystallites, t, is shown in Table 1. This is determined by Scherer's formula t = κλ/Bcosθ, where k is a shape factor (k = 0.9 for spherical crystals), λ (Å) is the wavelength, θ is the diffraction angle of the peak, and B (rad) is the line broadening at the full width at half maximum (FWHM) values of the peaks [16].

Table 1. Crystallite size of Ni/MPT (a) and Co/MPT (b) catalysts.

(a) Sample	Crystallite Size (nm) 2θ = 44.38 (111) h	Crystallite Size (nm) 2θ = 51.67 (200) h
40% Ni	26	19
30% Ni	29	24
20% Ni	27	24

(b) Sample	Crystallite Size (nm)								
	2θ = 41.58 (100) h	2θ = 41.78 (100) c	2θ = 44.60 (002) h	2θ = 44.37 (111) c	2θ = 47.30 (101) h	2θ = 47.30 (101) c	2θ = 62.71 (102) h	2θ = 75.37 (110) h	2θ = 75.37 (220) c
20% Co	39		28		22		20	17	
30% Co	41		34		23		50	21	
40% Co	30		20		15		48	17	

The efficiency of the prepared catalysts toward both partial reactions of water electrolysis was evaluated in an aqueous solution of 25 wt.% KOH at room temperature. The influence of the metal content on the catalytic efficiency was investigated using linear sweep voltammetry. It was previously found that Ni/MPT more significantly catalyzes the HER, while Co/MPT is much more efficient in the OER, showing remarkable stability at high anodic potentials [17]. Figure 5 presents the cathodic and anodic polarization curves obtained on electrodes with Ni/MPT and Co/MPT, respectively at different metal: support ratios.

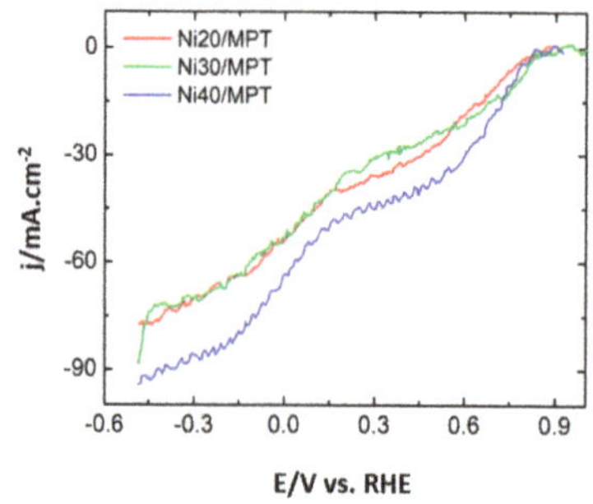

(a)

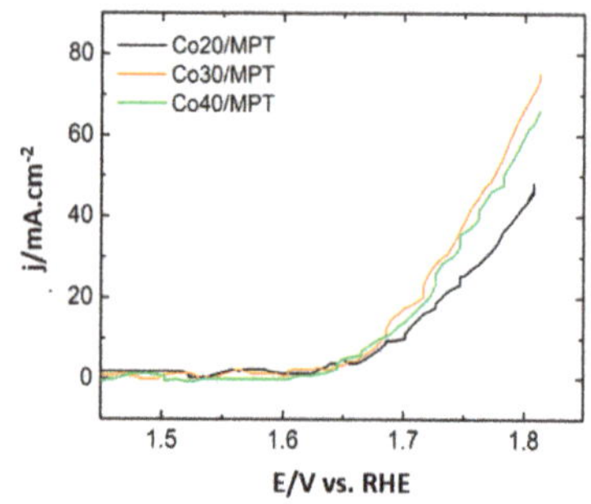

(b)

Figure 5. Polarization curves on Ni/MPT (**a**) and Co/MPT (**b**) with varying metal content recorded in 25% KOH at room temperature; potential scan rate: 1 mV s^{-1}.

The obtained results show that the cathodic partial reaction is most intensive on the catalyst with 40 wt.% metal (Ni40/MPT), while the rate of the oxygen evolution on the anode is very similar for Co30/MPT and Co40/MPT and proceeds with higher overvoltage on Co20/MPT. The registered different trend of the catalytic efficiency with the changing metal: support ratio for both metals could be explained with the more homogeneously dispersed Ni compared to Co particles. The cobalt particles at metal content above 30% wt. form bigger clusters, thus decreasing the active surface area available for the electrochemical reaction [18].

The role of the catalytic support is illustrated in Figure 6, where the intensity of the partial electrode reaction on the best performing samples and on non-supported metal nanoparticles is compared. It is seen that HER on the cathode with unsupported Ni nanoparticles proceeds with lower rate compared to Ni dispersed on MPT. The opposite effect is observed for the anode catalysts where the pure Co nanoparticles demonstrate distinct superior OER performance.

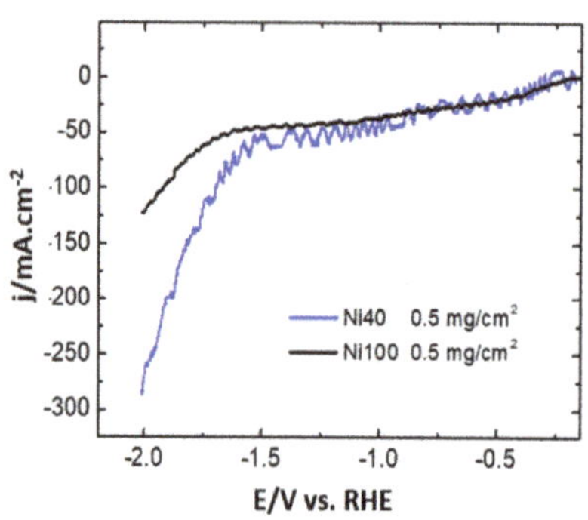

(a)

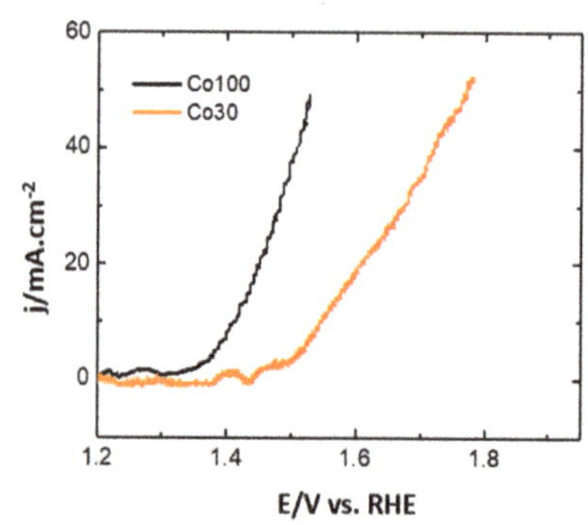

(b)

Figure 6. Polarization curves demonstrating the interactive role of Magneli phase titania as interactive catalytic support; (**a**) HER on Ni40/MPT and on non-supported Ni, (**b**) OER on Co30/MPT and on non-supported Co; potential scan rate: 1 mV s^{-1}.

It is well known that the general efficiency of electrocatalysts is determined by two major groups of factors: electronic factors affecting the surface–intermediate bonds strength and geometric factors related to the real active surface area of the catalyst [19,20]. The catalysts under study have similar surface area—they were deposited on the same substrate and XRD data showed similarity in their crystallites size. Therefore, the results obtained cannot be explained with the influence of geometry factors. On the other hand, since the partial electrode reactions are heterogeneous catalytic reactions, their rate depends on the strength of intermediate species–electrode surface chemisorptive bonds, which in turn, is influenced by the electronic density of states in the catalyst atoms. According to the Brewer theory [21], the increase in the electronic density of states increases the availability of metal valence electrons for the formation of chemisorptive bonds, and the result is the weakening of the metal-electron donor bonds such as M-H and strengthening of metal-electron acceptor bonds such as M-O. The decrease in the electronic density of states has an opposite effect. The optimal electrocatalytic activity is achieved at a neither too strong nor too weak level of adsorption bonding, but at a moderate level.

In the case under study, both Ni and Co are hyper-d-metals with electronic configurations $[Ar]3d^8 4s^2$ and $[Ar]3d^7 4s^2$, respectively, and their interatomic interactions with the hypo-d-electron Ti from the support with $[Ar]3d^3 4s^2$ structure should lead to decrease in the electronic density of states. As a result, the strength of the Ni-H bond should decrease, having a positive effect on the efficiency of HER. In contrast, Co-OH and Co-O bonds strength should increase, thus increasing OER overpotential. Based on these considerations and the results from the polarization tests (Figures 5 and 6), it could be suggested with great probability that the registered increase in the efficiency of the composite Ni/MPT catalysts and decrease in Co/MPT performance compared to the pure metals are due to electronic interactions between the hyper-d-electron metals and the hypo-d-electron support. To reveal the exact mechanism of the registered synergetic effects, further focused research is needed that would be outside the scope of this study.

Based on the performed preliminary electrochemical screening of the synthesized catalysts in aqueous alkaline electrolytes, Ni40/MPT and Co30/MPT samples were chosen for further testing in the electrolysis cell with an anion exchange membrane. The fact that AEMECs can run efficiently with non-noble catalysts is one of their most important benefits. Another major advantage is the much lower cost of OH-conductive membranes compared to the proton exchange membranes. On the other hand, the poor thermal and chemical stability of the available commercial anion-exchange membranes as well as their relatively low ionic conductivity are still unsolved issues. The commercially available OH-conductive membranes are mainly based on aromatic and aliphatic resins in ionomeric quaternized ammonia. At elevated temperatures (above 50 °C) and in alkaline environments, these resins are very prone to chemical degradation, leading to a rapid deterioration in performance for all MEA [22,23]. This problem can be solved by using anion-conductive membranes based on the chemically stable polybenzimidazoles (PBIs). The preparation and properties of membranes based on para-polybenzimidazole (p-PBI) doped with 50% aqueous KOH have previously been reported [24–27].

For the purpose of current research, five MEAs with different types of AEM and different catalysts integrated into the cathode and anode were prepared and investigated. Their composition is given in Table 2.

Table 2. Composition of the membrane electrode assemblies under study.

MEA	Cathode	Membrane	Anode
MEA1	Pt	Fumapem	Pt
MEA2	Pt	PBI	Pt
MEA3	Ni/MPT	PBI	Pt
MEA4	Pt	PBI	Co/MPT
MEA5	Ni/MPT	PBI	Co/MPT

In Figure 7c, the cyclic voltamograms of MEA1 and MEA2 are compared with Fumapem and PBI-based membranes, respectively which contain commercial carbon-supported platinum catalyst in both electrodes.

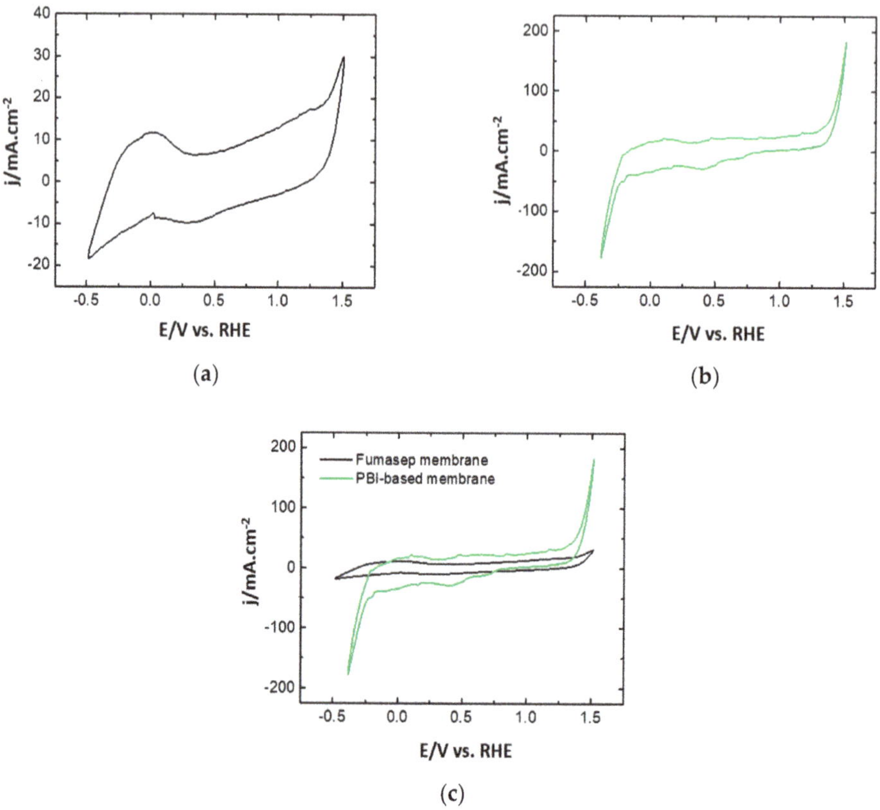

Figure 7. Influence of membrane type on electrochemical behavior of MEAs with commercial electrodes (0.5 mgPt cm^{-2}) and different electrolyte membranes: (**a**) Fumapem—MEA1; (**b**) PBI-based—MEA2; (**c**) both; potential scan rate: 100 mV s^{-1}.

The surface reactions inherent to Pt can be distinguished on both curves in the relevant hydrogen and oxygen CV branches: adsorption and desorption of hydrogen, coverage of the electrode by oxygen-containing intermediates followed by oxygen evolution and reduction at the reverse potential scan. However, the potential values at which these reactions occur and the measured currents fundamentally differ. The potential window of electrochemical water splitting on MEA1 is above 2.0 V (Figure 7a). The partial reaction of oxygen evolution starts at 1.45 V, while the start-off potential of hydrogen evolution on the cathode is rather high, about −0.55 V. Since the catalytic properties of Pt with respect to these reactions are well known, it can be assumed that the overly wide potential window and high HER overvoltage are due to the low ionic conductivity of the polymer electrolyte. The MEA2 containing the laboratory-prepared PBI membrane shows much better performance (Figure 7b) than MEA1. The characteristic current peaks are clearly illustrated, and the measured currents are considerably higher. The cathode reaction of hydrogen evolution begins at a potential of −0.25 V and its intensity increases rapidly reaching almost 200 mA cm^{-2} at −0.4 V. The oxygen evolution reaction starts at a potential of 1.4 V, and at 1.5 V, the current density reaches value of 180 mA cm^{-2}. The potential window in which electrochemical water splitting occurs is 1.7 V, i.e., about 300 mV narrower than that of MEA1. For a better visualization of the observed differences, the CVs of both MEAs are compared in Figure 7c.

Similar conclusions can be made when comparing the polarization characteristics of MEA1 and MEA2 presented in Figure 8a,b. Clearly, ohmic losses are significantly lower for the laboratory-prepared membrane than for the commercial Fumasep FA. The oxygen evolution reaction on MEA2 starts at a potential of 1.4 V, which is approximately 200 mV lower overvoltage compared to the same electrode integrated in MEA1 with a commercial membrane. The obtained results clearly show that the laboratory-prepared anion-conducting PBI membrane is much more appropriate as an electrolyte in AEMWE. Therefore, this membrane is further used to assemble MEAs with the synthesized composite MPT-supported catalysts.

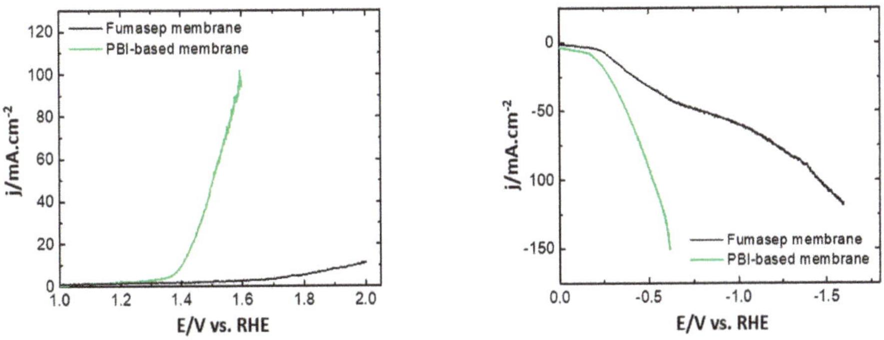

(a) (b)

Figure 8. Influence of the type of membrane on the intensity of: (**a**) OER and (**b**) HER.

The performance of the pre-selected catalysts in AEMWE cell was investigated by LSV and chronoamperometry. Figure 9a,b show the cathodic and anodic polarization curves of MEA3 (0.5 mg cm^{-2} Ni40/MPT at the cathode) and MEA4 (0.5 mg cm^{-2} Co30/MPT at the anode), respectively. In both figures, the results obtained for MEA2 (0.4 mg cm^{-2} Pt/C on both electrodes) are also presented for comparison. The curves were recorded under identical conditions with a potential scan rate of 1 mV s^{-1}.

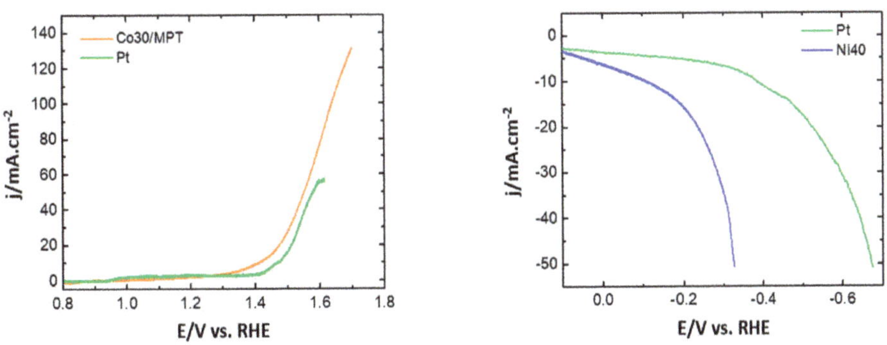

(a) (b)

Figure 9. Polarization curves of MEAs with synthesized MPT-supported catalysts (0.5 mgMe cm^{-2}) and with Pt/C catalyst (0.4 mgPt cm^{-2}) in AEMEC; (**a**) HER; (**b**) OER; potential scan rate 1 mV s^{-1}.

The hydrogen evolution reaction on the nickel-based catalyst reaches a current density of 50 mA cm^{-2} at a potential of −0.3 V, while for Pt, this value is −0.7 V. The obtained results confirm the superior performance of the synthesized composite Ni/MPT compared to Pt/C. Taking into consideration that the rate-determining partial electrode reaction in alkaline water electrolysis is the hydrogen evolution of the cathode, Ni/MPT can be considered as a very promising HER catalyst for water splitting in AEMEC.

The oxygen evolution reaction on Pt/C catalyst starts at a potential of 1.4 V, while on the Co30/MPT—about 50 mV earlier. The current densities observed at a potential of 1.6 V are 55 mA cm^{-2} and 80 mA cm^{-2}, respectively. At higher potentials, the current density of MEA2 reaches a limiting value, indicating a change in the OER mechanism due to transport limitations or other factors impeding the reaction. In contrast, the polarization curve on the Co/MPT composite catalyst preserves the linear trend, reaching 140 mA cm^{-2} at potential of 1.7 V. The reason for the observed discrepancy in the performance of both MEA can be explained by their different compositions. The commercial Pt catalysts in MEA2 are supported on carbon black XC72, which at high potentials of the OER, is not electrochemically stable and is prone to oxidation. This leads to the loss of active catalyst surfaces and a decrease in efficiency. Other factors such as transport limitations might be also involved. In contrast, MPT support is very stable under these experimental conditions, which explains why no signs of degradation or loss of OER intensity are observed on the MEA4 polarization curve.

In the final part of this study, an MEA5, with newly developed catalysts integrated in both electrodes (Ni40/MPT on the cathode and Co30/MPT on the anode), was fabricated and its performance at a cell voltage of 2.0 V was followed in a long-term chronoamperometric test. The obtained results are presented in Figure 10. For comparison, the curve recorded on MEA2 under identical conditions is also shown. It is seen that both MEAs demonstrate stable electrochemical behavior for the whole test duration. After some initial oscillations, the current density stabilized and almost did not change until the end of the 240 h experiment. The measured current density on the membrane electrode assembly with the commercial Pt electrodes (MEA2) was comparatively high at the beginning of the test (130–140 mA cm^{-2}). However, the intensity of the electrolysis rapidly decreased with time, reaching a constant current density of 22–25 mA cm^{-2}.

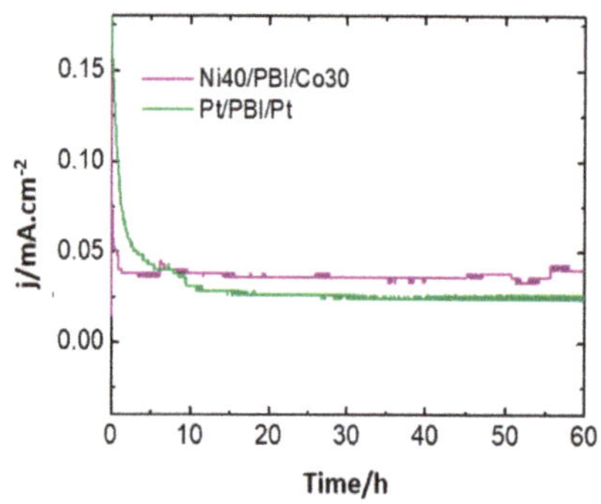

Figure 10. Chronoamperometric curves of MEA2 and MEA5 in AEMEC; cell voltage: 2.0 V.

The drop in the rate of the process on MEA5 was not as sharp, since it started at a current density of about 50 mA cm^{-2} lower and reached stability at values in the range of 38–40 mA cm^{-2}, which is about 1.5 times higher than on MEA2. These results once again confirm the stable performance of the novel composite MPT-supported Ni and Co catalysts and justify current research that is focusing on membrane thermal stability, the increase in operating temperature, scaling up of the MEA, and design of a short electrolysis stack.

3. Materials and Methods

The Magneli phase titania (MPT) chosen as catalytic support has excellent electrical conductivity approaching that of a metal combined with high corrosion resistance both in acid and in alkaline media [28–30]. In addition, its chemical composition and hypo-d-electron character enable interaction with hyper-d-electron elements form the transition series (Ni, Co, Fe, Mn, etc.), which facilitates the formation of intermediate species on the electrode surface and speeds up the kinetics of the electrode reactions. This synergetic effect is known as strong metal–support interaction (SMSI effect) and is well documented in the literature [31–34].

In this study, composite catalysts were deposited using a sol–gel process as nanoparticles using acetylacetonate precursors (M(($C_5H_7O_2$)n)m, (M = Ni, Co) (Alpha Acer) on Magneli Phases Titanium Oxide with the general formula Ti_nO_{2n-1}, (commercially available from Ti-dynamics Co., Ltd., Changsha, China, designated MPT). According to the manufacturer's technical specification, MPT has an area of 7–8 m^2 g^{-1}, very good electrical conductivity and corrosion resistance in both acidic and alkaline environments. The content of the metal part (Co or Ni) varied in the range of 0–40 wt.%. The test samples were denoted MX/MPT where M is Co or Ni and X is the metal content in wt.%. In the first step of the synthesis, the carrier, precursor, and a defined amount of isopropanol were mixed using a magnetic stirrer and then treated in an ultrasonic bath by heating at 60 °C to obtain a fine gel. In the second step, the mixture was heated to 240 °C at a rate of 2 °C min^{-1} in a reducing atmosphere, and then the temperature was maintained for 2 h, followed by gradual cooling to room temperature. The obtained powder was homogenized in a mortar and stored in a dry oxygen-free atmosphere. Unsupported nanoparticles were also synthesized and investigated to evaluate and better understand the expected metal–support interactions. The structural and phase composition of the prepared materials were examined using X-ray diffraction (XRD). Diffraction data were collected using X-ray diffractometer Philips ADP15 with Cu-Kα radiation λ = 1.54178 Å) at a constant rate of 0.02° s^{-1} over an angle range from 4° to 80° 2θ. The metal content in the composite catalysts was determined using energy dispersing X-ray analysis while the morphology and surface structure were investigated by scanning electron microscopy. The SEM analysis were carried out using JEOL JSM 6390 electron microscope (JEOL, Tokyo, Japan) (images in secondary and back-scattered electrons), equipped with INCA Oxford elemental detector (Oxford Instruments, London, UK).

The electrochemical performance of the MPT-supported catalysts was investigated firstly in aqueous alkaline solutions (25 wt.% KOH). To prepare the test electrodes, a catalytic ink was obtained by mixing the powder catalyst with isopropanol, which was then spread on the surface of a commercial gas-diffusion layer (Freudenberg FCCTKG H2315). The screening of the synthesized catalysts was performed in a standard electrochemical cell against a Pt wire as a counter electrode and Ag/AgCl reference electrode by means of linear sweep voltammetry (LSV). The best performing samples were integrated in membrane electrode assemblies using two types of anion exchange membranes—Fumasep FAA-3 with a thickness of 50 μm (commercial product of Fumapem) and a custom-made membrane, based on polybenzimidazole (PBI) doped with KOH under controlled laboratory conditions.

The characteristics of the MEA strongly depend on variety of factors, including the properties of the catalysts, the electrical conductivity and corrosion resistance of the catalytic support, OH conductivity, and the mechanical properties of the polymer electrolyte. In addition, the architecture of the MEA influences gas permeability and hydrophilicity, thus having a serious impact on the transport of reagents to and from the reaction zone where energy conversion takes place, which is another key factor for the efficiency of electrolysis [35,36]. The method and conditions for MEA preparation are also important.

The membrane electrode assemblies investigated in this study had a working area of 0.5 cm^{-2}. Both electrodes were prepared following the already described procedure for the electrodes tested in the aqueous electrolyte. The polymer electrolyte membrane was activated to provide a high OH conductivity, as the exact procedure depended on the type of the membrane used. The commercial Fumasep membrane was activated according to the technical specifications of the manufacturer. It includes immersion for 24 h in 1 M KOH and subsequent rinsing with distilled water. The laboratory PBI-based membrane was provided in an activated form and had a proven high conductivity, which, depending on the temperature and the content of doped KOH (%), was in the range from 298 S cm^{-1} to 1140 S cm^{-1} [37–39]. The MEA assembling requires binding material to attach the electrodes on both sides of the membrane. For this purpose, an OH-conducting fine gel was prepared by chemical dissolution of a piece of the membrane, which was spread over the catalytic layer of the gas diffusion electrodes and then dried at room temperature for 30 min. The final step was the mechanical clamping of MEA between the current collectors of the test cell, applying an experimentally determined optimal pressure of 8 N m^{-1}.

For the purpose of the research undertaken, five MEAs with different types of membrane and different catalysts integrated into the cathode and anode were assembled. Their performance under electrochemical water splitting conditions was investigated in a custom-made test cell shown in Figure 11. The cell has four hermetically sealed chambers—two for both working electrodes and two for the respective reference electrodes.

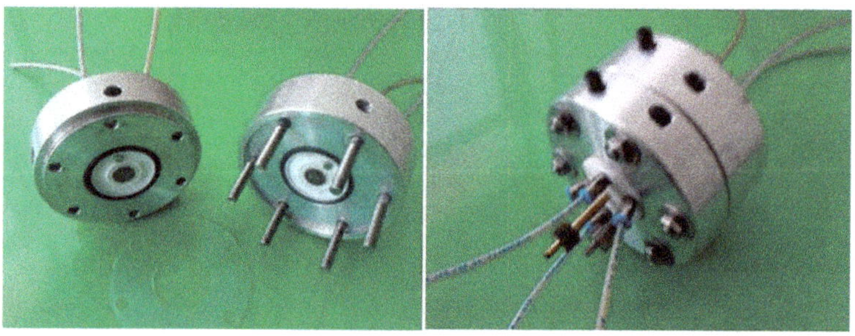

Figure 11. Pictures of the custom-made test cell.

The cell design allows simultaneous measurement of the anode and cathode polarization as well as monitoring the cell voltage. The current collectors of the cell are made from stainless steel and have flow-field channels to deliver the reagent water and to take off the reaction products (H_2 and O_2). All experiments reported in this study were carried out at room temperature using Gamry Instruments 1010E (Gamry Instruments, Warminster, PA, USA).

4. Conclusions

The newly synthesized composite non-noble metal catalysts supported in Magneli phase titania demonstrate good efficiency toward electrochemical water splitting in alkaline aqueous media combined with a high chemical stability. The combination of hyper-d-electron transition metals and hypo-d-electron MPT causes the SMSI effect, thus influencing the catalytic performance. The MEAs with the integrated Ni40/MPT cathode, and Co30/MPT anode show better performance than MEAs with Pt/C electrodes under identical experimental conditions. The previously proven high OH conductivity of the doped p-PBI membrane is confirmed when it is integrated in MEA and tested during long-term AEMEC operation. Further research is required to optimize catalyst loading as well as investigate the performance of the studied advanced materials at elevated temperatures and accelerated durability tests.

Author Contributions: Conceptualization, E.S. and G.B.; methodology, E.S., K.M.-D. and B.M.; validation, G.B. and B.M.; formal analysis, G.B.; investigation, G.B.; data curation, B.M.; writing—K.M.-D.;

writing—review and editing, E.S. All authors have read and agreed to the published version of the manuscript.

Funding: This research was funded by Bulgarian Science Fund, project КП-06-ОПР04/3. The experiments were performed using equipment of project № BG05M2OP001-1.002-0014 "Center of competence HITMOBIL—Technologies and systems for generation, storage and consumption of clean energy", funded by Operational Program "Science and Education for Smart Growth" 2014–2020 and co-funded by the EU from the European Regional Development Fund.

Data Availability Statement: The data presented in this article are available on request from the corresponding author because they are part of the ongoing study.

Acknowledgments: The authors express their gratitude to the Laboratory of Conjugated Polymers at the Institute of Polymers, Bulgarian Academy of Sciences, for the provision of the PBI-type polymer electrolyte membrane.

Conflicts of Interest: The authors declare no conflicts of interest. The funders had no role in the design of the study; in the collection, analyses, or interpretation of data; in the writing of the manuscript; or in the decision to publish the results.

References

1. Kuroda, Y.; Nishimoto, T.; Mitsushima, S. Self-repairing hybrid nanosheet anode catalysts for alkaline water electrolysis connected with fluctuating renewable energy. *Electrochim. Acta* **2019**, *323*, 134812. [CrossRef]
2. Chade, D.; Berlouis, L.; Infield, D.; Nielsen, T.; Mathiesen, T. Deactivation mechanisms of atmospheric plasma spraying Raney nickel electrodes. *J. Electrochem. Soc.* **2016**, *163*, F308. [CrossRef]
3. Todoroki, N.; Wadayama, T. Electrochemical stability of stainless-steel-made anode for alkaline water electrolysis: Surface catalyst nanostructures and oxygen evolution overpotentials under applying potential cycle loading. *Electrochem. Commun.* **2021**, *122*, 106902. [CrossRef]
4. Zeng, K.; Zhang, D. Recent progress in alkaline water electrolysis for hydrogen production and applications. *Prog. Energy Combust. Sci.* **2010**, *36*, 307. [CrossRef]
5. Eiler, K.; Suriñach, S.; Sort, J.; Pellicer, E. Mesoporous Ni-rich Ni–Pt thin flms: Electrodeposition, characterization and performance toward hydrogen evolution reaction in acidic media. *Appl. Catal. B Environ.* **2020**, *265*, 118597. [CrossRef]
6. Ayers, K.; Renner, J.; Danilovic, N.; Wang, J.; Zhang, Y.; Maric, R.; Yu, H. Pathways to ultra-low platinum group metal catalyst loading in proton exchange membrane electrolyzers. *Catal. Today* **2016**, *262*, 121. [CrossRef]
7. Zhong, W.; Tu, W.; Wang, Z.; Lin, Z.; Xu, A.; Ye, X.; Chen, D.; Xiao, B. Ultralow-temperature assisted synthesis of single platinum atoms anchored on carbon nanotubes for efficiently electrocatalytic acidic hydrogen evolution. *J. Energy Chem.* **2020**, *51*, 280. [CrossRef]
8. Suryanto, B.; Lu, X.; Zhao, C. Layer-by-layer assembly of transparent amorphous Co_3O_4 nanoparticles/graphene composite electrodes for sustained oxygen evolution reaction. *J. Mater. Chem. A* **2013**, *1*, 12726. [CrossRef]
9. Wang, D.; Chen, X.; Evans, D.; Yang, W. Well-dispersed Co_3O_4/Co_2MnO_4 nanocomposites as a synergistic bifunctional catalyst for oxygen reduction and oxygen evolution reactions. *Nanoscale* **2013**, *5*, 5312. [CrossRef]
10. Sa, Y.J.; Kwon, K.; Cheon, J.Y.; Kleitz, F.; Joo, S.H. Ordered mesoporous Co_3O_4 spinels as stable, bifunctional, noble metal-free oxygen electrocatalysts. *J. Mater. Chem. A* **2013**, *1*, 9992. [CrossRef]
11. Borisov, G.; Bachvarov, V.; Penchev, H.; Rashkov, R.; Slavcheva, E. Multi-metallic electrodeposited catalysts applicable for oxygen evolution reaction in AEM water electrolysis. *Mater. Lett.* **2021**, *286*, 129248. [CrossRef]
12. Borisov, G.; Bachvarov, V.; Rashkov, R.; Slavcheva, E. Advanced Alkaline Water Electrolysis Stack with Non-Noble Catalysts and Hybrid Electrical Connections of the Single Cells. *Catalysts* **2024**, *14*, 179. [CrossRef]
13. Velikokhatnyi, O.; Kumtal, P. Ab Initio Study of Doped Tin Niobates and Tantalates as Prospective Catalyst Supports for Water Electrolysis. *ECS Trans.* **2010**, *28*, 37. [CrossRef]
14. Santos, J.S.; Araújo, P.S.; Pissolitto, Y.B.; Lopes, P.P.; Simon, A.P.; Sikora, M.S.; Strixino, F.T. The Use of Anodic Oxides in Practical and Sustainable Devices for Energy Conversion and Storage. *Materials* **2021**, *14*, 383. [CrossRef] [PubMed]
15. Borisov, G.; Penchev, H.; Maksimova-Dimitrova, K.; Ublekov, F.; Lefterova, E.; Sinigersky, V.; Slavcheva, E. Alkaline water electrolysis facilitated via non-precious monometallic catalysts combined with highly KOH doped polybenzimidazole membrane. *Mater. Lett.* **2019**, *240*, 144. [CrossRef]
16. Wang, G.; Liu, Y.; He, W.; Ye, J. Conduction and sintering mechanism of high electrical conductivity Magneli phase Ti_4O_7. *Ceram. Int.* **2023**, *49*, 27117. [CrossRef]
17. Rani, N.; Chahal, S.; Chauhan, A.; Kumar, P.; Shukla, R.; Singh, S. X-ray Analysis of MgO Nanoparticles by Modified Scherer's Williamson-Hall and Size-Strain Method. *Mater. Today Proc.* **2019**, *12*, 543. [CrossRef]
18. Maksimova, K.; Lefterova, E.; Slavcheva, E. Nanostructured nickel and cobalt supported on Magneli—Phase titania—Preparation, properties and catalytic efficiency toward alkaline water electrolysis. *Nanosci. Nanotechnol.* **2015**, *15*, 40.
19. Zeradjanin, A.; Narangoda, P.; Masa, J.; Schlögl, R. What Controls Activity Trends of Electrocatalytic Hydrogen Evolution Reaction? Activation Energy versus Frequency Factor. *ACS Catal.* **2022**, *12*, 11597. [CrossRef]

20. Divanis, S.; Kutlusoy, T.; Boye, I.M.I.; Man, I.C.; Rossmeisl, J. Oxygen evolution reaction: A perspective on a decade of atomic scale simulations. *Chem. Sci.* **2020**, *11*, 2943. [CrossRef]
21. Koper, M.; Marc, M. Activity volcanoes for the electrocatalysis of homolytic and heterolytic hydrogen evolution. *J. Solid State Electrochem.* **2016**, *20*, 895. [CrossRef]
22. Hume-Rothery, W. The Engel-Brewer theories of metals and alloys. *Prog. Mater. Sci.* **1968**, *13*, 229. [CrossRef]
23. You, W.; Noonan, K.; Coates, G. Alkaline-stable anion exchange membranes: A review of synthetic approaches. *Prog. Polym. Sci.* **2020**, *100*, 101177. [CrossRef]
24. Dekel, D. Review of cell performance in anion exchange membrane fuel cells. *J. Power Sources* **2018**, *375*, 158. [CrossRef]
25. Abdi, Z.; Chiu, T.; Pan, Y.; Chien, Y. Anion exchange membranes based on ionic polybenzimidazoles crosslinked by thiolene reaction. *React. Funct. Polym.* **2020**, *156*, 104719. [CrossRef]
26. Hao, J.; Jiang, Y.; Gao, X.; Lu, X.; Xiao, Y.; Shao, Z.; Yi, B. Functionalization of polybenzimidazole-crosslinked poly(vinylbenzyl chloride) with two cyclic quaternary ammonium cations for anion exchange membranes. *J. Membr. Sci.* **2018**, *548*, 1–10. [CrossRef]
27. Guo, M.; Ban, T.; Wang, Y.; Wang, X.; Wang, Z.; Zhu, X. Polybenzimidazoles incorporating imidazole N-spirocyclic quaternary ammonium cation for anion exchange membranes water electrolysis. *J. Membr. Sci.* **2023**, *684*, 121903. [CrossRef]
28. Touni, A.; Grammenos, O.; Banti, A.; Karfaridis, D.; Prochaska, C.; Lambropoulou, D.; Pavlidou, E.; Sotiropoulos, S. Iridium oxide-nickel-coated titanium anodes for the oxygen evolution reaction. *Electrochim. Acta* **2021**, *390*, 138866. [CrossRef]
29. Fan, Y.; Feng, X.; Zhou, W.; Murakami, S.; Kikuchi, K.; Nomura, N.; Wangc, L.; Jianga, W.; Kawasakib, A. Preparation of monophasic titanium sub-oxides of Magnéli phase with enhanced thermoelectric performance. *J. Eur. Ceram. Soc.* **2018**, *38*, 507. [CrossRef]
30. Liu, H.; Xiao, H.; Qiao, Y.; Luo, M.; Wang, C.; Yang, L.; Zeng, C.; Chao, F. Preparation, characterization, and electrochemical behavior of a novel porous Magneli phase Ti_4O_7-coated Ti electrode. *Ceram. Int.* **2023**, *49*, 20564. [CrossRef]
31. Kim, M.; Choi, J.; Lee, W.; Ahn, Y.-Y.N.; Lee, H.; Cho, K.; Lee, J. Performance of Magneli phase Ti_4O_7 and Ti^{3+} self-doped TiO_2 as oxygen vacancy-rich titanium oxide anodes: Comparison in terms of treatment efficiency, anodic degradative pathways, and long-term stability. *Appl. Catal. B Environ.* **2023**, *337*, 122993. [CrossRef]
32. Slavcheva, E.; Nikolova, V.; Petkova, T.; Lefterova, E.; Dragieva, I.; Vitanov, T.; Budevski, E. Electrocatalytic activity of Pt and PtCo deposited on Ebonex by BH reduction. *Electrochim. Acta* **2005**, *50*, 5444. [CrossRef]
33. Figueiredo, W.; Prakash, R.; Vieira, C.; Lima, D.; Carvalho, V.; Soares, E.; Buchner, S.; Raschke, H.; Oscar, W.; Lopez, P.; et al. New insights on the electronic factor of the SMSI effect in Pd/TiO_2 nanoparticles. *Appl. Surf. Sci.* **2022**, *574*, 151647. [CrossRef]
34. Bertella, F.; Concepcióna, P.; Martínez, A. TiO_2 polymorph dependent SMSI effect in Co-Ru/TiO_2 catalysts and its relevance to Fischer-Tropsch synthesis. *Catal. Today* **2017**, *289*, 181. [CrossRef]
35. Rajalakshm, R.; Srividhy, G.; Viswanatha, C.; Ponpandian, N. The intriguing bifunctional effect of strong metal support interaction (SMSI) and hydrogen spillover effect (HSPE) for effective hydrogen evolution reaction. *Appl. Catal. B Environ.* **2023**, *339*, 12308. [CrossRef]
36. Knöppel, J.; Möckl, M.; Escalera-López, D.; Stojanovski, K.; Bierling, M.; Böhm, T.; Thiele, S.; Rzepka, M.; Cherevko, S. On the limitations in assessing stability of oxygen evolution catalysts using aqueous model electrochemical cells. *Nat. Commun.* **2021**, *12*, 2231.
37. Lagadec, M.; Grimaud, A. Water electrolysers with closed and open electrochemical systems. *Nat. Mater.* **2020**, *19*, 1140. [CrossRef] [PubMed]
38. Hassan, N.; Zheng, Y.; Kohl, P.; Mustain, W. KOH vs Deionized Water Operation in Anion Exchange Membrane Electrolyzers. *J. Electrochem. Soc.* **2022**, *169*, 044526. [CrossRef]
39. Penchev, H.; Borisov, G.; Petkucheva, E.; Ublekov, F.; Sinigersky, V.; Radev, I.; Slavcheva, E. Highly KOH doped para-polybenzimidazole anion exchange membrane and its performance in Pt/$TinO_{2n-1}$ catalyzed water electrolysis cell. *Mater. Lett.* **2018**, *221*, 128. [CrossRef]

inorganics

MDPI

Article

Hierarchically Electrodeposited Nickel/Graphene Coatings for Improved Corrosion Resistance of Ni Foam Flow Field in PEMFC

Yuzhen Xia [1], Qibin Zuo [1], Chuanfu Sun [1], Guilin Hu [1,*] and Baizeng Fang [2,*]

[1] School of Mechanical and Energy Engineering, Zhejiang University of Science and Technology, Hangzhou 310023, China; yuzhen.xia@zust.edu.cn (Y.X.); zuoqibin06@163.com (Q.Z.); 222301855023@zust.edu.cn (C.S.)

[2] School of Chemical Engineering and Energy Technology, Dongguan University of Technology, Dongguan 523808, China

* Correspondence: enehgl@163.com (G.H.); baizengfang@163.com (B.F.)

Abstract: Metal foams are promising materials for the flow fields of proton exchange membrane fuel cells (PEMFCs) because of excellent mass transport characteristics and high electronic conductivity. To resolve the corrosion problem in the acidic environment under high temperature, nickel/graphene (Ni/G) composite coatings with hierarchical structures were electrodeposited on the surface of Ni foam. The effect of grain size and the distribution in the double layer was discussed. It was found that Ni/G5-10, with larger inner size and middle outer size, exhibited the best corrosion performance. Meanwhile, the corrosion current in the Tafel plots and the steady current density in constant potential analysis was lower than that obtained under steady and gradient currents. Combined with the results of XRD, XPS, and SEM, it was proven that a uniform and dense protective film was produced during the two-step electrodeposition. Moreover, the ICR value was 8.820 mΩ·cm^2, which met the requirement of 2025 DOE.

Keywords: PEMFC; metal foam; flow field; electrodeposition; corrosion resistance; nickel/graphene composite

1. Introduction

With the increasing global demand for renewable energy and low-carbon technologies [1,2], fuel cells, as an efficient and environmentally friendly energy conversion technology, are receiving increasing attention. Fuel cells directly convert chemical energy into electrical energy during operation, with water as the only byproduct [3]. Among various types of fuel cells, proton exchange membrane fuel cells (PEMFCs) are favored for their high energy conversion efficiency, rapid start-up response, and low-temperature operation [4,5].

Among the main components of PEMFCs, the bipolar plate (BP) accounts for approximately 80% of the total mass and 45% of the cost, serving essential functions in mass transport, water management, and as a mechanical support in the stack [6,7]. Currently, the materials commonly used for bipolar plates (BPs) include graphite, composite materials, and metals. Graphite, as the first material developed for BPs, possesses good electrical conductivity and corrosion resistance; however, it suffers from low mechanical strength and processing difficulties [8,9]. Composite materials combine lightweight properties with excellent conductivity, making them suitable for high-performance applications [10–12]. Nonetheless, existing composite BPs face challenges such as high interfacial contact resistance and elevated costs. In contrast, metallic BPs offer advantages such as good mechanical properties, electrical conductivity, thermal conductivity, and ease of processing, making them a primary choice for BPs [9]. However, under the high-temperature and acidic operating conditions of PEMFCs, metallic materials are prone to corrosion, releasing corrosive metal ions that adversely affect fuel cell performance. Therefore, the introduction of protective coatings to enhance corrosion resistance is crucial [13,14].

Noble metal coatings exhibit excellent corrosion resistance and electrical conductivity; however, their high cost limits widespread application [15]. Consequently, non-noble metal coating materials such as tin (Sn) [16] and nickel (Ni) [17] are receiving increasing attention. For instance, Kim et al. [18] prepared a pure niobium (Nb) coating on a 316 L stainless steel substrate using pulsed direct current magnetron sputtering. Their study revealed that the Nb coating significantly improved the corrosion behavior of the substrate, increasing polarization resistance and reducing corrosion rate by more than 100 times and 50 times, respectively. Kumaravel et al. [19] applied nickel coatings on weathering steel via electroplating, finding that the Ni-coated samples exhibited the lowest corrosion rate when compared to copper and zinc coatings. Wang et al. [20] developed a Ni/Sn composite coating on nickel foam through electrodeposition, which showed no significant pitting after 8 h of corrosion testing compared to the uncoated samples.

Additionally, an increasing number of non-metal coatings are being used in corrosion-resistant coatings [19]. Among these, graphene is gaining traction as a protective coating due to its unique chemical inertness and impermeability [21]. Lee et al. [22] prepared tin, gold, and nanoscale graphene coatings on porous nickel foam, and through Tafel analysis at 80 °C, they found that the graphene-coated nickel foam exhibited the lowest corrosion current density (0.65 $\mu A/cm^2$) and a corrosion potential of −0.22 V. Compared to the other two materials, the performance of the graphene coating is superior. Graphene coatings were prepared on titanium-based bipolar plates, which demonstrated significantly enhanced corrosion resistance and stability in PEMFC environments [23,24]. Feng et al. [25] utilized chemical vapor deposition to create vertically aligned graphene coatings on titanium substrates, achieving a corrosion rate reduction in one to two orders of magnitude and a decrease in interfacial contact resistance (ICR) by approximately 100 times compared to uncoated samples.

Graphene–metal composite coatings not only addressed the limitations of single-layer coatings but also reduced costs and improved mechanical stability [26,27]. Rekha et al. [28] electrodeposited Cr–graphene composite coatings on low-carbon steel substrates and conducted corrosion studies using a 3.5% NaCl solution as the electrolytic medium. Their findings revealed an 80.4% reduction in corrosion current and a 72.8% decrease in ICR, significantly enhancing both corrosion resistance and conductivity compared to pure chromium coatings. Punith Kumar et al. [29] prepared Zn–graphene composite coatings through electrodeposition and conducted experiments in a 3.5% NaCl solution. The results indicated that the Zn–graphene coatings exhibited smaller grain sizes and fewer surface defects, leading to enhanced corrosion resistance. Abdul et al. [30] also developed Ni–graphene composite coatings through electrodeposition, investigating the effects of different deposition temperatures on coating performance, with optimal corrosion resistance achieved at 45 °C in a 3.5% NaCl solution. Sun et al. [31] prepared a Ni–graphene composite coating by direct current electrodeposition. Under the test condition of pH = 3, the coating showed good corrosion resistance and stability. Electrodeposition is widely used as a low-cost and easy-to-control coating method. In the process of electrodeposition, the larger the current, the smaller the particles.

Metal foam is used as the flow field to replace traditional BP plates with ribs and channels because of lightweight advantages and high mass transfer efficiency. However, it suffers more severe corrosion issues in the operating environment on account of high temperatures and the presence of acids. Meanwhile, due to the complex internal structure and high surface roughness, the electrodeposition mechanism is different from that on the flat plate. In this work, nickel/graphene (Ni/G) composite coatings with a hierarchical structure are employed by the electrodeposition method. While a larger grain size is obtained under a lower electrodeposition current, the coatings with various size distributions are compared. Their physical properties are analyzed along with their impact on corrosion resistance, ultimately identifying the coating with optimal performance.

2. Results and Discussions

2.1. XRD Analysis

The XRD spectrum of the Ni/G composite coating is shown in Figure 1. The spectrum exhibited three main diffraction peaks at 2θ values of 44.4°, 51.8°, and 76.3°, corresponding to the Ni (111), (200), and (220) planes, respectively, which closely matched the reference pattern for the face-centered cubic nickel. Each coating displayed a strong Ni (111) peak, indicating that during the deposition process, nickel tended to grow preferentially in the (111) orientation [32].

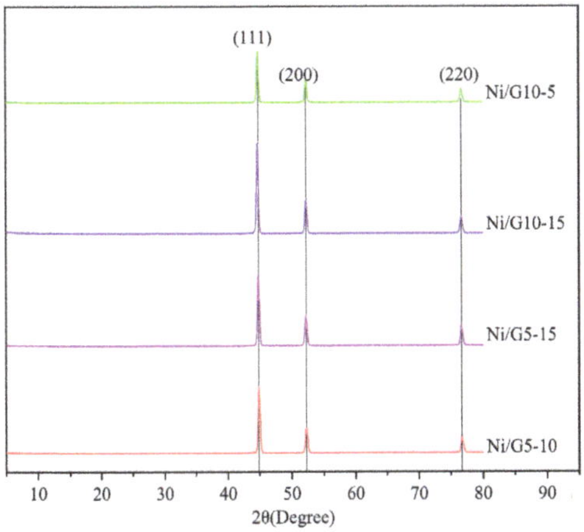

Figure 1. XRD patterns of composite coatings.

Using the Scherrer equation, we calculated the average crystallite sizes for Ni/G5-10, Ni/G5-15, Ni/G10-15, and Ni/G10-5 based on the Ni (111) peak, yielding values of 31.3 nm, 29.2 nm, 28.7 nm, and 31.3 nm, respectively, which was larger than that of Ni/G10 and Ni/G0-10 in our previous work [32]. As Ni/G5-10 had the same size as Ni/G10-5, we concluded that the grain size was not only related to the upper layer, but the average size of the two layers.

2.2. XPS Analysis

As to Ni/G5-10, the XPS spectrum of Ni 2p, C 1s, and O 1s is shown in Figure 2a. To accurately identify the chemical phases presented in the samples, peak fitting analysis was performed on the Ni 2p spectrum. As shown in Figure 2b, the main peaks of Ni 2p3/2 were located at approximately 856 eV, 857.5 eV, and 862.5 eV, which corresponded to the nickel compounds such as $Ni(OH)_2$, $NiSO_4$, and $NiCl_2$, indicating that nickel existed in a divalent state [33–35]. Additionally, a Ni(0) peak appeared around 852 eV, confirming the presence of metallic nickel in the coating [36]. The OH^{-1}, ${SO_4}^{-2}$, and Cl^{-1} ions might be derived from the plating solution used during coating preparation.

2.3. Tafel Analysis

The Tafel curves for the Ni/G composite coatings in pH 3 sulfuric acid are shown in Figure 3, with testing temperatures of 50 °C and 80 °C, respectively. The data for corrosion potential (E_{corr}), corrosion current (I_{corr}), and corrosion rate (V_{corr}) obtained from the Tafel analysis are presented in Table 1. E_{corr} is defined as the intersection of the anodic and cathodic polarization curves, indicating the self-corrosion potential of the coating. I_{corr} is a kinetic parameter obtained through extrapolation [37]. According to the corrosion rate equation, the corrosion rate coefficient is derived from the corrosion current.

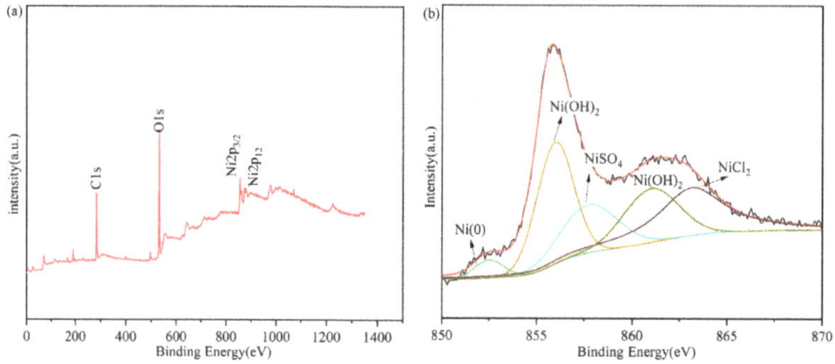

Figure 2. XPS of Ni/G5-10 (**a**) and Ni 2p (**b**) fitting spectrum.

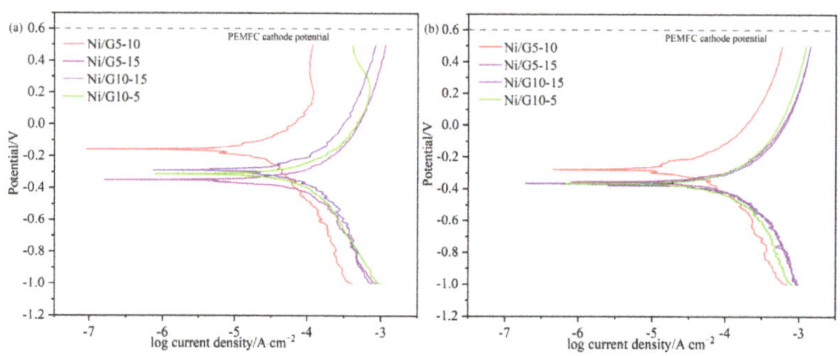

Figure 3. Tafel curves for Ni/G composite coatings tested at 50 °C (**a**) and 80 °C (**b**).

Table 1. Tafel polarization parameters for Ni/G composite coatings at 50 °C and 80 °C.

	I_{corr} (μA·cm^{-2})		E_{corr} (V)		V_{corr} ($\times 10^{-2}$ mm·y^{-1})	
	50 °C	80 °C	50 °C	80 °C	50 °C	80 °C
Ni/G5-10	26.14	43.80	−0.159	−0.278	0.283	0.473
Ni/G5-15	52.51	62.25	−0.348	−0.355	0.568	0.673
Ni/G10-15	66.27	71.70	−0.289	−0.366	0.716	0.775
Ni/G10-5	73.54	81.14	−0.312	−0.368	0.795	0.877

As shown in Table 1, with increasing temperature, all samples exhibited an increase in I_{corr}, a more negative shift in E_{corr}, and an increased V_{corr}, indicating that higher temperatures correlated with more severe corrosion and accelerated corrosion rates. At 50 °C, the gradient coatings Ni/G5-10, Ni/G5-15, and Ni/G10-15 outperformed Ni/G10-5. Ni/G5-10 exhibited the lowest I_{corr} and the most positive E_{corr}, with the values of 26.14 μA·cm^{-2} and −0.159 V, respectively. Compared to the values of Ni/G10-5, its I_{corr} decreased by 64.45% and E_{corr} improved by 24.46%. The same trend was observed at 80 °C.

2.4. Constant Potential Analysis

The stability of Ni/G coatings was tested at a constant potential of 0.6 V and 50 °C, with the current versus time changed shown in Figure 4. All samples quickly reached a stable current state within the first few minutes, exhibiting no significant fluctuations over 120 min and thus indicating a rapid self-repair process of the materials [38]. The current densities of Ni/G5-10, Ni/G5-15, Ni/G10-15, and Ni/G10-5 are 0.12, 0.38, 0.42, and 0.58 mA·cm^{-2}, respectively. Among them, Ni/G5-10 has a positive stable point voltage, indicating that the corrosion resistance is better, which has the same rule as Tafel analysis.

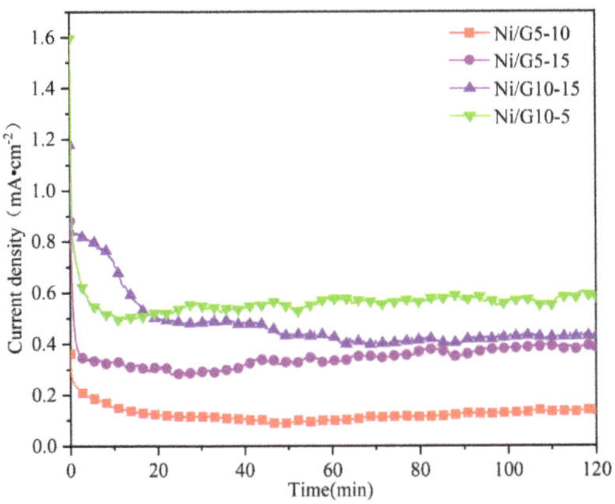

Figure 4. Constant potential curve for Ni/G composite coatings.

2.5. EIS Analysis

Figure 5 presents the results of all samples tested in a 0.5 M H_2SO_4 solution with 2 ppm HF at 50 °C. As shown in Figure 5a, the Nyquist plots for all coatings exhibit a semi-circular arc, indicating a similar corrosion mechanism across the samples. Among them, Ni/G5-10 displays the largest impedance semicircle radius. The Randles circuit was chosen as the equivalent circuit, as illustrated in Figure 5b, which consists of solution resistance (R_s), coating resistance (R_{coat}), and a constant phase element (C_{coat}). The fitted values are summarized in Table 2. Compared to the uncoated foam nickel (R_{coat} = 5.27 Ω·cm^2) [21], the Ni/G composite coatings exhibit significantly higher R_{coat} values, indicating better corrosion resistance. Notably, the gradient coating Ni/G5-10 has the highest R_{coat}, demonstrating the strongest corrosion resistance among the samples.

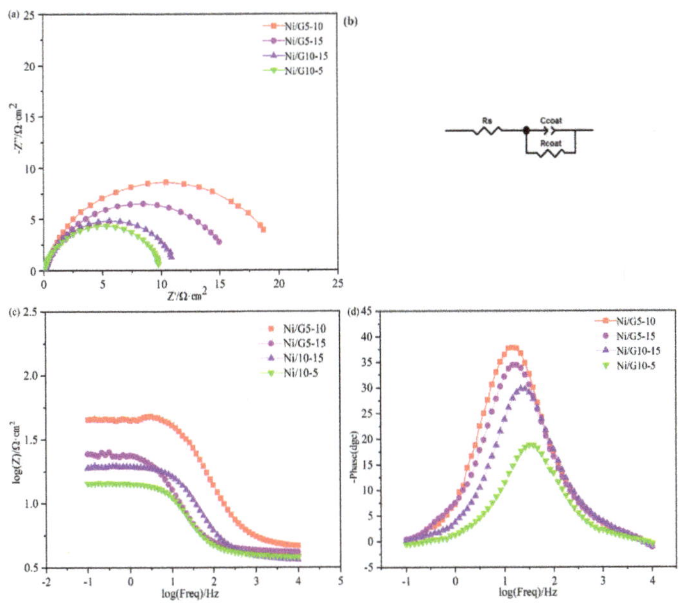

Figure 5. EIS analysis results for Ni/G composite coatings: (**a**) Nyquist impedance plots; (**b**) equivalent circuit; (**c**,**d**) Bode impedance plots.

Table 2. EIS fitting results for Ni/G composite coatings.

Samples	R_s ($\Omega \cdot cm^2$)	C_{coat}		R_{coat} ($\Omega \cdot cm^2$)
		$C_{coat\text{-}T}$	$C_{coat\text{-}P}$	
5–10	4.28	2.21×10^{-3}	0.87	19.79
5–15	3.72	1.06×10^{-3}	0.85	16.09
10–15	3.02	2.8×10^{-3}	0.88	12.26
10–5	8.35	1.05×10^{-3}	0.91	9.78

In Figure 5c,d, all Bode plots display high-frequency (10^2–10^4 Hz), mid-frequency (10^1–10^2 Hz), and low-frequency (10^{-1}–10^1 Hz) regions, which correspond to the corrosion processes at the coating surface, within the coating system, and at the metal/coating interface, respectively. In Figure 5c, Ni/G5-10 shows the highest log(Z) values across the entire frequency range, indicating the best corrosion resistance of the coating. In Figure 5d, the phase angles for Ni/G5-10, Ni/G5-15, and Ni/G10-15 are all greater than that of Ni/G10-5, which correlates with improved corrosion resistance.

2.6. SEM Analysis

The stability of the Ni/G composite coatings was studied by testing for 8 h at a constant potential of 0.6 V. The SEM images before and after the test are shown in Figure 6. As depicted in Figure 6a–d, the surface of the composite coatings is smooth and uniform, with no significant aggregation observed. In our previous work, the gain sizes of the coatings that were electrodeposited at 5, 10, and 15 mA·cm^{-2}, were found to be 37.0 nm, 33.2 nm, and 24.9 nm, respectively. In the schematic diagrams, colored particles were used to represent the grains in the hierarchical coatings. The graphene layer covered the surface of nickel ions, hindering further growth during the electrodeposition process while providing more nucleation sites for the reduction of nickel ions, resulting in smaller and more uniform composite coatings [39,40]. After the corrosion test, more sheet graphene was exposed, as shown in Figure 6a–d.

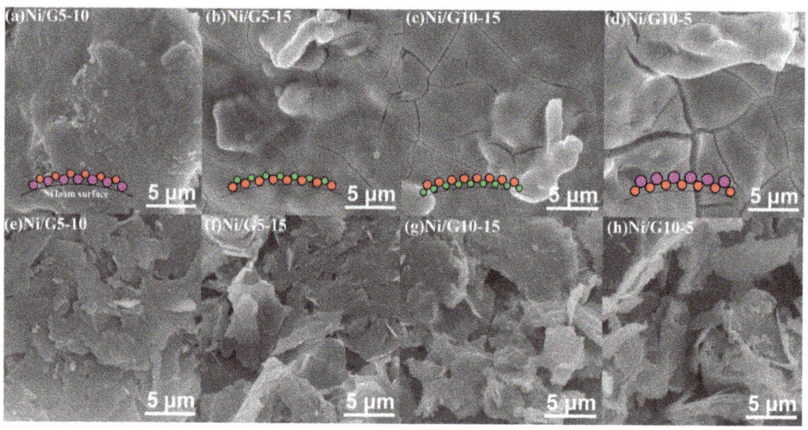

Figure 6. SEM images of Ni/G composite coatings and schematic diagrams with colored particles for hierarchical coatings before (**a–d**) and after (**e–h**) the 8h corrosion (purple, red, and green represents grains electrodeposited at 5, 10, and 15 mA·cm^{-2}, respectively).

2.7. EDS Analysis

The atomic composition of the main elements in the Ni/G5-10 composite coating before and after corrosion was analyzed using energy dispersive spectroscopy (EDS). As shown in Figure 7a, the presence of carbon confirmed the incorporation of graphene into the coating, which at the same time aided the reduction of nickel ions during the electrodeposition process, leading to the formation of metallic nickel [41]. Higher oxygen content might be attributed to the oxidation reaction, especially that of the nano-particles, occurring at the coating surface.

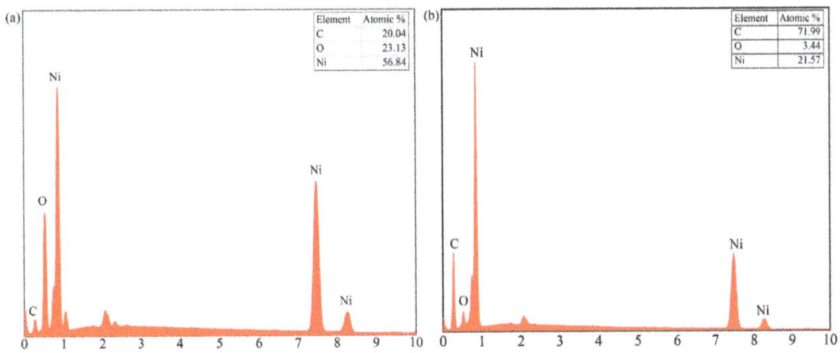

Figure 7. EDS of Ni/G5-10 before (**a**) and after (**b**) 8 h corrosion.

As shown in Figure 7b, after 8 h of corrosion, the carbon content in the coating increased significantly from 20.04% to 71.99%, while the nickel content decreased from 56.84% to 21.57%. During the corrosion process, metallic nickel gradually dissolved and converted to Ni^{2+} ions, which then entered into the solution. In contrast, graphene, with superior corrosion resistance and stability, played a protective role on the surface.

2.8. ICR Analysis

In addition to corrosion resistance, the interfacial contact resistance (ICR) is also a critical parameter for bipolar plates. A high ICR leads to additional resistance to current flow within the cell, resulting in reduced efficiency and negatively impacting the fuel cell's output power. In Figure 8a, the values of all the samples decreased with the pressure. Ni/G5-10 exhibited the lowest ICR value, which might be attributed to the formation of a uniform and dense coating. In Figure 8b, the value at a pressure of 150 $N{\cdot}cm^{-2}$ (the typical assembly pressure for PEMFCs) of Ni/G5-10 was 7.16 $m\Omega{\cdot}cm^2$, which met the 2025 DOE standard requirement (<10 $m\Omega{\cdot}cm^2$) [17].

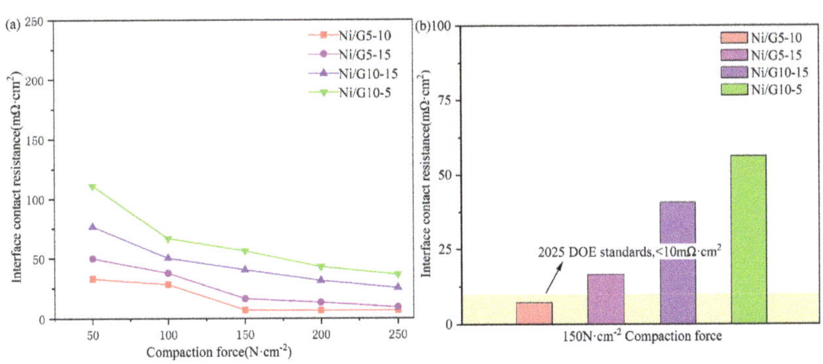

Figure 8. ICR of Ni/G composite coatings under various compaction force (**a**) and the values under 150 $N{\cdot}cm^{-2}$ (**b**).

3. Experimental

3.1. Electrodeposition

The foam nickel used in the experiment was obtained from Kunshan Green Creation Electronics Technology Co., Ltd. (Kunshan, China), with a thickness of 1 mm and a pore diameter around 600 μm. Prior to electrodeposition, the foam nickel underwent pretreatment and activation [42]. The Ni/G composite coating bath consisted of 12.9 g·L^{-1} $(SO_3NH_2)_2{\cdot}4H_2O$ (Shanghai Macklin Biochemical Technology Co., Ltd., Shanghai, China), 9.51 g·L^{-1} $NiCl_2{\cdot}6H_2O$ (Shanghai Macklin Biochemical Technology Co., Ltd., Shanghai, China), 40 g·L^{-1} H_3BO_3 (Shanghai Macklin Biochemical Technology Co., Ltd., Shanghai, China), 0.05 g·L^{-1} $C_{12}H_{25}SO_4Na$ (Shanghai Macklin Biochemical Technology Co., Ltd., Shanghai, China), 10 g·L^{-1} $C_6H_5Na_3O_7$ (Shanghai Macklin Biochemical Technology Co., Ltd., Shanghai, China), 0.048 g·L^{-1} graphene (Shanghai Macklin Biochemical Technology Co., Ltd., Shanghai, China) powder (97%), and 100 mL·L^{-1} C_5H_9NO. In the three-electrode system using a CHI660E (Shanghai Chenhua Co., Ltd., Shanghai, China), the foam nickel served as the cathode while a nickel plate (99.9%, Chen Shuo Co., Ltd., Fuzhou, China) acted as the anode. The electrodeposition was conducted at a temperature of 328 K for a duration of 20 min, under various current densities: 5, 10, and 15 mA·cm^{-2}. The hierarchical structure was obtained by depositing under two current densities with each current deposited for 10 min. The coated samples were named Ni/G5-10, Ni/G5-15, Ni/G10-15, and Ni/G10-5, respectively. Following electrodeposition, the samples were dried in an oven at 60 °C for 8 h.

3.2. Morphology Investigation

X-ray diffraction (XRD) was tested using an X-ray diffractometer model (Kuraray Co., Ltd., Tokyo, Japan) KYOWAGLAS-XA H-12 using Cu Kα radiation at operating voltages of 40 mA, 40 kV, and 2θ = 5~80°.

X-ray photoelectron spectroscopy (XPS) analysis was performed using Horiba LabRAM HR X-ray (HORIBA SAS Co., Ltd., Palaiseau, France) photoelectron spectroscopy.

Field emission scanning electron microscopy (SEM) was conducted using a Phenom Pro X model from Funa Scientific Instruments (Shanghai, China) Co., Ltd., with an applied voltage set between 10 and 15 kV. Prior to testing, the samples were gold-coated and then placed on the microscope sample stage for analysis.

The elements in the samples were identified and quantitatively analyzed by EDS detection. In this study, SU1510 (Hitachi High-tech Company, Tokyo, Japan) was used to test the foam metal sample. The voltage was set at 15.0 kV, the working distance was 15 mm, and the mode was selected as surface scan. The content and distribution of carbon, oxygen, and nickel were the main areas of study.

The experimental method of interface contact resistance (ICR) was the same as Liu et al.'s [24] The test sample was placed between two sheets of carbon paper, then between copper electrodes, and the two electrodes were fixed on the TS-C40L-B (Tycoway Automation Systems Co., Ltd., Shenzhen, China) press, and 3 mA constant current was output through the MP3030D (MAISHENG Co., Ltd., Shenzhen, China) voltage regulator power supply. At the same time, the press applied pressure from small to large, and the ICR value of each sample was obtained under different pressures.

3.3. Corrosion Resistance Measurements

Electrochemical testing was conducted in a traditional three-electrode configuration [22], with a platinum sheet (Chuxi Industry Co., Ltd., Shanghai, China) as the counter electrode, an Ag/AgCl (Chuxi Industry Co., Ltd., Shanghai, China) reference electrode, and a working electrode comprising a foam nickel with an area of 1 cm^2. The Tafel test was performed in H_2SO_4 solution with pH = 3, the voltage range was set to −1 to 0.5, and the scan rate was 1 mV·s^{-1}. A potentiostatic test was performed at 50 °C, with a voltage of 0.6 V and air pumped into the sulfuric acid solution to simulate the fuel cell cathode environment. The electrochemical impedance spectroscopy (EIS) test was performed at

50 °C and in a 2 ppm HF (Shanghai Macklin Biochemical Technology Co., Ltd., Shanghai, China) solution containing 0.5 mol·L^{-1} H_2SO_4 (Shanghai Macklin Biochemical Technology Co., Ltd., Shanghai, China), with a set frequency range from 10 kHz to 0.1 Hz and an amplitude voltage of 10 mV.

4. Conclusions

This study successfully fabricated Ni/G composite coatings with hierarchical structures on foam nickel using electrodeposition under different current densities. During the co-deposition of Ni and graphene, the effect of the grain size in the inner layer and the outer layer was discussed. It was found that Ni/G5-10, with larger inner size and a middle outer size, exhibited the best corrosion performance, with the lowest dynamic and steady-state corrosion currents, as well as the highest coating resistance. Compared with the coatings obtained under the steady and gradient current, the hierarchically electrodeposited coating improved the stability of the foam metals. Although the grain size was larger than that of the former coatings, the composite of NiO and graphene covered the surface of metal foam and formed a uniform and dense protective film. Moreover, its ICR value was 7.16 mΩ·cm^2, with a 11.8% reduction relative to the US DOE 2025 goal, resulting in lower ohmic resistance and less heat power in the fuel cell operation.

Author Contributions: Conceptualization, Y.X., G.H. and B.F.; Methodology, Q.Z. and C.S.; Formal analysis, Q.Z.; Investigation, Q.Z. and C.S.; Writing—original draft, Y.X. and Q.Z.; Writing—review & editing, Y.X.; Supervision, Y.X., G.H. and B.F.; Project administration, G.H. All authors have read and agreed to the published version of the manuscript.

Funding: This research received no external funding.

Data Availability Statement: The original contributions presented in the study are included in the article, further inquiries can be directed to the corresponding authors.

Conflicts of Interest: The authors declare no conflict of interest.

References

1. Gayen, D.; Chatterjee, R.; Roy, S. A review on environmental impacts of renewable energy for sustainable development. *Int. J. Environ. Sci. Technol.* **2024**, *21*, 5285–5310. [CrossRef]
2. Stergaard, P.A.; Duic, N.; Noorollahi, Y. Renewable energy for sustainable development. *Renew. Energy* **2022**, *199*, 1145–1152. [CrossRef]
3. Ahmed, H.; Adebayo, P.; Ahmed, M.; Arbab, A.I. Hydrogen fuel cell technology: Benefits, challenges, and future potential. *J. Energy Technol. Policy* **2023**, *13*, 2224–3232.
4. Aminudin, M.; Kamarudin, S.; Lim, B.; Majilan, E.; Masdar, M.; Shaari, N. An overview: Current progress on hydrogen fuel cell vehicles. *Int. J. Hydrogen Energy* **2023**, *48*, 4371–4388. [CrossRef]
5. Cai, F.; Cai, S.; Tu, Z. Proton exchange membrane fuel cell (PEMFC) operation in high current density (HCD): Problem, progress and perspective. *Energy Convers. Manag.* **2024**, *307*, 118348. [CrossRef]
6. Hermann, A.; Chaudhuri, T.; Spagnol, P. Bipolar plates for PEM fuel cells: A review. *Int. J. Hydrogen Energy* **2005**, *30*, 1297–1302. [CrossRef]
7. Liu, R.; Jia, Q.; Zhang, B.; Lai, Z.; Chen, L. Protective coatings for metal bipolar plates of fuel cells: A review. *Int. J. Hydrogen Energy* **2022**, *47*, 22915–22937. [CrossRef]
8. Kloess, J.P.; Wang, X.; Liu, J.; Shi, Z.; Guessous, L. Investigation of bio-inspired flow channel designs for bipolar plates in proton exchange membrane fuel cells. *J. Power Sources* **2009**, *188*, 132–140. [CrossRef]
9. Liu, Q.; Lan, F.; Zeng, C.; Chen, J.; Wang, J. A review of proton exchange membrane fuel cell's bipolar plate design and fabrication process. *J. Power Sources* **2022**, *538*, 231543. [CrossRef]
10. Kuan, Y.D.; Wang, C.K.; Yang, C.; Lee, P.C.; Siao, Y.S.; Lee, C.Y. Fuel cell stack design using carbon fiber composites. *Sens. Mater.* **2020**, *32*, 4233–4244. [CrossRef]
11. Gomez Sanchez, A.; Franco Luján, V.A.; Alfaro López, H.M.; Hernández-Sánchez, L.; Cruz-Martínez, H.; Medina, D.I. carbon material-reinforced polymer composites for bipolar plates in polymer electrolyte membrane fuel cells. *Polymers* **2024**, *16*, 671. [CrossRef] [PubMed]
12. Wenkai, L.; Zhiyong, X.; Haodong, Z. Current status of research on composite bipolar plates for proton exchange membrane fuel cells (PEMFCs): Nanofillers and structure optimization. *RSC Adv.* **2024**, *14*, 7172–7194. [PubMed]

13. Wang, X.Z.; Fan, H.Q.; Muneshwar, T.; Cadien, K.; Luo, J.L. Balancing the corrosion resistance and through-plane electrical conductivity of cr coating via oxygen plasma treatment. *J. Mater. Sci. Technol.* **2021**, *61*, 75–84.
14. McCay, K.; Kongstein, O.E.; Oedegaard, A.; Barnett, A.O.; Seland, F. Soldering a gas diffusion layer to a stainless steel bipolar plate using metallic tin. *Int. J. Hydrogen Energy* **2018**, *43*, 9006–9014.
15. Chanda, U.K.; Behera, A.; Roy, S.; Pati, S. Evaluation of Ni-Cr-P coatings electrodeposited on low carbon steel bipolar plates for polymer electrolyte membrane fuel cell. *Int. J. Hydrogen Energy* **2018**, *43*, 23430–23440.
16. Pukha, V.; Glukhov, A.; Belmesov, A.; Kabachkov, E.; Khodos, I.; Khadem, M.; Kim, D.E.; Karaseov, P. Corrosion-resistant nanostructured carbon-based coatings for applications in fuel cells based on bipolar plates. *Vacuum* **2023**, *218*, 112643.
17. Li, S.; Jin, R.; Li, S. High corrosion resistance and conductivity of Al_2O_3/CrN coating for metal bipolar plates in PEMFCs: Al_2O_3 hinders CrN columnar crystals growth. *Int. J. Hydrogen Energy* **2024**, *50*, 805–816. [CrossRef]
18. Kim, Y.S.; Lee, I.S.; Choi, J.Y.; Jun, S.; Kim, D.; Cha, B.C.; Kim, D.W. Corrosion behavior of niobium-coated 316L stainless steels as metal bipolar plates for polymer electrolyte membrane fuel cells. *Materials* **2021**, *14*, 4972. [CrossRef]
19. Chauhan, D.S.; Quraishi, M.; Ansari, K.; Saleh, T.A. Graphene and graphene oxide as new class of materials for corrosion control and protection: Present status and future scenario. *Prog. Org. Coat.* **2020**, *147*, 105741.
20. Wang, Z.; Xia, Y.; Lei, H.; Hu, G. Enhanced corrosion resistance of Ni/Sn nano-electrodeposited metal foam for flow field application in simulated PEMFC cathode environment. *Int. J. Hydrogen Energy* **2022**, *47*, 35412–35422. [CrossRef]
21. Ollik, K.; Lieder, M. Review of the application of graphene-based coatings as anticorrosion layers. *Coatings* **2020**, *10*, 883. [CrossRef]
22. Lee, Y.H.; Li, S.M.; Tseng, C.J.; Su, C.Y.; Lin, S.C.; Jhuang, J.W. Graphene as corrosion protection for metal foam flow distributor in proton exchange membrane fuel cells. *Int. J. Hydrogen Energy* **2017**, *42*, 22201–22207. [CrossRef]
23. Wang, J.; Min, L.; Fang, F.; Zhang, W.; Wang, Y. Electrodeposition of graphene nano-thick coating for highly enhanced performance of titanium bipolar plates in fuel cells. *Int. J. Hydrogen Energy* **2019**, *44*, 16909–16917. [CrossRef]
24. Liu, Y.; Min, L.; Zhang, W.; Wang, Y. High-performance graphene coating on titanium bipolar plates in fuel cells via cathodic electrophoretic deposition. *Coatings* **2021**, *11*, 437. [CrossRef]
25. Yu, F.; Wang, K.; Cui, L. Vertical-graphene-reinforced titanium alloy bipolar plates in fuel cells. *Adv. Mater.* **2022**, *34*, 2110565.
26. Pandey, U.; Singh, A.; Sharma, C. Development of anti-corrosive novel nickel-graphene oxide-polypyrrole composite coatings on mild steel employing electrodeposition technique. *Synth. Met.* **2022**, *290*, 117135. [CrossRef]
27. Mu, J.; Gao, F.; Cui, G.; Wang, S.; Tang, S.; Li, Z. A comprehensive review of anticorrosive graphene-composite coatings. *Prog. Org. Coat.* **2021**, *157*, 106321.
28. Rekha, M.; Kumar, M.P.; Srivastava, C. Electrochemical behaviour of chromium-graphene composite coating. *RSC Adv.* **2016**, *6*, 62083–62090.
29. Kumar, M.P.; Singh, M.P.; Srivastava, C. Electrochemical behavior of Zn-graphene composite coatings. *RSC Adv.* **2015**, *5*, 25603–25608.
30. Jabbar, A.; Yasin, G.; Khan, W.Q.; Anwar, M.Y.; Korai, R.M.; Nizam, M.N.; Muhyodin, G. Electrochemical deposition of nickel graphene composite coatings: Effect of deposition temperature on its surface morphology and corrosion resistance. *RSC Adv.* **2017**, *7*, 31100–31109.
31. Sun, C.; Hu, G.; Cao, L.; Pan, T.; Guo, C.; Xia, Y. Ni/graphene coating for enhanced corrosion resistance of metal foam flow field in simulated pemfc cathode environment. *ACS Omega* **2024**, *9*, 29797–29804. [CrossRef] [PubMed]
32. Yigit, O.; Dikici, B.; Senocak, T.C.; Ozdemir, N. One-step synthesis of nano-hydroxyapatite/graphene nanosheet hybrid coatings on Ti6Al4V alloys by hydrothermal method and their in-vitro corrosion responses. *Surf. Coat. Technol.* **2020**, *394*, 125858. [CrossRef]
33. Nesbitt, H.; Legrand, D.; Bancroft, G. Interpretation of Ni2p XPS spectra of Ni conductors and Ni insulators. *Phys. Chem. Miner.* **2000**, *27*, 357–366. [CrossRef]
34. Saldan, I.; Lewin, E.; Dobrovetska, O.; Karlsson, D.; Bilan, O.; Kuntyi, O. Surface analysis of nickel nanomaterials electrodeposited on graphite surface. *Micro Nano Lett.* **2019**, *14*, 1233–1237. [CrossRef]
35. Michał, G.; Mohammad, A.; Gregory, J. Surface science and electrochemical analysis of nickel foams. *ACS Appl. Mater. Interfaces* **2012**, *4*, 3012–3021.
36. Prieto, P.; Nistor, V.; Nouneh, K. XPS study of silver, nickel and bimetallic silver-nickel nanoparticles prepared by seed-mediated growth. *Appl. Surf. Sci.* **2012**, *258*, 8807–8813. [CrossRef]
37. Siddaiah, A.; Kumar, P.; Henderson, A.; Misra, M.; Menezes, P.L. Surface energy and tribology of electrodeposited Ni and Ni-graphene coatings on steel. *Lubricants* **2019**, *7*, 87. [CrossRef]
38. Wang, H.; Sweikart, M.A.; Turner, J.A. Stainless steel as bipolar plate material for polymer electrolyte membrane fuel cells. *J. Power Sources* **2003**, *115*, 243–251. [CrossRef]
39. Li, X.; Shen, Q.; Zhang, Y.; Wang, L.; Nie, C. Wear behavior of electrodeposited nickel/graphene composite coating. *Diam. Relat. Mater.* **2021**, *119*, 108589. [CrossRef]
40. Kumar, C.P.; Venkatesha, T.; Shabadi, R. Preparation and corrosion behavior of Ni and Ni-graphene composite coatings. *Mater. Res. Bull.* **2013**, *48*, 1477–1483. [CrossRef]

41. Algul, H.; Tokur, M.; Ozcan, S.; Uysal, M.; Cetinkaya, T.; Akbulut, H.; Alp, A. The effect of graphene content and sliding speed on the wear mechanism of nickel-graphene nanocomposites. *Appl. Surf. Sci.* **2015**, *359*, 340–348. [CrossRef]
42. Li, N.; Xu, H.; Li, X. Tribological properties and corrosion resistance of porous structure Ni-Mo/ZrO_2 Alloys. *Coatings* **2020**, *10*, 767. [CrossRef]

inorganics

MDPI

Article

Boosting C-C Coupling for Electrochemical CO_2 Reduction over Novel Cu-Cubic Catalysts with an Amorphous Shell

Hanlin Wang [1], Tian Wang [2], Gaigai Dong [1], Linbo Zhang [1], Fan Pan [1,3] and Yunqing Zhu [1,*]

1 School of Environmental Science and Engineering, Shaanxi University of Science and Technology, Xi'an 710021, China; yyh200003@163.com (H.W.); 220312090@sust.edu.cn (G.D.); 220312058@sust.edu.cn (L.Z.); skdpanfan@163.com (F.P.)

2 School of Environmental Engineering, Henan University of Technology, Zhengzhou 450001, China; wangtian@haut.edu.cn

3 School of Chemistry and Chemical Engineering, Shihezi University, Shihezi 832003, China

* Correspondence: zhuyunqing@sust.edu.cn

Abstract: Currently, the electrochemical reduction of carbon dioxide faces significant challenges, including poor selectivity for C_2 products and low conversion efficiency. An effective strategy for optimizing the reduction reaction pathway and enhancing catalytic performance involves manipulating highly unsaturated atomic sites on the catalyst's surface, thereby increasing the number of active sites. In this study, we employed sodium dodecylbenzenesulfonate (SDBS) as a surfactant in the electrodeposition method to synthesize copper cubes encapsulated with an amorphous shell (100 nm–250 nm) containing numerous defect sites on its surface. The electrocatalytic CO_2 reduction reactions in an H-type reactor showed that, compared to ED-Cu synthesized without additives, AS (amorphous shell)-Cu-5 exhibited a Faradaic efficiency value for ethylene that was 1.7 times greater than that of ED-Cu while significantly decreasing the Faradaic efficiency of hydrogen production. In situ attenuated total reflectance surface-enhanced infrared spectroscopy (ATR-SEIRAS) revealed that introducing an amorphous shell and abundant defects altered both the intermediate species and reaction pathways on the AS-Cu-5 catalyst's surface, favoring C_2H_4 formation. The density functional theory (DFT) calculations further confirmed that amorphous copper lowers the energy barrier required for C-C coupling, resulting in a marked enhancement in FE-C_2H_4. Therefore, additive-assisted electrodeposition presents a simple and rapid synthesis method for improving ethylene selectivity in copper catalysts.

Keywords: CO_2 electroreduction; Cu cube; amorphous Cu; C–C coupling; *COCHO intermediate

1. Introduction

The extensive reliance on fossil fuels poses a significant environmental challenge and energy crisis that the world confronts presently [1]. The electrochemical reduction of carbon dioxide (CO_2RR) presents a promising method for converting carbon dioxide into valuable products, such as ethanol and ethylene, establishing the method as green and sustainable due to CO_2 utilization [2–5]. Among the various carbonaceous products produced via CO_2RR, C_{2+} products are especially appealing due to their superior volumetric energy density and economic value compared to C_1 products. As a result, enhancing the selectivity for C_{2+} products has emerged as a key focus within the field of electrochemical CO_2 reduction research [6–8]. Currently, electrocatalysts for the synthesis of C_{2+} products are mostly Cu-based catalysts. In addition to copper-based catalysts, Abdinejad et al. [9]

used heterogeneous nickel-supported iron electrocatalysts, facilitating electrochemical carbon–carbon coupling to produce ethanol.

Nanomaterials have attracted extensive attention due to their unique characteristics, such as having surface, small-size, and quantum-size effects at the nanoscale, and exhibiting broad application prospects [10–13]. Due to these qualities, nanomaterials have been widely studied in the field of carbon dioxide reduction [14–16]. Among these catalysts, copper is one of the known heterogeneous catalysts capable of electrochemically reducing carbon dioxide to carbon couplings, such as ethylene and ethanol [7,17]. However, the lengthy and complex reaction pathways associated with copper-based catalysts result in poor selectivity for C_{2+} products, which hinders their further application. In recent years, researchers have employed various methods—including hydrothermal synthesis [18], wet chemical approaches [19], and thermal annealing [20]—to enhance both the catalytic performance and C_{2+} product selectivity of copper-based catalysts. These efforts have resulted in the development of advanced electrocatalysts, such as single-crystal copper [21], copper alloys [22], oxide-derived copper [23], and single-atom copper [24]. However, most of the current preparation processes for these copper-based catalysts remain relatively complicated, energy intensive, and time consuming. Therefore, the development of new methods for fabricating simple yet high-performance copper-based catalysts is essential for advancing electrochemical CO_2 reduction technologies aimed at producing multi-carbon compounds. Among these methods, electrodeposition has attracted significant interest from researchers due to its advantages: shorter processing times, ease of operation, and robust stability in loading processes. For instance, Chen et al. [25] employed polyvinylpyrrolidone (PVP) as an additive using a controlled grain growth method during electrodeposition to successfully synthesize a copper metal structure enriched with grain boundaries. This enhancement improved the adsorption of key intermediates (*CO) on the surface of copper and facilitated further reductions of carbon dioxide. Additionally, Chen et al. [26] prepared LA Cu_2O catalysts via direct electrodeposition on gas diffusion electrodes (GDEs) utilizing lactic acid (LA) as an additive. Compared to systems without additives, the enhanced electrochemically active surface area (ECSA) and the abundance of active sites—such as grain boundaries and defects—are primary factors contributing to the heightened CO_2RR activity observed in LA Cu_2O.

In addition, researchers have discovered that amorphous nanomaterials exhibit considerable potential and superior performance in catalysis. Recent studies have successfully synthesized these amorphous materials using various methods, including hydrothermal synthesis and chemical reduction [10,27]. It has been demonstrated that, compared to conventional crystalline materials, the abundant unsaturated sites on the surface of amorphous materials can function as effective adsorption sites for carbon dioxide molecules. This characteristic results in a significantly enhanced CO_2 adsorption capacity that facilitates the further activation of CO_2. Chen et al. [28] employed Raman spectroscopy to investigate intermediates during the initial reactions of CO_2 reduction utilizing both crystalline and amorphous Cu catalysts. The peak corresponding to CO_2 on amorphous copper was observed at −0.2 V, which is lower than the value of −0.25 V detected for crystalline copper. This observation indicates that CO_2 is more readily activated on amorphous Cu. Furthermore, Chen et al. [28] conducted simulations to evaluate the adsorption energy of CO on an amorphous copper model and found that, compared to Cu(111), amorphous Cu offers stronger active sites for binding CO, thereby promoting the generation of C_{2+} products. However, current research concerning non-crystalline copper primarily focuses on the strength of adsorption for either CO_2 or CO*, often neglecting a comprehensive exploration into the types or pathways of involved intermediates, particularly regarding coupling between intermediate species OCCO* and OCCHO*. In addition, C-C coupling barriers within ethylene production pathways remain insufficiently reported.

Therefore, this study investigates the changes in the reaction pathways and C-C coupling energy barriers of amorphous copper catalysts through additive-assisted electrodeposition. In this research study, sodium dodecylbenzenesulfonate (SDBS) was added to an electrolyte solution to successfully synthesize copper cubic nanocatalysts with an amorphous shell via electrodeposition. Typically, at the initial stage of electrodeposition, Cu^{2+} ions are usually reduced to Cu^0 and form an ordered crystal structure. However, by adjusting the microscopic environment of the electrolyte, the formation process of crystalline copper can be effectively regulated. When the anionic surfactant SDBS is added to the electrolyte, it will adsorb on the surface of copper. The adsorption of surfactants can alter the potential and electric field distribution on the electrode surface, thereby forming an electrical double layer (EDL) near the Cu surface [29]. The surfactants have charges on their molecules, which can interact electrostatically with metal ions, thereby promoting the formation of amorphous structures by hindering the ordered arrangement of Cu crystal forms. Therefore, the addition of SDBS to the electrolyte will hinder the deposition process of copper and promote the formation of amorphous structures by inhibiting the ordered arrangement of Cu crystal forms. From ATR-SEIRAS, key intermediate species *OCCO and *OCCHO were observed on the AS-Cu-5 electrode with an amorphous shell and abundant defects, which are critical intermediates for synthesizing C_{2+} products. This contrasts with the only existing intermediate *CHO found on ED-Cu. DFT calculations further demonstrate that amorphous materials exhibit lower energy barriers during OC-CHO coupling; this is crucial for facilitating further C−C coupling, resulting in C_{2+} products.

2. Results

2.1. Structure and Morphology

A copper catalyst was prepared using carbon paper as the support in an acidic solution of Cu^{2+} with or without the addition of sodium dodecylbenzenesulfonate (SDBS) for constant current deposition. This process enabled the in situ formation of a cubic nanostructured copper catalyst on the carbon paper. Without altering other conditions, we varied the amount of SDBS added in order to investigate changes in the morphology and structure of the catalyst. The catalysts prepared with SDBS concentrations of 3 mM, 5 mM, and 7 mM were designated AS-Cu-3, AS-Cu-5, and AS-Cu-7, respectively. In contrast, the material synthesized under additive-free conditions was labeled ED-Cu. Scanning electron microscopy (SEM) analysis revealed that AS-Cu-3, AS-Cu-5, and AS-Cu-7 exhibited rough copper surfaces adorned with numerous small copper cubes (Figure 1a–c). When utilizing a concentration of 3 mM SDBS, it was observed that the average particle size of these cubic structures was approximately 250 nm. Increasing the concentration to 5 mM resulted in stacked cubes with diameters ranging from 120 to 150 nm attached to the roughened surface. Conversely, at an SDBS concentration of 7 mM, the particle size further decreased to about 100 nm. Notably, there is an inverse relationship between increasing the amounts of additives and decreasing cube sizes. In stark contrast to this behavior observed for samples containing SDBS additives, only a rough surface was detected in ED-Cu without any formation of cube-like structures (Figure 1d). Thus, we can conclude that introducing SDBS indeed alters copper's surface morphology by facilitating the generation of cubical-shaped copper-based nanomaterials. Moreover, the varying concentrations of SDBS serve as effective means for further modulating the dimensions of these copper cubes. Studies have shown that varying the deposition time can yield catalysts with different morphologies. Therefore, we fixed the additive concentration at 5 mM and performed electrodeposition for various durations, subsequently characterizing the samples using SEM. As illustrated in the presented figures, no copper cubes were formed prior to a deposition time of 10 min; however, copper cubes began to form after a deposition period of 15 min. After extending the deposition time to 20 min, excessive sedimentation and copper stacking resulted in a three-dimensional structure (Figures S1–S4). In addition, SEM tests were performed for two materials reacting at the same time under the same conditions (Figures S5 and S6). In Figures S5 and S6, we can observe that AS-Cu-5 and ED-Cu both showed a tendency to disperse after CO_2RR. On the one hand, the copper cube in AS-Cu-5 disappears, forming a coral-like catalyst. On the other hand, many copper nanoparticles appear on the irregular surface of ED-Cu.

Figure 1. The SEM images of (**a**) AS−Cu−3, (**b**) AS−Cu−5, (**c**) AS−Cu−7, and (**d**) ED−Cu. (**e**–**g**) The HRTEM images of AS−Cu−5 and FFT patterns. (**h**,**i**) The HRTEM images of ED−Cu and FFT patterns.

We further collected samples prepared through additive-assisted electrodeposition on carbon paper for TEM analyses. The TEM results revealed that when SDBS was used as an additive during electrodeposition, a core–shell structure was formed, where an amorphous shell was enveloped by a rich polycrystalline interior. The average thickness of this amorphous shell was approximately 7 nm, while within it, Cu(111) planes could be observed along with numerous lattice defects (Figure 1e). Additionally, FTIR analyses confirmed both the presence of the external amorphous shell (Figure 1f) and the internal crystalline structure characterized by Cu(111) planes (Figure 1g). For comparative purposes, we also conducted TEM testing on polycrystalline copper ED-Cu catalysts synthesized without any additives. The ED-Cu electrodes exhibited uniform Cu(111) lattice orientations (Figures 1h and S7), with subsequent FTIR analyses corroborating their crystalline structure (Figure 1i). In summary, utilizing sodium dodecyl benzenesulfonate as an auxiliary agent during electrodeposition allowed us to synthesize controllable-sized copper cubes featuring both an amorphous copper shell and abundant structural defects.

To eliminate the performance differences arising from the valence state and crystal structure of copper, we conducted X-ray diffraction (XRD) and X-ray photoelectron spectroscopy (XPS) tests on ED-Cu, AS-Cu-3, AS-Cu-5, and AS-Cu-7. In the XRD testing results, all catalysts exhibited three peaks at angles of 43.3°, 50.4°, and 74.1° (Figure 2a), corresponding to Cu(111), Cu(200), and Cu(220), respectively. The diffraction peak observed at 54.21° is attributed to the carbon paper substrate, indicating that the introduction of SDBS did not alter the crystal structure of the catalysts. Further analysis was performed using XPS to investigate the elemental composition and valence states present in the AS-Cu-5 and ED-Cu electrodes. The XPS spectra predominantly revealed elements such as C, O, F, Cu, Na, and S in AS−Cu−5 (Figure S8). Notably, C and F originate from carbon paper that is typically modified with polytetrafluoroethylene. In the XPS spectrum of Cu 2p regions, as shown in Figure 2b, two peaks were identified at binding energies centered around 932.1 eV and 952.1 eV; these correspond to $Cu^{0/1+}$ for Cu $2p^{3/2}$ and Cu $2p^{1/2}$, respectively. Additionally, the peak with a binding energy of approximately 933.88 eV corresponds to Cu^{2+} in conjunction with a vibrational peak at around 945.00 eV, confirming the trace presence of cupric ions on the surface [30,31]. The ED-Cu electrode exhibited a similar XPS pattern (Figure S9). In addition, the atomic percentage of Cu in ED-Cu and AS-Cu-5 electrodes was quantitatively analyzed according to the Cu2p peak intensity, and the results show that there is no obvious difference in the Cu content of ED-Cu and AS-Cu-5 electrodes (Table S1). Overall analyses from both XRD and XPS measurements indicate that incorporating additives does not influence either the crystal structure or valence state of copper within these catalysts. Therefore, the enhancement observed in CO_2RR efficacy cannot be attributed to changes in the copper's crystalline form or its oxidation states. In previous studies, it has been demonstrated that amorphous Cu structures exhibit high activity levels for the conversion of carbon dioxide to carbon monoxide (CO), with a greater number of CO molecules adsorbing on the catalyst's surface [28,32]. This characteristic facilitates the further dimerization of CO into multi-carbon products. To ensure that enhanced performances in the CO_2 reduction reaction (CO_2RR) observed in this experiment are indeed attributable to the introduction of an amorphous shell, it is essential to eliminate any influence from additives. Therefore, we conducted Fourier transform infrared spectroscopy tests on AS-Cu-5 and ED-Cu materials. As illustrated in Figure 2c, the spectra of the two materials are largely similar. Peaks at approximately 1200 cm^{-1} and 1068 cm^{-1} correspond to S-O bonds within the $SO_4{}^{-2}$ groups. The characteristic peak of the S=O bond mainly appears within 1320–1380 cm^{-1}, which is due to sulfonic acid groups, while a peak at around 1500 cm^{-1} can be attributed to aromatic ring vibrations. Notably, we clearly observe the characteristic peak of S-O, but no significant S=O characteristic peaks were observed in Figure 2c, thus ruling out any enhancements in performance due to functionalization by sulfonic acid groups. Although additives do not directly participate in the CO_2RR process, they may alter the selectivity toward ethylene production by inducing abundant defects and promoting the formation of an amorphous copper shell.

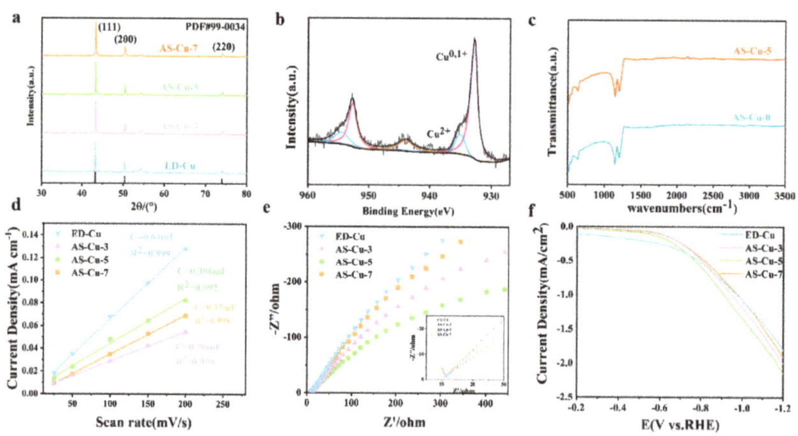

Figure 2. (**a**) XRD of AS–Cu–3, AS–Cu–5, AS–Cu–7, and ED-Cu; (**b**) XPS spectra of AS–Cu–5; (**c**) FTIR spectra of AS–Cu–3, AS–Cu–5, AS–Cu–7, and ED–Cu; (**d**) double-layer capacitance measurement of AS–Cu–3, AS–Cu–5, AS–Cu–7, and ED-Cu; (**e**) Nyquist plots over AS–Cu–3, AS–Cu–5, AS–Cu–7, and ED–Cu; (**f**) LSV of AS–Cu–3, AS–Cu–5, AS–Cu–7, and ED-Cu in CO_2–saturated 0.1 M KCl.

It is generally known that electrodes with larger surface areas can provide more catalytic active sites, thereby better facilitating CO_2 reduction reactions [33–35]. Therefore, we conducted a study on the electrochemically active surface area of four materials: ED-Cu, AS-Cu-3, AS-Cu-5, and AS-Cu-7. Notably, after the addition of SDBS as an auxiliary for electrodeposition, all three materials exhibited varying degrees of reduction in their active surfaces. Among these four materials, ED-Cu presented the largest double-layer capacitance value (0.64 $mF \cdot cm^{-2}$) rather than AS-Cu-5 (0.396 $mF \cdot cm^{-2}$), which demonstrated the best catalytic performance (Figure 2d). Moreover, differing deposition times altered the morphology of the catalysts; thus, we further investigated their electrochemical surface areas under different deposition durations. As shown in Figure S10, AS-Cu-5 deposited for 20 min exhibited the highest double-layer capacitance value (0.652 mF·cm–2), while AS-Cu-5 at 5 min exhibited the lowest capacitance value (0.27 $mF \cdot cm^{-2}$). With an increase in deposition times, there was a corresponding increase in material electrochemical surface area. However, this did not result in enhanced reduction performances over time. Consequently, we propose that the high catalytic activity of AS-Cu-5 is not attributable to an increase in the electrochemically active area but is instead related to improvements in the intrinsic active sites. The lower resistance values are likely beneficial for charge transfers during electrochemical reactions and may represent one significant factor contributing to the superior catalytic activity [19]. In order to further investigate the electron transfer capabilities of these four electrodes, we conducted electrochemical impedance spectroscopy (EIS) on the materials. As shown in Figures 2e and S11, there is no significant difference in the charge transfer resistance among the four materials. Furthermore, after varying the deposition time, the impedance remains relatively consistent (Figure S12). Subsequently, analyses of the linear sweep voltammetry curves (Figure 2f) indicate that AS-Cu-5 does not exhibit a notable increase in current density compared to the other three electrodes.

2.2. *Activity for CO_2 Reduction*

We conducted electrochemical tests on the synthesized copper-based catalysts for CO_2 reduction reactions (CO_2RR) using a standard H-type electrolytic cell employing a constant potential method. After completing the reaction, we collected the gaseous products and analyzed them via gas chromatography. Firstly, through our characterizations, we discovered that varying concentrations of sodium dodecylbenzenesulfonate (SDBS) significantly influenced the morphology of the Cu catalyst, resulting in different copper cube sizes. Consequently, we further compared the CO_2RR performance of Cu catalysts fabricated via electrodeposition assisted by various amounts of SDBS (Figure 3a). From the results presented in the figure, it is evident that when SDBS was added at a concentration of 5 mM, the Faradaic efficiency for C_2H_4 reached its peak at 43.3%. In contrast, with addition levels of 3 mM and 7 mM SDBS yielding Faradaic efficiencies of 35.2% and 37.8% for C_2H_4, respectively, both were found to enhance ethylene production efficiency compared to electrodialyzed copper (ED-Cu) while also effectively suppressing hydrogen evolution reactions [19]. The effects of different deposition times on the performance of catalysts in CO_2 reduction reactions (CO_2RR) were further investigated. Figure 3a illustrates the influence of the deposition time of electrode materials on their performance under a fixed current density. It was observed that as the deposition time increased from 5 min to 15 min, the Faradaic efficiency for ethylene (C_2H_4) production gradually improved, reaching a maximum value of 43.3% at a deposition time of 15 min. However, with an extension of 20 min, the Faradaic efficiency for C_2H_4 dropped to 36.7% (as shown in Figure 3b). Therefore, we determined that the optimal electro-deposition time is 15 min. This enhancement can be attributed to cumulative effects during the deposition process, resulting in the gradual formation of a copper cubic structure that provides more opportunities for the C-C coupling of intermediates. Conversely, excessive or insufficient copper thicknesses may contribute to catalyst agglomeration, which negatively impacts reaction progression.

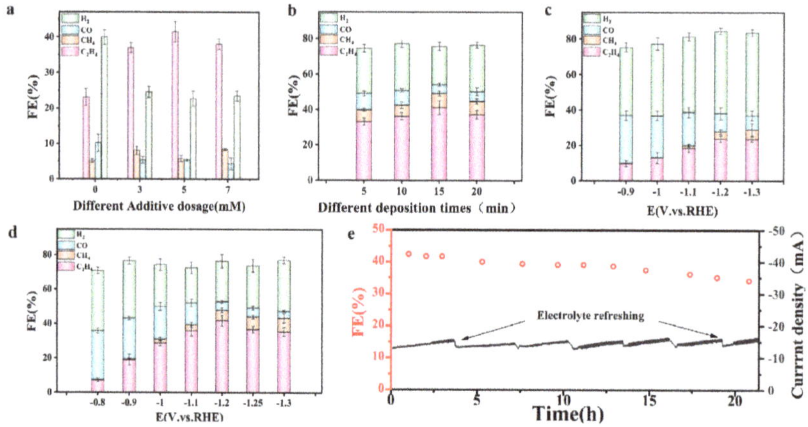

Figure 3. CO_2RR performance in CO_2−saturated KCL. (**a**) CO_2RR performance of different additive amounts. (**b**) CO_2RR performance at various deposition times. The FE at different applied potentials over (**c**) ED−Cu and (**d**) AS−Cu−5. (**e**) Stability test at −1.2 V vs. RHE of AS−Cu−5.

As anticipated, the four materials demonstrated the typical CO_2RR performance of polycrystalline copper in full-potential tests (Figures 3c,d, S13 and S14). Specifically, with increasing overpotential, the faradaic efficiency of C_2H_4 (FE-C_2H_4) gradually increased while that of CO (FE-CO) decreased. At −1.2 V. RHE, the FEC_2H_4 of ED-Cu was recorded at 24.04%, with FE-H_2 at 40.4%. Typically, after CO_2 adsorption on the catalyst and the subsequent formation of CO*, it can directly desorb as CO or participate in C-C coupling on the catalyst's surface to yield multi-carbon products. In contrast, AS-Cu-5 exhibited a higher FEC_2H_4 (43.3%) and lower FE-H_2 (23.7%). Additionally, we observed that FE-CO on ED-Cu was significantly greater than that on AS-Cu-5. The performance tests indicate that more carbon dioxide can be reduced on AS-Cu-5, resulting in the greater utilization of CO for dimerization reactions producing C_2H_4, whereas for ED-Cu, adsorbed CO* is released directly as CO. Across the entire test potential range, the reaction current density for AS-Cu-5 consistently surpassed that of ED-Cu (Figure S15), evidencing enhanced catalytic activity for activating CO_2 at the AS-Cu-5 electrode. In Figures S16 and S17, we compare the faradaic efficiencies of ethylene and methane between optimal additive-loaded AS-Cu-5 and additive-free electrodeposited ED-Cu. Notably, we observed a significant enhancement in both ethylene and methane faradaic efficiencies across all potentials with AS-Cu-5 compared to ED-Cu. Furthermore, during stability testing conducted with the AS-Cu-5 catalyst, we found that both the reaction's current density and the Faradaic efficiency of C_2H_4 were maintained within a consistent range throughout the reaction process (Figure 3e), indicating the stability of this catalyst.

3. Discussion

From the above research, we have identified that the incorporation of SDBS-assisted electrodeposition presents a simple yet efficient method for preparing copper catalysts capable of electrochemically reducing CO_2 to C_2H_4. Compared to catalysts produced without SDBS, those formed via this approach exhibit superior reduction performances and lower hydrogen evolution efficiencies. The currently known reduction processes result in the formation of rough copper surfaces characterized by increased roughness [36], abundant grain boundaries [37], and unsaturated coordination environments [38]. These unique properties may contribute to enhanced selectivity toward multi-carbon products. Furthermore, some studies emphasize that the presence of Cu^{1+} and Cu^{2+} enhances the selectivity for multi-carbon products [39,40]. However, in our study, following the addition of SDBS during assisted electrodeposition, there was no significant difference in the concentrations of Cu^{1+} and Cu^{2+} between AS-Cu-3, AS-Cu-5, and AS-Cu-7 electrodes and those prepared without additives (ED-Cu). To further elucidate the origin of variations in CO_2RR activity after adding SDBS to the catalyst preparation process, we compared AS-Cu-5—the most effective sample—to ED-Cu under identical CO_2RR conditions (same reactor setup, electrolyte composition, carbon dioxide flow rate, etc.). As shown in Figure S18, we conducted TEM tests on AS-Cu-5 after CO_2RR reactions, and the results showed that AS-Cu-5 still had stable amorphous shells. Analyzing XRD patterns post-reaction (Figure S19) revealed no changes in the crystal structure between AS-Cu-5 and ED-Cu before and after reactions. Moreover, XPS spectra for Cu 2p (Figure S20) indicated a reduction in peaks associated with Cu^{2+}, suggesting that both catalysts demonstrate an overall transition from higher valence states to lower ones during the reaction. In summary, it appears that neither the oxidation state nor morphology within these materials predominantly influences performance differences. Rather than these factors being central determinants affecting catalytic efficacy in CO_2RR activities, especially regarding C_2H_4 products, the enhancements observed in the materials derived through SDBS-assisted electrodeposition can be attributed primarily to exhibited amorphous shell characteristics along with rich defect structures; such features are conducive to elevating both reactivities toward C_2H_4 production and selectivity.

To further investigate how the amorphous shell and abundant defects enhance the performance of electrodes in CO_2 reduction reactions (CO_2RR), we employed in situ attenuated total reflectance surface-enhanced infrared absorption spectroscopy (ATR-SEIRAS) to monitor the types of adsorbates on the catalyst's surface, aiming to elucidate the reasons and mechanisms behind the improved CO_2RR performance. Firstly, a peak at 2077 cm^{-1} observed for ED-Cu corresponds to *CO (Figure S21); this peak diminishes as the potential increases from -0.2 V to -0.5 V, indicating the consumption of *CO. A similar peak at 2072 cm^{-1} corresponding to *CO is noted on the AS-Cu-5 catalyst (Figure S22). AS-Cu-5 exhibits stronger CO adsorption in a larger range of sites. The weak signal from *CO intermediates in both materials may be attributed to further rapid reduction processes involving *CO. Furthermore, the peaks observed at 1685 cm^{-1} and 1384 cm^{-1} for ED-Cu correspond to the *CHO and *COOH species, respectively (Figure 4a), with *CHO serving as a crucial intermediate in pathways resulting in CH_4 formation [41,42]. In addition to the peak related to *COOH, AS-Cu-5 also reveals intermediate species at 1575 cm^{-1} (*OCCHO) and 1535 cm^{-1} (*OCCO) [43], both of which are key intermediates for synthesizing C_2H_4. Notably, between -0.1 V and -0.4 V, the *OCCO intermediate demonstrates a significant advantage. However, beyond -0.5 V, *OCCHO becomes the predominant intermediate. Typically, CO_2 adsorbs on the catalyst's surface and undergoes hydrogenation to form *COOH, which can further be hydrogenated to generate HCOOH or activated relative to adsorbed *CO. Following the formation of *CO, there are four potential pathways: (1) direct desorption as CO; (2) hydrogenation to form *CHO, followed by additional hydrogenation to produce CH_4; (3) two *CO molecules coupling through the C-C linkage to yield *OCCO [44,45]; (4) coupling between *CO and *CHO, resulting in the formation of *OCCHO [43,46]. The latter two pathways typically contribute to the production of C_2H_4. Consequently, enhancing the concentration of surface-adsorbed intermediates *OCCO and *OCCHO on the catalyst is more conducive to generating multi-carbon products. It can thus be inferred that the formation of amorphous shells with abundant defects may alter the microenvironment on the catalyst's surface, modulating both reaction intermediates and pathways and thereby favoring C_2H_4 product generation during the electroreduction of CO_2 with AS-Cu-5 catalysts.

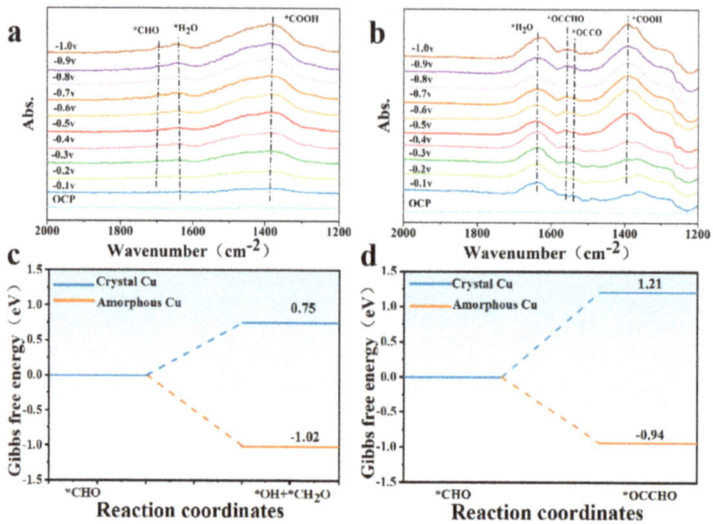

Figure 4. In situ ATR–SEIRAS spectra of (**a**) ED–Cu and (**b**) AS–Cu–5. (**c**) Free energy diagram of the process from CHO to $*CH_2O$ on crystal Cu and amorphous Cu. (**d**) Free energy diagram of C–C coupling on crystal Cu and amorphous Cu.

Afterward, DFT calculations were performed to gain a deeper understanding of the reaction mechanisms at the molecular level. Crystal Cu(111) (Figure S23) and amorphous copper-encapsulated Cu(111) structures (Figure S24) were constructed. The DFT calculations were informed by previous research studies concerning the pathways for methane production [47–49] (Figrues 4c and S25). The free energy of the electroreduction from CO_2 to CH_4 is calculated (Figures S26 and S27). Among these pathways, in addition to *CO protonation to *CHO being commonly regarded as a potentially rate-determining step, it is also suggested that the $*CH_2O$ and $*CH_3O$ species formed through electron coupling from *CHO are crucial intermediates that determine selectivity [50]. Therefore, one reason for the enhanced activity of methane relative to amorphous copper electrocatalysts can be attributed to the exothermic nature of the process from CHO to $*CH_2O$. Additionally, based on the prior literature, we hypothesized that the typical mechanism for CO_2 reduction to ethylene involves two adsorbed *CO molecules [51,52]. Specifically, one adsorbed *CO species can undergo hydrogenation to form *CHO while another undergoes dimerization, resulting in the formation of an intermediate compound known as *OCCHO [53] (Figures 4d and S28). The free energy of the electroreduction from CO_2 to C_2H_4 was calculated (Figures S29 and S30). The computational results indicate that, for this critical step in generating *OCCHO using amorphous catalysts, the energy barrier is significantly lower at −0.94 eV compared to that observed for crystalline copper catalysts. This finding suggests that there are relatively small energy barriers associated with ethylene production via amorphous copper and thus faster reaction rates compared to crystalline counterparts. This correlation aligns with in situ infrared spectroscopy data indicating higher concentrations of *OCCHO for AS-Cu-5 catalysts relative to ED-Cu catalysts, and it is consistent with experimental observations showing superior CO_2RR performances by AS-Cu-5 compared to ED-Cu.

4. Materials and Methods

4.1. Materials

Carbon fiber paper (TGP-H-60, 0.19 nm thickness) was purchased from Toray (Tokyo, Japan). A Nafion N-117 membrane (0.180 mm thick, ≥0.90 meg/g exchange capacity), copper sulfate pentahydrate ($CuSO_4 \cdot 5H_2O$, ≥99%, Tianjin Damao Chemical Reagent Factory, Tianjin, China), sulfuric acid (H_2SO_4, 96%~98%, Damao chemical reagent), sodium dodecylbenzenesulfonate (SDBS, Tianjin Tianli Chemical Reagent Co,. Ltd., Tianjin, China), potassium chloride (KCL, Tianjin Damao Chemical Reagent Factory, Tianjin, China), and potassium hydrogen carbonate ($KHCO_3$, 99.7–100.5%, Thermo Fisher Scientific, Shanghai, China) were used. High-purity carbon dioxide mixed gas (99.999%), argon gas (99.9993%), and high-purity nitrogen (99.9993%) were used as feedstock and carrier gases. All reagents were used as received.

4.2. Synthesis of Electrode

Synthesis of ED-Cu: The preparation of ED-Cu was modified according to the instructions of a previous report [26]. The carbon fiber paper (1×2 cm^2) was washed with ethanol and ultrapure water (18.2 MΩ) in turn. Electrochemical deposition was performed in 0.1 M $CuSO_4$ and 0.1 M H_2SO_4 solutions using a 50 mL electrolyzer. After thorough stirring, the carbon paper and Pt net with an area of 1 cm^2 are used as the cathode and anode, respectively. Electrodeposition was conducted in situ, and the process was carried out at −20 mA on the carbon paper for 15 min using a high-resolution DC power supply (DH1718E-6, Beijing Dahua Radio Instrument Co., Ltd., Beijing, China).

Synthesis of AS-Cu-X (X = 3, 5, 7): AS-Cu-5 was deposited under the same conditions as ED-Cu. In a typical synthesis stage, AS-Cu-X was electrodeposited in 0.1 M $CuSO_4$ and 0.1 M H_2SO_4 solutions using a 50 mL electrolyzer. Then, X (X = 3, 5, and 7) mM SDBS was added. The working reference and counter electrode are consistent with that of ED Cu. After thorough stirring, AS-Cu-X was deposited on the carbon fiber paper with a 1 cm^2 working area (1×1 cm^2) using a high-resolution DC power supply (DH1718E-6, Beijing Dahua Radio Instrument Co., Ltd., Beijing, China), which outputted a −20 mA current for the desired time.

4.3. Electrochemical Measurements

All electrochemical experiments were conducted on an electrochemical workstation (EnergyLab XM, Ametek Corporation, Berwin, IL, USA). Electrochemical CO_2RR was measured in a standard H-cell, with Ag/AgCl and Pt gauze (1×1 cm^2) as the reference and counter electrodes, respectively. An anion exchange membrane (Nafion 117) was used to separate two compartments. Then, a 0.1 M KCl solution and 0.1 M $KHCO_3$ solution were utilized as the cathode and anodic electrolytes, respectively. Prior to CO_2RR measurements, 99.995% CO_2 was continually bubbled into electrolytes (0.1 M KCl) with a flow rate of 20 sccm in order to reach saturation. LSV measurements in gas-saturated electrolytes were carried out within the potential range of 0 V to −1.4 V versus RHE at a sweep rate of 50 mV s^{-1} before experiments. Slight magnetic stirring was employed to acquire uniform electrolytes. All potential readings were measured against Ag/AgCl and then converted to RHE using the following equation: E (vs. RHE) = E (vs. Ag/AgCl) + 0.197 V + 0.0591 V × pH. Gaseous reduction products were analyzed using an online gas chromatograph (GC, Shimadzu 2014, Shimadzu Corporation, Kyoto, Japan). A bio-Logic VMP3 multichannel potentiostat/galvanostat with built-in EIS analysis was used for all electrochemical measurements under ambient pressure. The Faradic efficiency (FE) of each product was calculated using the following equation:

$$FE = \frac{Q_{product}}{Q_{Total}} \times 100\% = \frac{\text{amount of product} \times n \times F}{C} \times 100\%$$

where n represents the number of electrons transferred, F represents the Faradaic constant, and C represents the Coulomb number.

4.4. *In Situ ATR-SEIRAS Measurements*

In situ ATR-SEIRAS measurements: In situ ATR-SEIRAS measurements were carried out using Germany Bruker INVENIO S (Berlin) in a reactor, which consisted of two chambers with three electrodes. The Ag/AgCl electrode and Pt mesh were used as the reference and counter electrodes. Then, 0.1 M KCl solution and 0.1 M $KHCO_3$ solution were utilized as the cathode and anodic electrolytes, respectively. The flow rate of CO_2 for saturating the cathode electrolyte was 20 sccm. Different potentials were applied during in situ SERS measurements, and electrolysis was performed from 0 V vs. RHE to −1 V vs. RHE for 15 min, after which the signal was collected.

5. Conclusions

In summary, we successfully synthesized copper cubes designated as AS-Cu-5 through additive-assisted electrodeposition, which demonstrated enhanced performances compared to the ED-Cu produced via traditional electrodeposition without additives. Transmission electron microscopy (TEM) analyses indicate that the AS-Cu-5 catalysts exhibit an amorphous copper shell characterized by a significant presence of defects. Specifically, varying sizes of copper cubes can be obtained by manipulating the quantity of additives used. In an H-cell setup, the AS-Cu-5 electrode achieved a high Faradaic efficiency (FE) of 43.3% for C_2H_4 production while reaching 23.7% FE for hydrogen generation at −1.2 V versus RHE. In situ ATR-SEIRAS results reveal that ED-Cu displays *CHO intermediates; in contrast, AS-Cu-5 is associated with two distinct intermediates: *OCCO and *OCCHO—neither of which are present in ED-Cu samples. The density functional theory (DFT) calculations further substantiate that the amorphous structure of copper presents a lower energy barrier during the C–C coupling process. These findings suggest that the amorphous Cu shell modulates both the reaction pathway and product selectivity for CO_2 reduction reactions (CO_2RR). Overall, this study highlights a straightforward and effective strategy for fabricating Cu catalysts featuring an amorphous copper shell, resulting in improved activity and selectivity relative to multi-carbon products.

Supplementary Materials: The following supporting information can be downloaded at: https://www.mdpi.com/article/10.3390/inorganics13050130/s1, Figure S1. SEM images of AS-Cu-5 after a deposition period of 5 min. Figure S2. SEM images of AS-Cu-5 after a deposition period of 10 min. Figure S3. SEM images of AS-Cu-5 after a deposition period of 15 min. Figure S4. SEM images of AS-Cu-5 after a deposition period of 20 min. Figure S5. SEM images of AS-Cu-5 after CO2RR. Figure S6. SEM images of ED-Cu after CO2RR. Figure S7. The HRTEM images of ED-Cu. Figure S8. The full XPS spectrum of the as-prepared AS-Cu-5 catalyst. Figure S9. XPS spectra of AS-Cu-5. Figure S10. Double-layer capacitance measurement of AS-Cu-5 after different deposition time. Figure S11. Equivalent circuit used for fitting the data of Nyquist plots. The components contain solution resistance (Rd), double layer capacitance (CPE), electron transfer resistance (Rct) and Warburg-type impedance (Zw). Figure S12. Nyquist plots over the four electrodes AS-Cu-5 after different deposition time. Figure S13. The FE at different applied potentials over AS-Cu-3. Figure S14. The FE at different applied potentials over AS-Cu-7. Figure S15. Current density of C2H4 over ED-Cu and AS-Cu-5 at different applied potentials. Figure S16. The FE-C2H4 over ED-Cu and AS-Cu-5 at different applied potentials. Figure S17. The FE-CH4 over ED-Cu and AS-Cu-5 at different applied potentials. Figure S18. The HRTEM images of AS-Cu-5 after CO2RR. Figure S19. XRD of AS-Cu-5 and ED-Cu after CO2RR. Figure S20. XPS spectra of AS-Cu-5 and ED-Cu after CO2RR. Figure S21. In-situ ATR-SEIRAS spectras of ED-Cu. Figure S22. In-situ ATR-SEIRAS spectras of AS-Cu-5. Figure S23. The crystal Cu(111) structures. Figure S24. Amorphous copper-encapsulated Cu(111) structures. Figure S25. The simulated mechanism for the crystal (a) and amorphization (b) to the CH4 production pathways. Figure S26. Calculated free energy of crystal Cu for CH4. Figure S27. Calculated free energy of amorphous Cu for CH4. Figure S28. The simulated mechanism for the crystal (a) and amorphization (b) to the C2H4 production pathways. Figure S29. Calculated free energy of crystal Cu for C2H4. Figure S30. Calculated free energy of amorphous Cu for C2H4. Table S1. Atomic ratio of Cu from Cu2p signals of ED-Cu and AS-Cu-5 electrodes. Table S2. Comparison of the results of CO2 electroreduction to ethylene over various Cu-based composite electrocatalysts in H-cell. References [54–59] are cited in the Supplementary Materials.

Author Contributions: H.W.: Conceptualization (lead); Supervision (lead); funding acquisition; writing—original draft (lead); writing—review and editing (lead); T.W.: conceptualization; investigation; writing—review and editing; G.D.: writing—review and editing; L.Z.: resources and supervision; F.P.: software; Y.Z.: supervision. All authors have read and agreed to the published version of the manuscript.

Funding: This work was financial supported by the National Natural Science Foundation of China (No. 21876105), Shaanxi "Scientist & Engineer" Team (2023KXJ-131) and Xianyang Key S&T Special Projects (L2023-ZDKJ-QCY-SXGG-GY-007).

Institutional Review Board Statement: Not applicable.

Informed Consent Statement: Not applicable.

Data Availability Statement: The original contributions presented in this study are included in the article/Supplementary Materials. Further inquiries can be directed to the corresponding author. The Supporting Information is available free of charge. Details of Chemicals and materials, characterization, supplementary results such as XRD, SEM, TEM, ECR products distributions; EIS, LSV test and In situ ATR-SEIRAS spectra; models and calculated free energy for DFT simulation.

Conflicts of Interest: The authors declare no conflicts of interest. The funders had no role in the design of the study; in the collection, analyses, or interpretation of data; in the writing of the manuscript; or in the decision to publish the results.

References

1. Gattrell, M.; Gupta, N.; Co, A. Electrochemical reduction of CO_2 to hydrocarbons to store renewable electrical energy and upgrade biogas. *Energy Convers. Manag.* **2007**, *48*, 1255–1265. [CrossRef]
2. Ding, H.; Liu, H.; Chu, W.; Wu, C.; Xie, Y. Structural transformation of heterogeneous materials for electrocatalytic oxygen evolution reaction. *Chem. Rev.* **2021**, *121*, 13174–13212. [CrossRef] [PubMed]
3. Gao, D.; Arán-Ais, R.M.; Jeon, H.S.; Roldan Cuenya, B. Rational catalyst and electrolyte design for CO_2 electroreduction towards multicarbon products. *Nat. Catal.* **2019**, *2*, 198–210. [CrossRef]
4. Gao, D.; Sinev, I.; Scholten, F.; Arán-Ais, R.M.; Divins, N.J.; Kvashnina, K.; Timoshenko, J.; Roldan Cuenya, B. Selective CO_2 electroreduction to ethylene and multicarbon alcohols via electrolyte-driven nanostructuring. *Angew. Chem. Int. Ed.* **2019**, *58*, 17047–17053. [CrossRef]
5. Yang, Y.; He, A.; Yang, M.; Zou, Q.; Li, H.; Liu, Z.; Tao, C.; Du, J. Selective electroreduction of CO_2 to ethanol over a highly stable catalyst derived from polyaniline/$CuBi_2O_4$. *Catal. Sci. Technol.* **2021**, *11*, 5908–5916. [CrossRef]
6. Fu, W.; Liu, Z.; Wang, T.; Liang, J.; Duan, S.; Xie, L.; Han, J.; Li, Q. Promoting C_{2+} production from electrochemical CO_2 reduction on shape-controlled cuprous oxide nanocrystals with high-index facets. *ACS Sustain. Chem. Eng.* **2020**, *8*, 15223–15229. [CrossRef]
7. Nitopi, S.; Bertheussen, E.; Scott, S.B.; Liu, X.; Engstfeld, A.K.; Horch, S.; Seger, B.; Stephens, I.E.L.; Chan, K.; Hahn, C.; et al. Progress and perspectives of electrochemical CO_2 reduction on copper in aqueous electrolyte. *Chem. Rev.* **2019**, *119*, 7610–7672. [CrossRef]
8. Yang, P.-P.; Zhang, X.-L.; Gao, F.-Y.; Zheng, Y.-R.; Niu, Z.-Z.; Yu, X.; Liu, R.; Wu, Z.-Z.; Qin, S.; Chi, L.-P.; et al. Protecting copper oxidation state via intermediate confinement for selective CO_2 electroreduction to C_{2+} Fuels. *J. Am. Chem. Soc.* **2020**, *142*, 6400–6408. [CrossRef]

9. Abdinejad, M.; Farzi, A.; Möller-Gulland, R.; Mulder, F.; Liu, C.; Shao, J.; Biemolt, J.; Robert, M.; Seifitokaldani, A.; Burdyny, T. Eliminating redox-mediated electron transfer mechanisms on a supported molecular catalyst enables CO_2 conversion to ethanol. *Nat. Catal.* **2024**, *7*, 1109–1119. [CrossRef]
10. Tan, M.; Huang, B.; Su, L.; Jiao, X.; Feng, F.; Gao, Y.; Huang, Q.; Huang, Z.; Ge, Y. Amorphous nanomaterials: Emerging catalysts for electrochemical carbon dioxide reduction. *Adv. Energy Mater.* **2024**, *14*, 2402424. [CrossRef]
11. Li, Z.; Zhai, L.; Ge, Y.; Huang, Z.; Shi, Z.; Liu, J.; Zhai, W.; Liang, J.; Zhang, H. Wet-chemical synthesis of two-dimensional metal nanomaterials for electrocatalysis. *Natl. Sci. Rev.* **2022**, *9*, 142. [CrossRef] [PubMed]
12. He, T.; Wang, W.; Shi, F.; Yang, X.; Li, X.; Wu, J.; Yin, Y.; Jin, M. Mastering the surface strain of platinum catalysts for efficient electrocatalysis. *Nature* **2021**, *598*, 76–81. [CrossRef] [PubMed]
13. Cao, L.; Liu, W.; Luo, Q.; Yin, R.; Wang, B.; Weissenrieder, J.; Soldemo, M.; Yan, H.; Lin, Y.; Sun, Z.; et al. Atomically dispersed iron hydroxide anchored on Pt for preferential oxidation of CO in H_2. *Nature* **2019**, *565*, 631–635. [CrossRef] [PubMed]
14. Gao, F.-Y.; Hu, S.-J.; Zhang, X.-L.; Zheng, Y.-R.; Wang, H.-J.; Niu, Z.-Z.; Yang, P.-P.; Bao, R.-C.; Ma, T.; Dang, Z.; et al. High-curvature transition-metal chalcogenide nanostructures with a pronounced proximity effect enable fast and selective CO_2 Electroreduction. *Angew. Chem. Int. Ed.* **2020**, *59*, 8706–8712. [CrossRef]
15. Zhai, Y.; Han, P.; Yun, Q.; Ge, Y.; Zhang, X.; Chen, Y.; Zhang, H. Phase engineering of metal nanocatalysts for electrochemical CO_2 reduction. *eScience* **2022**, *2*, 467–485. [CrossRef]
16. Feijóo, J.; Yang, Y.; Fonseca Guzman, M.V.; Vargas, A.; Chen, C.; Pollock, C.J.; Yang, P. Operando high-energy-resolution X-ray spectroscopy of evolving Cu nanoparticle electrocatalysts for CO_2 reduction. *J. Am. Chem. Soc.* **2023**, *145*, 20208–20213. [CrossRef]
17. Ahmad, T.; Liu, S.; Sajid, M.; Li, K.; Ali, M.; Liu, L.; Chen, W. Electrochemical CO_2 reduction to C_{2+} products using Cu-based electrocatalysts: A review. *Nano Res. Energy* **2022**, *1*, 9120021. [CrossRef]
18. Tan, Z.; Peng, T.; Tan, X.; Wang, W.; Wang, X.; Yang, Z.; Ning, H.; Zhao, Q.; Wu, M. Controllable synthesis of leaf-like CuO nanosheets for selective CO_2 electroreduction to ethylene. *ChemElectroChem* **2020**, *7*, 2020–2025. [CrossRef]
19. Luo, H.; Li, B.; Ma, J.-G.; Cheng, P. Surface modification of nano-Cu_2O for controlling CO_2 electrochemical reduction to ethylene and syngas. *Angew. Chem. Int. Ed.* **2022**, *61*, e202116736. [CrossRef]
20. Keerthiga, G.; Chetty, R. Electrochemical reduction of CO_2 on flame annealed Cu. *ChemElectroChem* **2021**, *6*, 2887–2892. [CrossRef]
21. Hori, Y.; Takahashi, I.; Koga, O.; Hoshi, N. Selective formation of C_2 compounds from electrochemical reduction of CO_2 at a series of copper single crystal electrodes. *J. Phys. Chem. B* **2002**, *106*, 15–17. [CrossRef]
22. Jiao, J.; Lin, R.; Liu, S.; Cheong, W.-C.; Zhang, C.; Chen, Z.; Pan, Y.; Tang, J.; Wu, K.; Hung, S.-F.; et al. Copper atom-pair catalyst anchored on alloy nanowires for selective and efficient electrochemical reduction of CO_2. *Nat. Chem.* **2019**, *11*, 222–228. [CrossRef] [PubMed]
23. Zhao, K.; Liu, Y.; Quan, X.; Chen, S.; Yu, H. CO_2 electroreduction at low overpotential on oxide-derived Cu/carbons fabricated from metal organic framework. *Acs Appl. Mater. Interfaces* **2017**, *9*, 5302–5311. [CrossRef] [PubMed]
24. Guan, A.; Chen, Z.; Quan, Y.; Peng, C.; Wang, Z.; Sham, T.-K.; Yang, C.; Ji, Y.; Qian, L.; Xu, X.; et al. Boosting CO_2 electroreduction to CH_4 via tuning neighboring single-Copper sites. *ACS Energy Lett.* **2020**, *5*, 1044–1053. [CrossRef]
25. Chen, Z.; Wang, T.; Liu, B.; Cheng, D.; Hu, C.; Zhang, G.; Zhu, W.; Wang, H.; Zhao, Z.-J.; Gong, J. Grain-boundary-rich copper for efficient solar-driven electrochemical CO_2 reduction to ethylene and ethanol. *J. Am. Chem. Soc.* **2020**, *142*, 6878–6883. [CrossRef]
26. Chen, L.; Chen, J.; Fan, L.; Chen, J.; Zhang, T.; Chen, J.; Xi, S.; Chen, B.; Wang, L. Additive-assisted electrodeposition of Cu on gas diffusion electrodes enables selective CO_2 reduction to multicarbon products. *ACS Catal.* **2023**, *13*, 11934–11944. [CrossRef]
27. Li, S.-J.; Bao, D.; Shi, M.-M.; Wulan, B.-R.; Yan, J.-M.; Jiang, Q. Amorphizing of Au Nanoparticles by CeO–RGO Hybrid Support towards Highly Efficient Electrocatalyst for N_2 Reduction under Ambient Conditions. *Adv. Mater.* **2017**, *29*, 1700001. [CrossRef]
28. Chen, C.; Yan, X.; Wu, Y.; Zhang, X.; Liu, S.; Zhang, F.; Sun, X.; Zhu, Q.; Zheng, L.; Zhang, J.; et al. Oxidation of metallic Cu by supercritical CO_2 and control synthesis of amorphous nano-metal catalysts for CO_2 electroreduction. *Nat. Commun.* **2023**, *14*, 1092. [CrossRef]
29. Yang, X.; Zhou, Q.; Wei, S.; Guo, X.; Chimtali, P.J.; Xu, W.; Chen, S.; Cao, Y.; Zhang, P.; Zhu, K.; et al. Anion additive integrated electric double layer and solvation shell for aqueous zinc ion battery. *Small Methods* **2024**, *8*, 2301115. [CrossRef]
30. Lim, C.F.C.; Harrington, D.A.; Marshall, A.T. Effects of mass transfer on the electrocatalytic CO_2 reduction on Cu. *Electrochim. Acta* **2017**, *238*, 56–63. [CrossRef]
31. Yuan, X.; Chen, S.; Cheng, D.; Li, L.; Zhu, W.; Zhong, D.; Zhao, Z.-J.; Li, J.; Wang, T.; Gong, J. Controllable Cu^0-Cu^+ sites for electrocatalytic reduction of carbon dioxide. *Angew. Chem. Int. Ed.* **2021**, *60*, 15344–15347. [CrossRef] [PubMed]
32. Vennekoetter, J.-B.; Sengpiel, R.; Wessling, M. Beyond the catalyst: How electrode and reactor design determine the product spectrum during electrochemical CO_2 reduction. *Chem. Eng. J.* **2019**, *364*, 89–101. [CrossRef]
33. Geng, Z.; Kong, X.; Chen, W.; Su, H.; Liu, Y.; Cai, F.; Wang, G.; Zeng, J. Oxygen vacancies in ZnO nanosheets enhance CO_2 electrochemical reduction to CO. *Angew. Chem. Int. Ed.* **2018**, *57*, 6054–6059. [CrossRef]
34. Jia, S.; Zhu, Q.; Wu, H.; Chu, M.; Han, S.; Feng, R.; Tu, J.; Zhai, J.; Han, B. Efficient electrocatalytic reduction of carbon dioxide to ethylene on copper–antimony bimetallic alloy catalyst. *Chin. J. Catal.* **2020**, *41*, 1091–1098. [CrossRef]

35. Wu, T.; Han, M.Y.; Xu, Z.J. Size effects of electrocatalysts: More than a variation of surface area. *ACS Nano* **2022**, *16*, 8531–8539. [CrossRef]
36. Hoang, T.T.H.; Ma, S.; Gold, J.I.; Kenis, P.J.A.; Gewirth, A.A. Nanoporous copper films by additive-controlled electrodeposition: CO_2 reduction catalysis. *ACS Catal.* **2017**, *7*, 3313–3321. [CrossRef]
37. Zhang, J.; Wang, Y.; Li, Z.; Xia, S.; Cai, R.; Ma, L.; Zhang, T.; Ackley, J.; Yang, S.; Wu, Y.; et al. Grain boundary-derived Cu^+/Cu^0 interfaces in CuO nanosheets for low overpotential carbon dioxide electroreduction to ethylene. *Adv. Sci.* **2022**, *9*, e2200454. [CrossRef]
38. Chou, T.-C.; Chang, C.-C.; Yu, H.-L.; Yu, W.-Y.; Dong, C.-L.; Velasco-Vélez, J.-J.; Chuang, C.-H.; Chen, L.-C.; Lee, J.-F.; Chen, J.-M.; et al. Controlling the oxidation state of the Cu electrode and reaction intermediates for electrochemical CO_2 reduction to ethylene. *J. Am. Chem. Soc.* **2020**, *142*, 2857–2867. [CrossRef]
39. Li, C.W.; Kanan, M.W. CO_2 reduction at low overpotential on Cu electrodes resulting from the reduction of thick Cu_2O Films. *J. Am. Chem. Soc.* **2012**, *134*, 7231–7234. [CrossRef]
40. Yue, K.; Qin, Y.; Huang, H.; Lv, Z.; Cai, M.; Su, Y.; Huang, F.; Yan, Y. Stabilized Cu^0-Cu^{1+} dual sites in a cyanamide framework for selective CO_2 electroreduction to ethylene. *Nat. Commun.* **2024**, *15*, 7820. [CrossRef]
41. Deng, B.; Huang, M.; Li, K.; Zhao, X.; Geng, Q.; Chen, S.; Xie, H.; Dong, X.; Wang, H.; Dong, F. The Crystal plane is not the key factor for CO_2-to-methane electrosynthesis on reconstructed Cu_2O microparticles. *Angew. Chem. Int. Ed.* **2021**, *134*, e202114080. [CrossRef]
42. Wang, F.; Zhang, W.; Wan, H.; Li, C.; An, W.; Sheng, X.; Liang, X.; Wang, X.; Ren, Y.; Zheng, X.; et al. Recent progress in advanced core-shell metal-based catalysts for electrochemical carbon dioxide reduction. *Chin. Chem. Lett.* **2022**, *33*, 2259–2269. [CrossRef]
43. Chhetri, M.; Wan, M.; Jin, Z.; Yeager, J.; Sandor, C.; Rapp, C.; Wang, H.; Lee, S.; Bodenschatz, C.J.; Zachman, M.J.; et al. Dual-site catalysts featuring platinum-group-metal atoms on copper shapes boost hydrocarbon formations in electrocatalytic CO_2 reduction. *Nat. Commun.* **2023**, *14*, 3075. [CrossRef]
44. Wang, J.; Qin, Y.; Jin, S.; Yang, Y.; Zhu, J.; Li, X.; Lv, X.; Fu, J.; Hong, Z.; Su, Y.; et al. Customizing CO_2 Electroreduction by Pulse-Induced Anion Enrichment. *J. Am. Chem. Soc.* **2023**, *145*, 26213–26221. [CrossRef]
45. Guo, S.; Liu, Y.; Huang, Y.; Wang, H.; Murphy, E.; Delafontaine, L.; Chen, J.L.; Zenyuk, I.V.; Atanassov, P. Promoting electrolysis of carbon monoxide toward acetate and 1-propanol in flow electrolyzer. *ACS Energy Lett.* **2023**, *8*, 935–942. [CrossRef]
46. Yao, Y.; Shi, T.; Chen, W.; Wu, J.; Fan, Y.; Liu, Y.; Cao, L.; Chen, Z. A surface strategy boosting the ethylene selectivity for CO_2 reduction and in situ mechanistic insights. *Nat. Commun.* **2024**, *15*, 1257. [CrossRef]
47. Zhao, J.; Zhang, P.; Yuan, T.; Cheng, D.; Zhen, S.; Gao, H.; Wang, T.; Zhao, Z.-J.; Gong, J. Modulation of $^*CH_xO$ adsorption to facilitate electrocatalytic reduction of CO_2 to CH_4 over Cu-based catalysts. *J. Am. Chem. Soc.* **2023**, *145*, 6622–6627. [CrossRef]
48. Huang, F.; Chen, X.; Sun, H.; Zeng, Q.; Ma, J.; Wei, D.; Zhu, J.; Chen, Z.; Liang, T.; Yin, X.; et al. Atmosphere induces tunable oxygen vacancies to stabilize single-atom copper in ceria for robust electrocatalytic CO_2 reduction to CH_4. *Angew. Chem. Int. Ed.* **2024**, *64*, e202415642. [CrossRef]
49. Hung, S.-F.; Xu, A.; Wang, X.; Li, F.; Hsu, S.-H.; Li, Y.; Wicks, J.; Cervantes, E.G.; Rasouli, A.S.; Li, Y.C.; et al. A metal-supported single-atom catalytic site enables carbon dioxide hydrogenation. *Nat. Commun.* **2022**, *13*, 819. [CrossRef]
50. Zhi, X.; Vasileff, A.; Zheng, Y.; Jiao, Y.; Qiao, S.-Z. Role of oxygen-bound reaction intermediates in selective electrochemical CO_2 reduction. *Energ. Environ. Sci.* **2021**, *14*, 3912–3930. [CrossRef]
51. Chen, S.; Zheng, X.; Zhu, P.; Li, Y.; Zhuang, Z.; Wu, H.; Zhu, J.; Xiao, C.; Chen, M.; Wang, P.; et al. Copper atom pairs stabilize *OCCO dipole toward highly selective CO_2 electroreduction to C_2H_4. *Angew. Chem. Int. Ed.* **2024**, *63*, e202411591. [CrossRef] [PubMed]
52. Zhang, Y.C.; Zhang, X.L.; Wu, Z.Z.; Niu, Z.Z.; Chi, L.P.; Gao, F.Y.; Yang, P.P.; Wang, Y.H.; Yu, P.C.; Duanmu, J.W.; et al. Facet-switching of rate-determining step on copper in CO_2-to-ethylene electroreduction. *Proc. Natl. Acad. Sci. USA* **2024**, *121*, e2400546121. [CrossRef]
53. Liu, W.; Zhai, P.; Li, A.; Wei, B.; Si, K.; Wei, Y.; Wang, X.; Zhu, G.; Chen, Q.; Gu, X.; et al. Electrochemical CO_2 reduction to ethylene by ultrathin CuO nanoplate arrays. *Nat. Commun.* **2022**, *13*, 1877. [CrossRef] [PubMed]
54. Huang, J.; Mensi, M.D.; Oveisi, E.; Mantella, V.; Buonsanti, R. Structural Sensitivities in Bimetallic Catalysts for Electrochemical CO2 Reduction Revealed by Ag-Cu Nanodimers. *J. Am. Chem. Soc.* **2019**, *141*, 2490–2499. [CrossRef]
55. Hou, L.; Han, J.; Wang, C.; Zhang, Y.; Wang, Y.; Bai, Z.; Gu, Y.; Gao, Y.; Yan, X. Ag nanoparticle embedded Cu nanoporous hybrid arrays for the selective electrocatalytic reduction of CO_2 towards ethylene. *Inorg. Chem. Front.* **2020**, *7*, 2097–2106. [CrossRef]
56. Ning, H.; Mao, Q.; Wang, W.; Yang, Z.; Wang, X.; Zhao, Q.; Song, Y.; Wu, M. N-doped reduced graphene oxide supported Cu_2O nanocubes as high active catalyst for CO_2 electroreduction to C_2H_4. *J. Alloys Compd.* **2019**, *785*, 7–12. [CrossRef]
57. Feng, Y.; Li, Z.; Liu, H.; Dong, C.; Wang, J.; Kulinich, S.A.; Du, X. Laser-Prepared CuZn Alloy Catalyst for Selective Electrochemical Reduction of CO_2 to Ethylene. *Langmuir* **2018**, *34*, 13544–13549. [CrossRef]

58. Mao, M.-J.; Zhang, M.-D.; Meng, D.-L.; Chen, J.-X.; He, C.; Huang, Y.-B.; Cao, R. Imidazolium-Functionalized Cationic Covalent Triazine Frameworks Stabilized Copper Nanoparticles for Enhanced CO_2 Electroreduction. *Chemcatchem* **2020**, *12*, 3530–3536. [CrossRef]
59. Chen, Y.; Fan, Z.; Wang, J.; Ling, C.; Niu, W.; Huang, Z.; Liu, G.; Chen, B.; Lai, Z.; Liu, X.; et al. Ethylene Selectivity in Electrocatalytic CO_2 Reduction on Cu Nanomaterials: A Crystal Phase-Dependent Study. *J. Am. Chem. Soc.* **2020**, *142*, 12760–12766. [CrossRef]

inorganics

MDPI

Article

Insights into Pd-Nb@In_2Se_3 Electrocatalyst for High-Performance and Selective CO_2 Reduction Reaction from DFT

Lin Ju [1,*], Xiao Tang [2], Yixin Zhang [1], Mengya Chen [1], Shuli Liu [1] and Chen Long [1]

[1] School of Physics and Electric Engineering, Anyang Normal University, Anyang 455000, China; 221101089@stu.aynu.edu.cn (Y.Z.); ciky-825@stu.aynu.edu.cn (M.C.); 01958@aynu.edu.cn (S.L.); longchen@aynu.edu.cn (C.L.)
[2] College of Science, Nanjing Forestry University, Nanjing 210037, China; xiaotang@njfu.edu.cn
* Correspondence: julin@aynu.edu.cn

Abstract: The electrochemical CO_2 reduction reaction (eCO_2RR), driven by renewable energy, represents a promising strategy for mitigating atmospheric CO_2 levels while generating valuable fuels and chemicals. Its practical implementation hinges on the development of highly efficient electrocatalysts. In this study, a novel dual-metal atomic catalyst (DAC), composed of niobium and palladium single atoms anchored on a ferroelectric α-In_2Se_3 monolayer (Nb-Pd@In_2Se_3), is proposed based on density functional theory (DFT) calculations. The investigation encompassed analyses of structural and electronic characteristics, CO_2 adsorption configurations, transition-state energetics, and Gibbs free energy changes during the eCO_2RR process, elucidating a synergistic catalytic mechanism. The Nb-Pd@In_2Se_3 DAC system demonstrates enhanced CO_2 activation compared to single-atom counterparts, which is attributed to the complementary roles of Nb and Pd sites. Specifically, Nb atoms primarily drive carbon reduction, while neighboring Pd atoms facilitate oxygen species removal through proton-coupled electron transfer. This dual-site interaction lowers the overall reaction barrier, promoting efficient CO_2 conversion. Notably, the polarization switching of the In_2Se_3 substrate dynamically modulates energy barriers and reaction pathways, thereby influencing product selectivity. Our work provides theoretical guidance for designing ferroelectric-supported DACs for the eCO_2RR.

Keywords: electrochemical CO_2 reduction reaction; density functional theory; dual-metal atomic catalyst; ferroelectric

1. Introduction

Carbon dioxide (CO_2), a principal climate-altering gas, drives critical environmental challenges including rising global temperatures, climatic instability, ocean acidification, and ecological carbon cycle disturbances [1,2]. Urgent action is required to achieve net-zero anthropogenic carbon dioxide equivalent emissions by 2050, as mandated by the Intergovernmental Panel on Climate Change (IPCC) to cap global temperature increases at 1.5 °C above preindustrial baselines [3]. This imperative is becoming increasingly critical given the persistent reliance on emission-generating technologies and escalating global energy demands. Addressing the dual challenge of atmospheric CO_2 mitigation and sustainable utilization has emerged as a pivotal research frontier [4–7]. Carbon management encompasses three interconnected approaches to mitigating anthropogenic CO_2 emissions. First, energy generation processes are being optimized through advanced combustion

 https://doi.org/10.3390/inorganics13050146

technologies, such as oxy-fuel combustion with pure oxygen injection and homogeneous charge compression ignition (HCCI) engines using biofuel blends [8,9]. Second, modern carbon capture technologies are advancing beyond conventional geological storage in saline aquifers or depleted oil fields by transforming captured CO_2 into value-added products like construction aggregates, agricultural bio-fertilizers, and energy-efficient syngas, leveraging advanced separation processes such as pressure swing adsorption (PSA) and membrane systems [10,11]. Third, the chemical conversion pathways, particularly electrocatalytic CO_2 reduction reaction (eCO_2RR), utilize multi-electron transfer processes ($2e^-$ to $12e^-$) with tailored catalysts (e.g., Cu for hydrocarbons, Ag/Zn for CO) to produce value-added fuels and chemicals like CO, formate, methane, ethylene, and ethanol, while enabling renewable energy integration and high-pressure reactor designs for industrial scalability [12,13].

Among the above approaches, the eCO_2RR stands out for its dual capacity to mitigate emissions and generate valuable chemical feedstocks [14–16]. And this technology has four key advantages. First, the eCO_2RR leverages renewable electricity from solar, wind, or tidal energy to convert CO_2 into chemicals, enabling carbon-neutral cycles while reducing reliance on fossil fuels [17]. Second, the process generates high-value chemicals like ethylene (feedstock for plastics) [18], methane (fuel) [19], and formate (industrial reagent) [20], directly replacing fossil-derived equivalents in existing supply chains. Third, unlike thermocatalytic methods requiring high temperatures/pressures, the eCO_2RR operates efficiently at room temperature and atmospheric pressure, reducing energy demands and infrastructure complexity. Forth, modular flow cell designs and membrane electrode assemblies (MEAs) allow scalable deployment for industrial adoption [21].

The eCO_2RR process involves intricate multi-stage reaction sequences: (1) CO_2 adsorption at electrode interfaces; (2) proton=coupled electron transfer (PCET) steps, which forms C–H bonds with broken C=O bonds; and (3) detachment of synthesized compounds from active sites through desorption mechanisms and their subsequent transport into the bulk reaction medium via dissolution–diffusion pathways [22]. This sequential process becomes particularly consequential in water-containing systems where the parasitic hydrogen evolution reaction (HER) actively competes with the CO_2RR for catalytic resources and electron transfer, fundamentally altering the thermodynamic landscape and resulting in suboptimal product distributions [23].

In the process of reducing CO_2 with metal electrodes, there are many challenges, including the impact of the poor stability of the electrodes. Many metal electrodes, such as palladium (Pd), indium (In), and tin (Sn), have stability issues when reducing CO_2. Take Pd as an example; CO has strong adsorption energy on its surface, which causes the electrode to be easily poisoned and deactivated, greatly shortening its service life [24]. Although In and Sn electrodes can achieve the reduction of CO_2, they often require a higher voltage to drive the reaction [25–27], which not only causes a large waste of energy but may also trigger side reactions due to high voltage, further reducing the selectivity and efficiency of the reaction. At the same time, the stability of the electrode deteriorates further under high voltage, accelerating the electrode's wear and tear. These problems seriously restrict the practical application of metal electrodes in the field of CO_2 reduction [26]. Effective eCO_2RR implementation hinges on catalytic systems being optimized across three critical dimensions, namely conversion efficiency under applied potentials, target product selectivity over competing reactions, and long-term structural and functional stability. These properties are intrinsically linked to atomic-scale catalyst architecture, particularly the coordination environment of active sites.

Since the conceptualization of single-atom catalysts (SACs) in 2011, these atomically dispersed metal systems have revolutionized catalytic research through maximum metal utilization efficiency (approaching 100%) [28–33], well-defined active site geometries for mechanistic studies [34–36], and enhanced selectivity profiles compared to nanoparticle analogues. However, SAC implementation faces two fundamental constraints: low active site density due to synthesis limitations and inadequate multi-intermediate stabilization in complex eCO_2RR pathways [37,38]. This has prompted researchers to turn their attention to bimetallic atomic catalyst (DAC) systems with greater structural flexibility.

DACs exhibit distinct advantages over isolated SACs, combining the inherent merits of SACs with supplementary synergistic benefits: (1) DACs substantially elevate metal loading capacities while expanding the population of accessible reactive centers, a critical factor for scalable catalytic applications [39]; (2) DACs permit dynamic spatial arrangements of active sites, enabling coordinated interactions essential for multi-step CO_2 reduction pathways [40,41]; (3) electronic modulation between dissimilar metal atoms allows precise regulation of intermediate binding strengths, creating tunable platforms for reaction pathway optimization and selectivity control [40,42–47]; (4) independent yet coexisting metallic centers enable parallel reaction channels, permitting simultaneous production of distinct value-added compounds through site-specific catalytic processes [48]; (5) and in DACs, one metallic component can modify support interactions that subsequently influence the catalytic behavior of adjacent metal sites, creating cascading activation mechanisms [30,49,50]. It is anticipated that further studies will concentrate on enhancing the performance of bimetallic atomic site catalysts.

Contemporary DACs predominantly utilize two-dimensional nitrogen-doped carbon substrates, as exemplified by Fe-Co [51], Ni-Fe [52], and Co-Pt [53] dual-atom configurations, which have been fabricated with dual-solvent, pyrolysis, and cyclic voltammetry methods, respectively. However, these conventional DACs exhibit inherent limitations in post-synthesis catalytic tunability due to their inability to leverage substrate property modifications for performance enhancement. Recent advancements in ferroelectric material engineering provide a promising solution. Substantial theoretical and experimental evidence demonstrates that polarization direction modulation in ferroelectrics not only governs molecular adsorption/desorption dynamics but also enables precise control over surface catalytic activity and product selectivity [54–56]. Capitalizing on this principle, we engineered a novel ferroelectric DAC system by anchoring niobium (Nb) and palladium (Pd) single atoms onto an α-In_2Se_3 monolayer substrate (Nb-Pd@In_2Se_3). The α-In_2Se_3 monolayer is a van der Waals material exhibiting coupled in-plane and out-of-plane ferroelectric polarization at room temperature [56–58]. It features a quintuple atomic-layer structure (Se-In-Se-In-Se) with spontaneous polarization arising from asymmetric Se atom displacements, enabling bidirectional electric field control of both polarization axes. This unique architecture allows for the dynamic regulation of metal–support interactions, addressing the static catalytic behavior inherent to traditional carbon-based DACs. The theoretical analysis revealed that the dual-metal atomic configuration significantly enhances CO_2 activation capability during the eCO_2RR process. In the synergistic catalytic mechanism, the Nb atomic sites predominantly function as catalytic centers for carbon reduction, while the adjacent Pd counterparts facilitate oxygen species removal through a complementary catalytic pathway. Notably, polarization switching of the ferroelectric substrate modulates both the energy barriers along the reaction coordinates and the dominant reaction pathways, ultimately dictating the selectivity of terminal products.

2. Results and Discussion

2.1. Geometric Structures of Nb-Pd@In_2Se_3

The 2D ferroelectric α-phase In_2Se_3 monolayer was chosen as the substrate material. Based on our previous work [56], we selected Pd and Nb transition metals as candidates for loaded metallic dual atoms, since they can stably form Pd@In_2Se_3 and Nb@In_2Se_3

SACs. According to the position of the intermediate Se layer offset (see Figure 1a,b), the α-In_2Se_3 monolayer has two stable polarization phases, namely polarization upward (P↑) and downward (P↓). We refer to the configuration where the Pd-Nb dual-atom pair is doped on the outer surface of two layers with a Se spacing of 3.06 Å as the Pd-Nb@P↓-In_2Se_3. Moreover, the configuration where the Pd-Nb dual-atom pair is doped on the outer surface of two layers of Se with a spacing of 3.84 Å is called Pd-Nb@P↑-In_2Se_3. As for the SACs [59–61], strong binding strength is essential for the stability of DACs. At the energetically most favorable site, the binding energy of the Nb-Pd dual-atom pair in Pd-Nb@P↓-In_2Se_3 and Pd-Nb@P↑-In_2Se_3 were −15.69 and −12.27 eV/unit, respectively. The low binding energies indicate that there is a strong interaction between the metal atoms and the substrate, as evidenced by the strong covalent Pd–Se and Nb-Se bonds (see Figure 1a,b), which are caused by the strong hybridization of Se *p* and Pd/Nb *d* orbitals near the Fermi level (see Figure S1a,b).

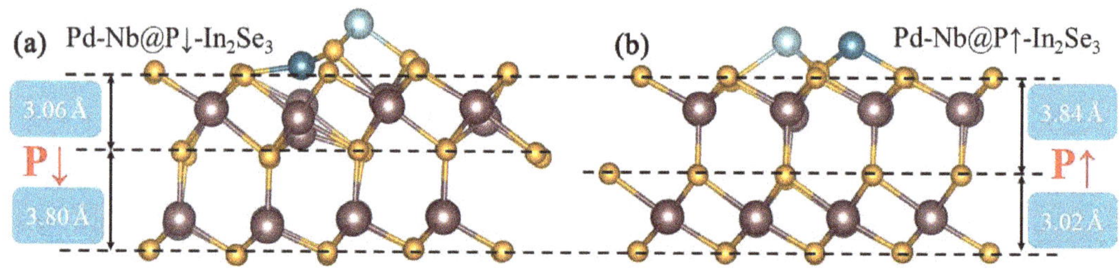

Figure 1. Side views of the optimized configuration of (**a**) Pd-Nb@P↓-In_2Se_3 and (**b**) Pd-Nb@P↑-In_2Se_3 DACs.

To further demonstrate the structural stability of Pd-Nb@In_2Se_3, ab initio molecular dynamics (AIMD) calculations were carried out on these two configuration at a temperature of 1000 K, with a time step of 1 fs and a total time length of 10 ps (see Figure 2a,b). During the AIMD simulations, the geometrical structure did not change significantly, and the total energy of the system tended to be stable, indicating that these configurations have good thermal stability. Overall, both the binding energy and molecular dynamics simulation results show the promising feasibility of fabricating Pd-Nb@In_2Se_3 DACs.

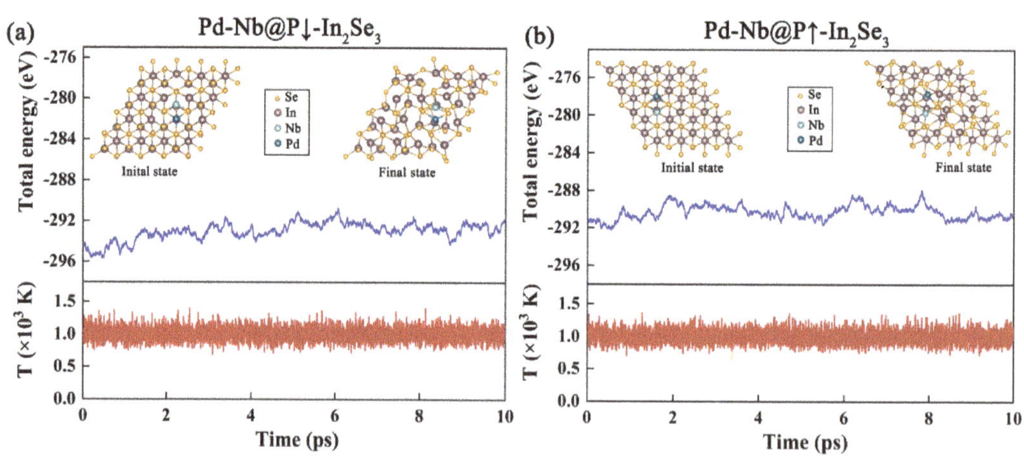

Figure 2. The ab initio molecular dynamics simulations of (**a**) Pd-Nb@P↓-In_2Se_3 and (**b**) Pd-Nb@P↑-In_2Se_3 configurations at 1000 K, showing the fluctuations in the total energy (upper) and temperature (lower) over a time step of 1 fs and a duration of 10 ps.

2.2. Activation of CO_2 on Pd-Nb@In_2Se_3 DACs

The effective activation of CO_2 molecules is a key step for the subsequent reduction reaction. The linear structure of CO_2 molecules contributes to their pronounced chemical inertness, primarily due to the symmetry-induced cancellation of dipole moments despite the presence of polar C=O bonds. This linear geometry stabilizes the molecule thermodynamically, as the two C=O bonds exhibit dissociation energies of approximately 750 $kJ \cdot mol^{-1}$. Additionally, the large energy gap (13.7 eV) between the highest occupied molecular orbital (HOMO) and lowest unoccupied molecular orbital (LUMO) creates a significant kinetic barrier to electron transfer, necessitating substantial energy input for chemical transformations. In catalytic systems employing isolated atomic site catalysts (SACs), CO_2 adsorption typically occurs via either oxygen– or carbon–metal (O-M or C-M) interactions at the active metal centers. However, the limited electron donating capacity and geometric constraints of isolated atomic sites restrict the extent of CO_2 activation. Consequently, SACs often require elevated temperatures and pressures to achieve moderate reaction rates, further reducing reaction efficiency and selectivity toward multi-carbon products.

Dual-site catalysts can demonstrate enhanced CO_2 adsorption via coordination involving both metallic dual atoms and support atoms, requiring lower activation energy than the SAC to activate CO_2 [62]. As shown in Table S1, by separately comparing the adsorption energy of CO_2 on the Pd and Nb sites of Pd-Nb@In_2Se_3 DACs (see Figure S2), we conclude that the adsorption of CO_2 on the Pd site had more negative adsorption energy in both P↑ and P↓ configurations, which means the CO_2 tends to adsorb on the Pd site of Pd-Nb@In_2Se_3 DACs. The ∠OCO values of CO_2 adsorbed on Pd-Nb@P↓-In_2Se_3 and Pd-Nb@P↑-In_2Se_3 were significantly reduced to 131.38° and 125.18° from an initial value of 180°, respectively. Based on the structural analysis, it could be stated that the Pd-Nb@In_2Se_3 catalyst could effectively activate CO_2 molecules. Moreover, according to our previous report [56] that the ∠OCO values of CO_2 adsorbed on Pd@P↓-In_2Se_3 and Pd@P↑-In_2Se_3 were 149.2° and 151.2°, respectively, it can be concluded that the synergistic effect of a bimetallic atomic site causes a more obvious deformation of adsorbed CO_2 molecules (17.82° and 26.02° for P↓ and P↑, respectively), indicating a higher activation of CO_2. The similar synergistic effects in the Pt-Ni@CeO_2 bimetallic system were extensively validated through comprehensive characterization and computational studies [62]. Temperature-programmed reduction (TPR) and desorption (TPD) analyses revealed that Pt-Ni synergy significantly enhances the generation of reactive oxygen intermediates (e.g., surface peroxides and hydroxyl radicals) and strengthens metal–support interactions via charge transfer mechanisms. These modifications collectively optimize CO_2 chemisorption capacity, as evidenced by the higher adsorption efficiency compared to the monometallic counterparts (Pt@CeO_2 and Ni@CeO_2). Density functional theory (DFT) simulations further demonstrated a 35–40% reduction in CO_2 activation energy (E_a = 10.1 kcal mol^{-1} for Pt-Ni@CeO_2 vs. 32.9 and 36.5 kcal mol^{-1} for Ni@CeO_2 and Pt@CeO_2, respectively), which is attributed to electronic structure modulation at Pt-Ni-CeO_2 interfacial sites.

In order to explore the activation mechanism, we employed a charge density difference calculation, which demonstrated that there is both a loss and gain of electrons on the adsorbed CO_2 molecule and the bimetallic atomic sites (see Figure 3a,b). This is because the empty Pd/Nb *d* orbital can accept electrons from the highest occupied molecular orbitals ($1\pi_g$ orbital) of the CO_2 molecule, while the occupied Pd/Nb *d* orbital can back-donate electrons to the lowest unoccupied molecular orbitals ($2\pi_u$ orbital) of the CO_2 molecule (see Figure 3c). The synergy of electron donation and backdonation-dependent *d* orbital occupation ensures that CO_2 can be efficiently activated [56].

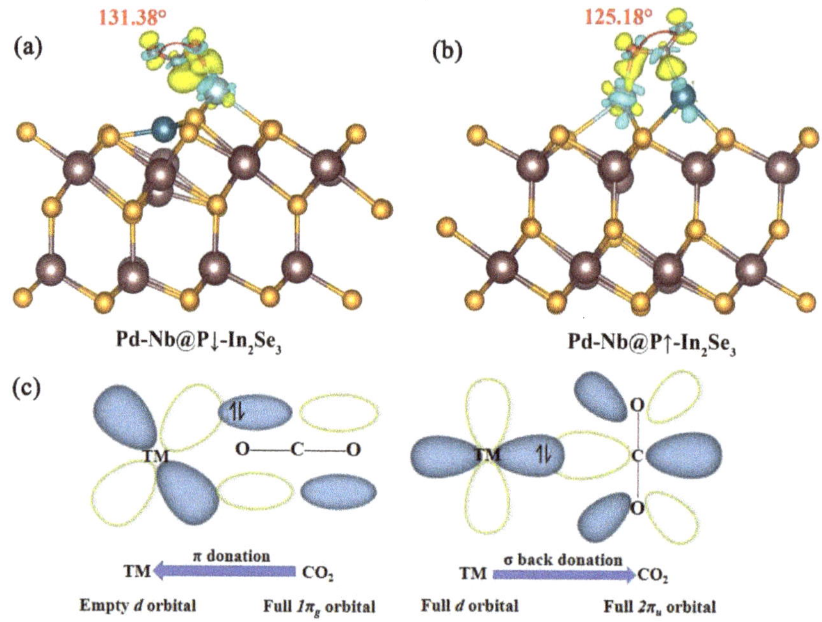

Figure 3. The charge differential density plots of CO_2 adsorbed on (**a**) Pd-Nb@P↓-In_2Se_3 and (**b**) Pd-Nb@P↑-In_2Se. The charge accumulation is indicated by areas that are yellow. The charge depletion is indicated by areas that are cyan. The value of the isosurface is 0.01 e Å^{-3}. (**c**) Simplified schematic diagrams of CO_2 activation on transition metals.

As mentioned before, the CO_2 adsorbed on Pd-Nb@P↑-In_2Se_3 (∠OCO = 125.18°) had a more obvious deformation than that absorbed on Pd-Nb@P↓-In_2Se_3 (∠OCO = 131.38°), which means Pd-Nb@P↑-In_2Se_3 has higher CO_2 activation. In order to investigate the reason for this, we calculated the Pd and Nb *d* orbitals in Pd-Nb@P↑-In_2Se_3 and Pd-Nb@P↓-In_2Se_3 DACs, respectively. Then, based on the partial electronic orbitals (see Figure 4a), we obtained the *d* band center of the Pd and Nb atoms for both kinds of Pd-Nb@In_2Se_3 DACs. As displayed in Figure 4b, both *d* band centers of the Pd (−2.09 eV) and Nb (0.11 eV) atoms in Pd-Nb@P↑-In_2Se_3 were closer to the Fermi level than the ones of Pd (−2.68 eV) and Nb (0.44 eV) atoms in Pd-Nb@P↓-In_2Se_3. And it has been reported that the exceptional electrocatalytic performance of Co_2 anchored graphdiyne in nitrogen fixation originates from the presence of localized electronic states adjacent to the Fermi level, which optimize charge transfer dynamics during the reaction [63]. Therefore, in the case of P↑, the proximity of the *d* band center of Pd and Nb atoms to the Fermi level induces a high CO_2 activation performance. Note that the different *d* band centers of bimetallic atomic sites in Pd-Nb@In_2Se_3 under different polarization states implies that both the reaction path and the limiting potential of the subsequent CO_2 hydrogenation process may be affected by the polarization switch.

2.3. eCO_2RR on Pd-Nb@In_2Se_3 DACs

Considering the complexity of the CO_2 reaction (see Figure S3), we have summarized the minimum-energy reaction paths of the CO_2RR over Pd-Nb@In_2Se_3 catalysts (see Figure 5). The CO_2RR pathways over Pd-Nb@In_2Se_3 catalysts exhibit stark polarization-dependent selectivity, governed by the structural and electronic asymmetry induced by the ferroelectric In_2Se_3 substrate's polarization orientation (P↓ and P↑). The CO_2 reduction reaction begins with the adsorption of CO_2 onto the Pd-Nb@P↓-In_2Se_3 catalyst surface, forming a CO_2 intermediate (CO_2 + * → CO_2*). This adsorbed CO_2 undergoes a proton-coupled electron transfer (PCET) step, where one hydrogen atom binds to an O atom (CO_2* + H^+ + e^- → OCHO*), leading to the formation of an OCHO intermediate (bent configuration with O–CH–O bonding). Subsequently, a second protonation occurs at the adjacent oxygen atom (OCHO* + H^+ + e^- → HOCHO*), stabilizing the HOCHO intermediate (adsorbed formate-like structure). Finally, the HOCHO intermediate desorbs from the catalyst surface (HOCHO* → HOCHO + *), releasing formic acid (HOCHO) as the end product. Due to the high energy barrier associated with stabilizing the second protonation step, the rate-determining step (PDS) is identified as the transformation of OCHO* → HOCHO*, requiring an applied potential of −0.45 V vs. RHE. In contrast, in the CO_2RR on a Pd-Nb@P↑-In_2Se_3 catalyst, the reaction course is significantly different from that on Pd-Nb@P↓-In_2Se_3, which undergoes an alternative protonation pathway. The first PCET step preferentially protonates the carbon atom, forming the OCOH intermediate (CO_2* + H^+ + e^- → OCOH*) (linear configuration with O–C–O–H bonding). Further reduction cleaves the C–O bond. A second protonation occurs at the hydroxyl group of OCOH intermediate. The hydroxyl group binds a hydrogen ion to generate H_2O and desorb (OCOH* + H^+ + e^- → CO* + H_2O), producing a CO intermediate adsorbed on the Pd site of the catalyst. Sequential hydrogenation steps follow: CO* + H^+ + e^- → CHO* (formyl), CHO* + H^+ + e^- → CH_2O* (formaldehyde-like) and CH_2O* + H^+ + e^- → CH_{2*} + OH* via C–O bond scission. The critical C–O bond scission step (CH_2O* + H^+ + e^- → CH_2* + OH*) in the methane production pathway over the Pd-Nb@P↓-In_2Se_3 catalyst is driven by the synergistic interplay between the Pd and Nb dual active sites. The CH_2 intermediate absorbs the Pd site, and the OH intermediate absorbs the Nb site. CH_2* undergoes progressive hydrogenation (CH_2* + OH* + H^+ + e^- → CH_3* + OH*), followed by a critical step where CH_3 reacts with a co-adsorbed OH group to form CH_3 and H_2O (CH_3* + OH* + H^+ + e^- → CH_3* + H_2O*). Then, the H_2O molecule desorbs from the Nb site (CH_3* + H_2O* → CH_3* + H_2O). This step is identified as the PDS, with a limiting potential of −0.66 V vs. RHE, reflecting the substantial energy required for H_2O dissociation. The final hydrogenation of CH_3 yields methane (CH_4) (CH_3* + H^+ + e^- → CH_4*), which desorbs from the catalyst surface (CH_4* → CH_4 + *). Consequently, the U_l, catalytic pathway, and the ultimate products during the eCO_2RR, are governed by polarization orientation and are adjustable via ferroelectric switching. Moreover, compared to the U_l of Pd@In_2Se_3 SACs for the eCO_2RR (−0.77 V under the P↓ state and −0.87 V under the P↑ state) [56], each corresponding Pd-Nb@In_2Se_3 DAC for the eCO_2RR has a smaller U_l, indicating that the synergistic effect of the bimetallic atomic site could effectively lower the eCO_2RR barrier. According to the reduction reaction path of CH_2O* → CH_2* + OH* → CH_3* + OH* → CH_3*, we found that carbon-containing intermediates (e.g., CH*, CH_2*) tend to undergo hydrogenation reduction at Pd atomic sites, while oxygen-containing intermediates (e.g., OH*) prefer to be hydrogenated and reduced into water at Nb atomic sites (see Figure S4). Therefore, the synergistic effect of Nb-Pd dual-site metal atoms can be described as such: Pd sites act as functional sites for carbon reduction, whereas Nb sites serve as functional sites for oxygen removal. Notably, the U_l of Pd@In_2Se_3 SACs for the eCO_2RR (−0.45 V under the P↓ state and −0.66 V under the P↑ state) are markedly superior to those of conventional eCO_2RR catalysts, such as Cu (211) (−0.8 V) [64], Co@Cu (−0.87 V) [65], and WC (0001) (−1.0 V) [66]. Since the equilibrium potential (E°) for the formation of HCOOH and CH_4 from CO_2, respectively, are −0.25 V vs. RHE and 0.17 V vs. RHE at pH = 0 [67], the corresponding overpotentials are −0.2 V vs. RHE and −0.83 V vs. RHE, which are lower than the ones of many eCO_2RR electrocatalysts synthesized experimentally, such as Pd_2 (−0.85 V vs. RHE for generating CO) [68], NiCu-NC (−1.07 V vs. RHE for generating CO) [69], FeNi-NSC (−1.0 V vs. RHE for generating CO) [70],

Cu/Ni-NC (−0.9 V vs. RHE for generating CO) [71], and Cu-N_2 (−0.84 V vs. RHE for generating CH_4) [72]. Therefore, the Pd-Nb@In_2Se_3 DAC is a promising candidate for a tunable and efficient eCO_2RR, particularly in achieving high selectivity for final products through controlled polarization modulation.

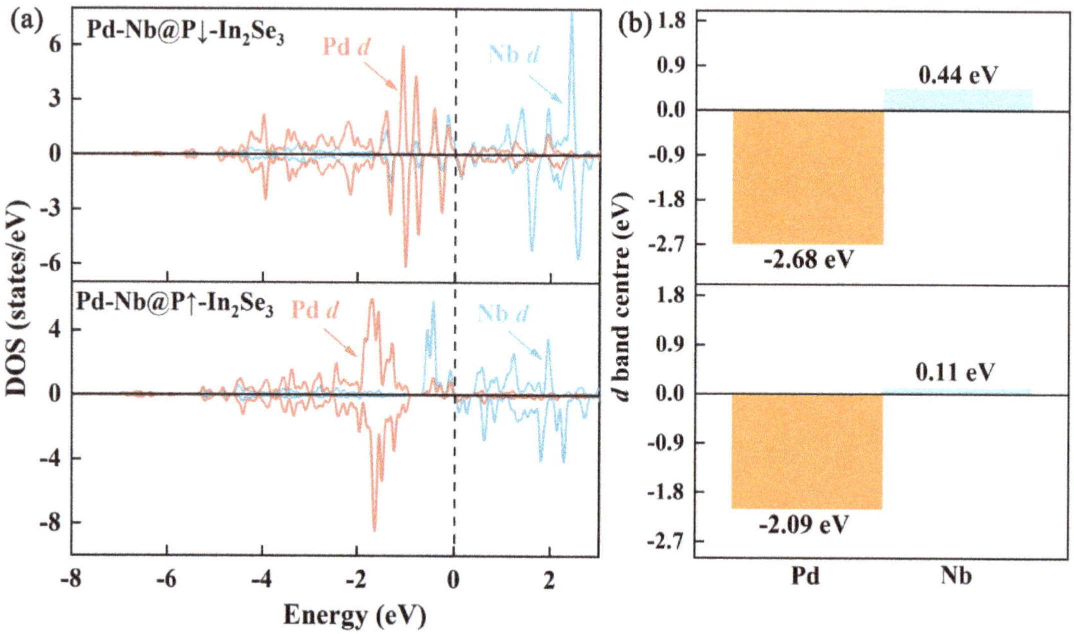

Figure 4. (**a**) Pd *d* orbitals (in light red) from Pd-Nb@P↓-In_2Se_3 and Pd-Nb@P↑-In_2Se_3 DACs. Nb *d* orbitals (in light blue) from Pd-Nb@P↓-In_2Se_3 and Pd-Nb@P↑-In_2Se_3 DACs. (**b**) The *d* band center of Pd-Nb@In_2Se_3, with orange columns for Pd atoms and blue columns for Nb atoms.

2.4. Selectivity for HER vs. eCO_2RR on Pd-Nb@In_2Se_3 DACs

The HER is a prevalent parasitic process in electrochemical systems. It poses a significant challenge to the Faradaic efficiency of the eCO_2RR by diverting critical proton and electron resources from the electrolyte. This competition arises through the HER's multi-step mechanisms—Volmer adsorption, Heyrovsky/Tafel recombination, and H_2 desorption—which operate synergistically to deplete reactive intermediates (H*) and impose additional energy barriers for CO_2 reduction pathways. Particularly in acidic media, the HER's thermodynamic favorability and low activation overpotential exacerbate its dominance over the eCO_2RR, resulting in difficulty in achieving an efficiency of the eCO_2RR that meets the needs of industrial production [73,74]. To check if the eCO_2RR is more favorable, we calculated the Gibbs free energy changes (ΔG) at the first hydrogenation step of the eCO_2RR ($* + CO_2 \rightarrow CO_2*$) and the HER ($* + H^+ + e^- \rightarrow H*$). As displayed in Figure 6, on the Pd-Nb@P↓-In_2Se_3 DAC, the ΔG for the formation of H* is 0.61 eV, while that of CO_2* is −0.22 eV. As for the case of the Pd-Nb@P↑-In_2Se_3 DAC, the ΔG for the formation of H* is −0.11 eV, while that of CO_2* is −1.04 eV. According to the Brønsted–Evans–Polanyi relation, reactions with lower ΔG have smaller reaction barriers and consequently are kinetically more favorable. The Gibbs free energy changes indicate that the Pd-Nb@In_2Se_3 DACs have a higher selectivity toward the CO_2RR, instead of the HER. These results demonstrate the feasibility of using Pd-Nb@In_2Se_3 catalysts as cathodes for the eCO_2RR with a high Faradaic efficiency.

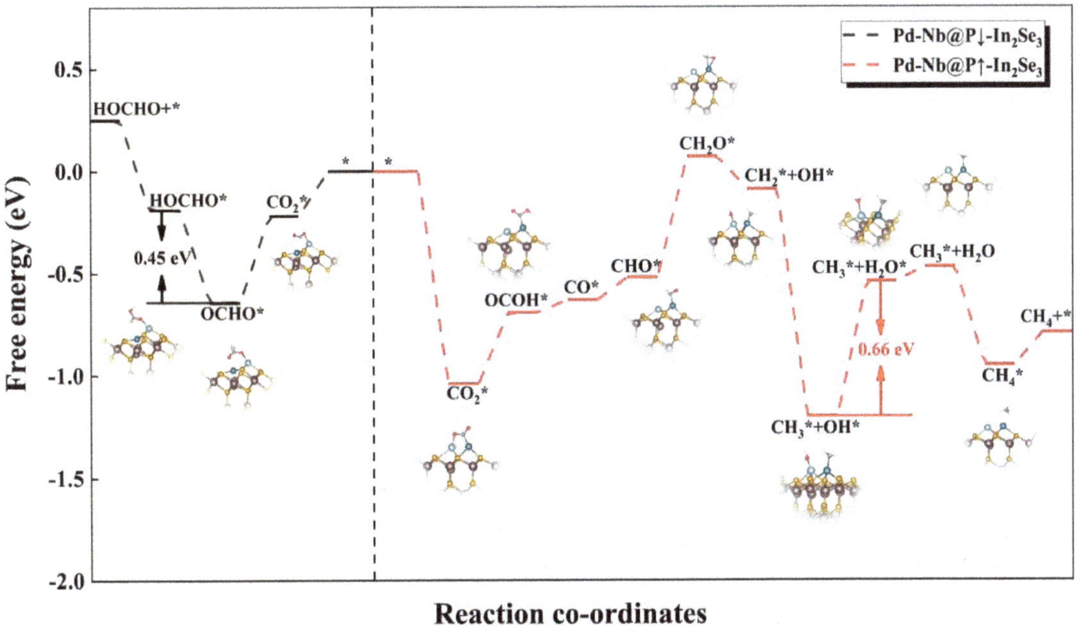

Figure 5. The free-energy profile for the eCO_2RR along the minimum energy path at 0 V (vs. RHE). Energy-minimized structural arrangements of transient species are displayed in the supplementary panels. Gold-shaded region delineates the catalytic process occurring on the Pd-Nb@P↓-In_2Se_3 DAC, while the cyan-tinted zone corresponds to the electrochemical transformation happening on the Pd-Nb@P↑-In_2Se_3 DAC. * stands for the adsorption site at the surface of catalyst.

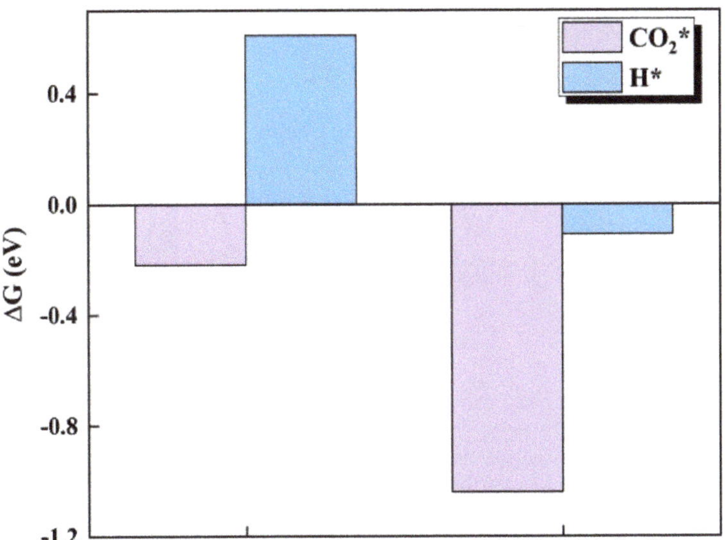

Figure 6. The blue bar indicates the H* formation reaction energy barrier of the optimal active site in the catalyst, and the purple bar indicates the CO_2 adsorption reaction energy barrier. A negative reaction barrier indicates a spontaneous exothermic reaction. * stands for the adsorption site at the surface of catalyst.

3. Methods

All density functional theory (DFT) calculations were performed using the Vienna Ab Initio Simulation Package (VASP) (Hanger team, University of Vienna, version 5.3) [75,76]. The calculations were based on the electron exchange correlation of the PBE generalized function calculation under the generalized gradient approximation (GGA) [77–79]. The van der Waals (vdW) interactions were described by applying Grimme's DFT-D3 method [80], with the colinear spin turned off, the truncation energy set to 500 eV, and the convergence thresholds for force and total energy set to 10^{-2} eV/Å and 10^{-5} eV. The catalytic system was constructed by distributing one Pd atom and one Nb atom on adjacent hexagonal hollow sites of an 4 × 4 α-In_2Se_3 supercell. For the sampling of the dimensional Brillouin zone, 1 × 1 × 1 gamma-centered Monkhorst-pack k-mesh was chosen for performing structural relaxation and frequency calculations. To minimize periodic image interactions, a vacuum layer exceeding 20 Å was introduced along the non-periodic direction. For Gibbs free energy calculations of the eCO_2RR, solvent effects were accounted for by using the implicit water model in VASPsol (version 5.3) [55,81]. Charge transfer analysis was performed using the Bader charge partitioning method.

The binding energy for the Nb-Pd dual-atom pair ($E_{b\text{-}(Nb\text{-}Pd)}$) added onto the substrate was calculated using the following equation [63,82]:

$$E_{b\text{-}(Nb\text{-}Pd)} = E_{tot} - E_{sub} - E_{Nb} - E_{Pd} \quad (1)$$

where E_{sub} and E_{tot} are the total energies of the substrate without and with the added Nb-Pd dual-atom pair, respectively. E_{Nb} and E_{Pd} are the total energy of the isolated Nb and Pd atoms.

The charge density difference ($\Delta\rho$) for the CO_2 molecule adsorbed on Pd-Nb@P↓-In_2Se_3 and Pd-Nb@P↑-In_2Se_3 was calculated using the following equation:

$$\Delta\rho = \rho_{CO_2\text{-}Pd\text{-}Nb\text{-}sub} - \rho_{CO_2} - \rho_{Pd\text{-}Nb\text{-}sub} \quad (2)$$

where $\rho_{CO_2\text{-}Pd\text{-}Nb\text{-}sub}$ and $\rho_{Pd\text{-}Nb\text{-}sub}$ are the charge density of Pd-Nb@In_2Se_3 with and without CO_2 molecules, respectively. ρ_{CO_2} is the charge density of the CO_2 molecule.

In order to obtain the individual orbital components, we employed a code named "splitdos.dos" to process the output file of density of states (DOS). The average *d* band (*d* band center) shifts were calculated for the surface metal atoms for both the total *d* partial DOS and the orbital-resolved *d* partial DOS. The *d* band center (ε_d) was calculated as [83]

$$\varepsilon_d = \frac{\int n_d(\varepsilon)\varepsilon ds}{\int n_d(\varepsilon)d\varepsilon} \quad (3)$$

where ε is the electronic energy of states, and $n_d(\varepsilon)$ is the electronic density of states.

4. Conclusions

Our study presents a theoretical investigation of a novel Pd-Nb@In_2Se_3 DAC for the eCO_2RR, leveraging DFT calculations to elucidate its structural, electronic, and catalytic properties. The Pd-Nb@In_2Se_3 system, anchored on a ferroelectric α-In_2Se_3 monolayer, combines Nb and Pd single atoms to exploit synergistic interactions, while the substrate's switchable polarization enables dynamic control over catalytic activity and selectivity. Structural analyses confirm the stability of Pd-Nb@In_2Se_3, with binding energies of −15.69 eV/unit (P↓ polarization) and −12.27 eV/unit (P↑ polarization) and ab initio molecular dynamics simulations demonstrating thermal stability at 1000 K. The CO_2 adsorption studies revealed enhanced activation on Pd sites, where the linear CO_2 molecule bends significantly, which is attributed to electron donation/backdonation between CO_2 π orbitals and the Pd/Nb *d* orbitals. The *d* band centers of Pd and Nb atoms, which are closer to the Fermi level in the P↑ configuration, correlate with stronger CO_2 activation, emphasizing the role of electronic structure in catalytic performance. Reaction pathway analyses showed polarization-dependent product selectivity. The P↓ polarization favors formic acid production via a U_l of −0.45 V vs. RHE, while the P↑ polarization directs the reaction toward methane with a U_l of −0.66 V vs. RHE. The dual-site mechanism involves Pd atoms driving carbon reduction and Nb atoms facilitating oxygen removal, synergistically lowering energy barriers compared to their SAC counterparts. Additionally, Gibbs free energy calculations indicate superior selectivity for the CO_2RR over the HER, with CO_2 adsorption energies significantly more favorable than H* formation. The ferroelectric substrate's polarization switching dynamically modulates metal–support interactions, adjusting *d* orbital occupancy and intermediate binding strengths, thereby controlling reaction pathways and product outcomes. This work not only advances the understanding of dual-metal catalysis and polarization–catalysis coupling but also provides a strategy to design tunable electrocatalysts for sustainable energy conversion and CO_2 utilization.

Supplementary Materials: The following supporting information can be downloaded at: https://www.mdpi.com/article/10.3390/inorganics13050146/s1, Table S1: Adsorption energy of CO_2 molecules adsorbed at the adsorption sites of Pd-Nb@In_2Se_3; Table S2: Zero-point energy correction (E_{ZPE}) and entropy contribution (*TS*, T = 298.15 K) of molecules; Table S3: Zero-point energy correction (E_{ZPE}), entropy contribution (*TS*, T = 298.15 K), total energy (*E*), and the Gibbs free energy (*G*) of molecules and adsorbates along the reaction pathway on Pd-Nb@P↓-In_2Se_3, where * represents the adsorption site; Table S4: Zero-point energy correction (E_{ZPE}), entropy contribution (*TS*, *T* = 298.15 K), total energy (*E*), and the Gibbs free energy (*G*) of molecules and adsorbates along the reaction pathway on Pd-Nb@P↑-In_2Se_3, where * represents the adsorption site; Figure S1: Pd *d* orbitals and Nb *d* orbitals from Pd-Nb@In_2Se_3; Figure S2: The selected adsorption sites in Pd-Nb@In_2Se_3; Figure S3: The search process for the minimum energy reaction pathways of the CO_2 reduction reactions on Pd-Nb@In_2Se_3; Figure S4: Three configurations on the optimal CO_2RR reduction path of Pd-Nb@P↑-In_2Se_3; The calculation details of the reaction free energy change. References [84,85] are cited in the supplementary materials.

Author Contributions: Conceptualization, X.T.; data curation, Y.Z.; formal analysis, L.J., X.T., M.C., S.L. and C.L.; funding acquisition, L.J.; investigation, L.J. and X.T.; software, Y.Z.; supervision, L.J.; validation, Y.Z. and M.C.; writing—original draft, L.J., X.T., Y.Z. and S.L. All authors have read and agreed to the published version of the manuscript.

Funding: This study was funded by the Young Scientist Project of Henan Province (Grant No. 225200810103), the Program for Science & Technology Innovation Talents in Universities of Henan Province (Grant No. 24HASTIT013), the College Students Innovation Fund of Anyang Normal University (Grant No. 202410479017), and the Natural Science Foundation of Henan Province (Grant No. 232300420128).

Institutional Review Board Statement: Not applicable.

Informed Consent Statement: Not applicable.

Data Availability Statement: The original contributions presented in this study are included in the article/supplementary material. Further inquiries can be directed to the corresponding author.

Conflicts of Interest: The authors declare no conflict of interest.

References

1. Obama, B. The irreversible momentum of clean energy. *Science* **2017**, *355*, 126–129. [CrossRef] [PubMed]
2. Lüthi, D.; Le Floch, M.; Bereiter, B.; Blunier, T.; Barnola, J.-M.; Siegenthaler, U.; Raynaud, D.; Jouzel, J.; Fischer, H.; Kawamura, K. High-resolution carbon dioxide concentration record 650,000–800,000 years before present. *Nature* **2008**, *453*, 379–382. [CrossRef] [PubMed]
3. Metz, B.; Davidson, O.; De Coninck, H.; Loos, M.; Meyer, L. Intergovernmental panel on climate change. In *IPCC Special Report on Carbon Dioxide Capture and Storage*; Cambridge University Press: New York, NY, USA, 2005; 442p.
4. Chu, S.; Cui, Y.; Liu, N. The path towards sustainable energy. *Nat. Mater.* **2017**, *16*, 16–22. [CrossRef] [PubMed]
5. Yuan, Z.; Zhu, X.; Jiang, Z. Recent Advances of Constructing Metal/Semiconductor Catalysts Designing for Photocatalytic CO_2 Hydrogenation. *Molecules* **2023**, *28*, 5693. [CrossRef]
6. Zhu, X.; Zong, H.; Pérez, C.J.V.; Miao, H.; Sun, W.; Yuan, Z.; Wang, S.; Zeng, G.; Xu, H.; Jiang, Z. Supercharged CO_2 photothermal catalytic methanation: High conversion, rate, and selectivity. *Angew. Chem.* **2023**, *135*, e202218694. [CrossRef]
7. Zhu, X.; Zhou, E.; Tai, X.; Zong, H.; Yi, J.; Yuan, Z.; Zhao, X.; Huang, P.; Xu, H.; Jiang, Z. g-C3N4 S-Scheme Homojunction through Van der Waals Interface Regulation by Intrinsic Polymerization Tailoring for Enhanced Photocatalytic H_2 Evolution and CO_2 Reduction. *Angew. Chem. Int. Ed.* **2025**, *64*, e202425439. [CrossRef]
8. Yap, D.; Megaritis, A. Applying Forced Induction to Bioethanol HCCI Operation with Residual Gas Trapping. *Energy Fuels* **2005**, *19*, 1812–1821. [CrossRef]
9. Kang, Z.; Wu, Z.; Zhang, Z.; Deng, J.; Hu, Z.; Li, L. Study of the Combustion Characteristics of a HCCI Engine Coupled with Oxy-Fuel Combustion Mode. *SAE Int. J. Engines* **2017**, *10*, 908–916. [CrossRef]
10. Berstad, D.; Straus, J.; Gundersen, T. CO_2 capture and enhanced hydrogen production enabled by low-temperature separation of PSA tail gas: A detailed exergy analysis. *Energies* **2024**, *17*, 1072. [CrossRef]
11. Reynolds, S.P.; Ebner, A.D.; Ritter, J.A. Stripping PSA cycles for CO_2 recovery from flue gas at high temperature using a hydrotalcite-like adsorbent. *Ind. Eng. Chem. Res.* **2006**, *45*, 4278–4294. [CrossRef]
12. Hoffert, M.I.; Caldeira, K.; Benford, G.; Criswell, D.R.; Green, C.; Herzog, H.; Jain, A.K.; Kheshgi, H.S.; Lackner, K.S.; Lewis, J.S. Advanced technology paths to global climate stability: Energy for a greenhouse planet. *Science* **2002**, *298*, 981–987. [CrossRef] [PubMed]
13. Duan, X.; Xu, J.; Wei, Z.; Ma, J.; Guo, S.; Wang, S.; Liu, H.; Dou, S. Metal-free carbon materials for CO_2 electrochemical reduction. *Adv. Mater.* **2017**, *29*, 1701784. [CrossRef] [PubMed]
14. Qiao, J.; Liu, Y.; Hong, F.; Zhang, J. A review of catalysts for the electroreduction of carbon dioxide to produce low-carbon fuels. *Chem. Soc. Rev.* **2014**, *43*, 631–675. [CrossRef]
15. Centi, G.; Quadrelli, E.A.; Perathoner, S. Catalysis for CO_2 conversion: A key technology for rapid introduction of renewable energy in the value chain of chemical industries. *Energy Environ. Sci.* **2013**, *6*, 1711–1731. [CrossRef]
16. Ross, M.B.; De Luna, P.; Li, Y.; Dinh, C.-T.; Kim, D.; Yang, P.; Sargent, E.H. Designing materials for electrochemical carbon dioxide recycling. *Nat. Catal.* **2019**, *2*, 648–658. [CrossRef]
17. Wang, G.; Zhong, S.; Xiong, X.; Li, J.; Wang, F.; Huo, L.; Wu, D.; Han, X.; Wang, Z.; Chen, Q. Plasma induced grain boundaries to boost electrochemical reduction of CO_2 to formate. *J. Energy Chem.* **2024**, *95*, 636–643. [CrossRef]
18. Wei, Z.; Wang, W.; Shao, T.; Yang, S.; Liu, C.; Si, D.; Cao, R.; Cao, M. Constructing Ag/Cu_2O interface for efficient neutral CO_2 electroreduction to C_2H_4. *Angew. Chem. Int. Ed.* **2025**, *64*, e202417066. [CrossRef]
19. Yue, M.; Xie, W.; Zhong, Z.; Li, M.; Zhang, T.; Shao, M.; Li, H.; Wang, Q. Isolating and stabilizing active copper species in layered double hydroxide to enhance electrocatalytic CO_2 reduction to CH_4. *J. Catal.* **2025**, *443*, 115959. [CrossRef]
20. Dong, J.; Liu, Y.; Pei, J.; Li, H.; Ji, S.; Shi, L.; Zhang, Y.; Li, C.; Tang, C.; Liao, J. Continuous electroproduction of formate via CO_2 reduction on local symmetry-broken single-atom catalysts. *Nat. Commun.* **2023**, *14*, 6849. [CrossRef]
21. Chatterjee, S.; Dutta, I.; Lum, Y.; Lai, Z.; Huang, K.-W. Enabling storage and utilization of low-carbon electricity: Power to formic acid. *Energy Environ. Sci.* **2021**, *14*, 1194–1246. [CrossRef]
22. Li, M.; Idros, M.N.; Wu, Y.; Burdyny, T.; Garg, S.; Zhao, X.S.; Wang, G.; Rufford, T.E. The role of electrode wettability in electrochemical reduction of carbon dioxide. *J. Mater. Chem. A* **2021**, *9*, 19369–19409. [CrossRef]
23. Möller, T.; Ju, W.; Bagger, A.; Wang, X.; Luo, F.; Ngo Thanh, T.; Varela, A.S.; Rossmeisl, J.; Strasser, P. Efficient CO_2 to CO electrolysis on solid Ni–N–C catalysts at industrial current densities. *Energy Environ. Sci.* **2019**, *12*, 640–647. [CrossRef]
24. Guo, S.; Liu, Y.; Murphy, E.; Ly, A.; Xu, M.; Matanovic, I.; Pan, X.; Atanassov, P. Robust palladium hydride catalyst for electrocatalytic formate formation with high CO tolerance. *Appl. Catal. B Environ.* **2022**, *316*, 121659. [CrossRef]
25. Detweiler, Z.M.; White, J.L.; Bernasek, S.L.; Bocarsly, A.B. Anodized indium metal electrodes for enhanced carbon dioxide reduction in aqueous electrolyte. *Langmuir* **2014**, *30*, 7593–7600. [CrossRef]
26. Ikeda, S.; Takagi, T.; Ito, K. Selective formation of formic acid, oxalic acid, and carbon monoxide by electrochemical reduction of carbon dioxide. *Bull. Chem. Soc. Jpn.* **1987**, *60*, 2517–2522. [CrossRef]

27. Hori, Y.; Wakebe, H.; Tsukamoto, T.; Koga, O. Electrocatalytic process of CO selectivity in electrochemical reduction of CO_2 at metal electrodes in aqueous media. *Electrochim. Acta* **1994**, *39*, 1833–1839. [CrossRef]
28. Ding, S.; Hülsey, M.J.; Pérez-Ramírez, J.; Yan, N. Transforming energy with single-atom catalysts. *Joule* **2019**, *3*, 2897–2929. [CrossRef]
29. Su, X.; Yang, X.-F.; Huang, Y.; Liu, B.; Zhang, T. Single-atom catalysis toward efficient CO_2 conversion to CO and formate products. *Acc. Chem. Res.* **2018**, *52*, 656–664. [CrossRef]
30. Liu, K.; Fu, J.; Zhu, L.; Zhang, X.; Li, H.; Liu, H.; Hu, J.; Liu, M. Single-atom transition metals supported on black phosphorene for electrochemical nitrogen reduction. *Nanoscale* **2020**, *12*, 4903–4908. [CrossRef]
31. Lu, F.; Yi, D.; Liu, S.; Zhan, F.; Zhou, B.; Gu, L.; Golberg, D.; Wang, X.; Yao, J. Engineering platinum–oxygen dual catalytic sites via charge transfer towards highly efficient hydrogen evolution. *Angew. Chem.* **2020**, *132*, 17865–17871. [CrossRef]
32. Nguyen, T.; Salehi, M.; Le, Q.; Seifitokaldani, A.; Dinh, C. Fundamentals of electrochemical CO_2 reduction on single-metal-atom catalysts. *ACS Catal.* **2020**, *10*, 10068–10095. [CrossRef]
33. Sun, J.-F.; Wu, J.-T.; Xu, Q.-Q.; Zhou, D.; Yin, J.-Z. CO_2 electrochemical reduction using single-atom catalysts. Preparation, characterization and anchoring strategies: A review. *Environ. Chem. Lett.* **2020**, *18*, 1593–1623. [CrossRef]
34. Asset, T.; Garcia, S.T.; Herrera, S.; Andersen, N.; Chen, Y.; Peterson, E.J.; Matanovic, I.; Artyushkova, K.; Lee, J.; Minteer, S.D. Investigating the nature of the active sites for the CO2 reduction reaction on carbon-based electrocatalysts. *ACS Catal.* **2019**, *9*, 7668–7678. [CrossRef]
35. Zhang, Q.; Guan, J. Single-atom catalysts for electrocatalytic applications. *Adv. Funct. Mater.* **2020**, *30*, 2000768. [CrossRef]
36. Zhang, Y.; Xia, B.; Ran, J.; Davey, K.; Qiao, S.Z. Atomic-level reactive sites for semiconductor-based photocatalytic CO_2 reduction. *Adv. Energy Mater.* **2020**, *10*, 1903879. [CrossRef]
37. Luo, G.; Jing, Y.; Li, Y. Rational design of dual-metal-site catalysts for electroreduction of carbon dioxide. *J. Mater. Chem. A* **2020**, *8*, 15809–15815. [CrossRef]
38. Ouyang, Y.; Shi, L.; Bai, X.; Li, Q.; Wang, J. Breaking scaling relations for efficient CO_2 electrochemical reduction through dual-atom catalysts. *Chem. Sci.* **2020**, *11*, 1807–1813. [CrossRef]
39. Li, Y.; Chen, C.; Cao, R.; Pan, Z.; He, H.; Zhou, K. Dual-atom Ag_2/graphene catalyst for efficient electroreduction of CO_2 to CO. *Appl. Catal. B Environ.* **2020**, *268*, 118747. [CrossRef]
40. Li, Y.; Su, H.; Chan, S.H.; Sun, Q. CO_2 Electroreduction Performance of Transition Metal Dimers Supported on Graphene: A Theoretical Study. *ACS Catal.* **2015**, *5*, 6658–6664. [CrossRef]
41. Li, X.; Sun, Y.; Xu, J.; Shao, Y.; Wu, J.; Xu, X.; Pan, Y.; Ju, H.; Zhu, J.; Xie, Y. Selective visible-light-driven photocatalytic CO_2 reduction to CH_4 mediated by atomically thin $CuIn_5S_8$ layers. *Nat. Energy* **2019**, *4*, 690–699. [CrossRef]
42. Li, S.; Nagarajan, A.V.; Alfonso, D.R.; Sun, M.; Kauffman, D.R.; Mpourmpakis, G.; Jin, R. Boosting CO_2 electrochemical reduction with atomically precise surface modification on gold nanoclusters. *Angew. Chem. Int. Ed.* **2021**, *60*, 6351–6356. [CrossRef] [PubMed]
43. Wen, G.; Lee, D.U.; Ren, B.; Hassan, F.M.; Jiang, G.; Cano, Z.P.; Gostick, J.; Croiset, E.; Bai, Z.; Yang, L. Orbital interactions in Bi-Sn bimetallic electrocatalysts for highly selective electrochemical CO_2 reduction toward formate production. *Adv. Energy Mater.* **2018**, *8*, 1802427. [CrossRef]
44. Lin, L.; Li, H.; Yan, C.; Li, H.; Si, R.; Li, M.; Xiao, J.; Wang, G.; Bao, X. Synergistic catalysis over iron-nitrogen sites anchored with cobalt phthalocyanine for efficient CO_2 electroreduction. *Adv. Mater.* **2019**, *31*, 1903470. [CrossRef]
45. Hu, K.; Li, Z.; Bai, L.; Yang, F.; Chu, X.; Bian, J.; Zhang, Z.; Xu, H.; Jing, L. Synergetic subnano Ni-and Mn-Oxo clusters anchored by chitosan oligomers on 2D g-C_3N_4 boost photocatalytic CO_2 reduction. *Sol. Rrl* **2021**, *5*, 2000472. [CrossRef]
46. Bok, J.; Lee, S.Y.; Lee, B.-H.; Kim, C.; Nguyen, D.L.T.; Kim, J.W.; Jung, E.; Lee, C.W.; Jung, Y.; Lee, H.S. Designing atomically dispersed Au on tensile-strained Pd for efficient CO_2 electroreduction to formate. *J. Am. Chem. Soc.* **2021**, *143*, 5386–5395. [CrossRef]
47. Ding, T.; Liu, X.; Tao, Z.; Liu, T.; Chen, T.; Zhang, W.; Shen, X.; Liu, D.; Wang, S.; Pang, B. Atomically precise dinuclear site active toward electrocatalytic CO_2 reduction. *J. Am. Chem. Soc.* **2021**, *143*, 11317–11324. [CrossRef]
48. He, Q.; Liu, D.; Lee, J.H.; Liu, Y.; Xie, Z.; Hwang, S.; Kattel, S.; Song, L.; Chen, J.G. Electrochemical conversion of CO_2 to syngas with controllable CO/H_2 ratios over Co and Ni single-atom catalysts. *Angew. Chem. Int. Ed.* **2020**, *59*, 3033–3037. [CrossRef]
49. Darif, B.; Ojala, S.; Pirault-Roy, L.; Bensitel, M.; Brahmi, R.; Keiski, R.L. Study on the catalytic oxidation of DMDS over Pt-Cu catalysts supported on Al_2O_3, $AlSi_{20}$ and SiO_2. *Appl. Catal. B Environ.* **2016**, *181*, 24–33. [CrossRef]
50. Winter, L.R.; Gomez, E.; Yan, B.; Yao, S.; Chen, J.G. Tuning Ni-catalyzed CO_2 hydrogenation selectivity via Ni-ceria support interactions and Ni-Fe bimetallic formation. *Appl. Catal. B Environ.* **2018**, *224*, 442–450. [CrossRef]
51. Wang, J.; Huang, Z.; Liu, W.; Chang, C.; Tang, H.; Li, Z.; Chen, W.; Jia, C.; Yao, T.; Wei, S.; et al. Design of N-Coordinated Dual-Metal Sites: A Stable and Active Pt-Free Catalyst for Acidic Oxygen Reduction Reaction. *J. Am. Chem. Soc.* **2017**, *139*, 17281–17284. [CrossRef]
52. Ren, W.; Tan, X.; Yang, W.; Jia, C.; Xu, S.; Wang, K.; Smith, S.C.; Zhao, C. Isolated Diatomic Ni-Fe Metal–Nitrogen Sites for Synergistic Electroreduction of CO_2. *Angew. Chem. Int. Ed.* **2019**, *58*, 6972–6976. [CrossRef] [PubMed]

53. Zhang, L.; Fischer, J.M.T.A.; Jia, Y.; Yan, X.; Xu, W.; Wang, X.; Chen, J.; Yang, D.; Liu, H.; Zhuang, L.; et al. Coordination of Atomic Co–Pt Coupling Species at Carbon Defects as Active Sites for Oxygen Reduction Reaction. *J. Am. Chem. Soc.* **2018**, *140*, 10757–10763. [CrossRef] [PubMed]
54. Li, D.; Zhao, M.H.; Garra, J.; Kolpak, A.M.; Rappe, A.M.; Bonnell, D.A.; Vohs, J.M. Direct in situ determination of the polarization dependence of physisorption on ferroelectric surfaces. *Nat. Mater.* **2008**, *7*, 473–477. [CrossRef]
55. Ju, L.; Shang, J.; Tang, X.; Kou, L. Tunable Photocatalytic Water Splitting by the Ferroelectric Switch in a 2D $AgBiP_2Se_6$ Monolayer. *J. Am. Chem. Soc.* **2019**, *142*, 1492–1500. [CrossRef]
56. Ju, L.; Tan, X.; Mao, X.; Gu, Y.; Smith, S.; Du, A.; Chen, Z.; Chen, C.; Kou, L. Controllable CO_2 electrocatalytic reduction via ferroelectric switching on single atom anchored In_2Se_3 monolayer. *Nat. Commun.* **2021**, *12*, 5128. [CrossRef]
57. Ding, W.; Zhu, J.; Wang, Z.; Gao, Y.; Xiao, D.; Gu, Y.; Zhang, Z.; Zhu, W. Prediction of intrinsic two-dimensional ferroelectrics in In_2Se_3 and other III_2-VI_3 van der Waals materials. *Nat. Commun.* **2017**, *8*, 14956. [CrossRef]
58. Cui, C.; Hu, W.J.; Yan, X.; Addiego, C.; Gao, W.; Wang, Y.; Wang, Z.; Li, L.; Cheng, Y.; Li, P.; et al. Intercorrelated In-Plane and Out-of-Plane Ferroelectricity in Ultrathin Two-Dimensional Layered Semiconductor In_2Se_3. *Nano Lett.* **2018**, *18*, 1253–1258. [CrossRef]
59. Gao, G.; Jiao, Y.; Waclawik, E.R.; Du, A. Single Atom (Pd/Pt) Supported on Graphitic Carbon Nitride as an Efficient Photocatalyst for Visible-Light Reduction of Carbon Dioxide. *J. Am. Chem. Soc.* **2016**, *138*, 6292–6297. [CrossRef]
60. Li, X.; Bi, W.; Zhang, L.; Tao, S.; Chu, W.; Zhang, Q.; Luo, Y.; Wu, C.; Xie, Y. Single-Atom Pt as Co-Catalyst for Enhanced Photocatalytic H_2 Evolution. *Adv. Mater.* **2016**, *28*, 2427–2431. [CrossRef]
61. Vile, G.; Albani, D.; Nachtegaal, M.; Chen, Z.; Dontsova, D.; Antonietti, M.; Lopez, N.; Perez-Ramirez, J. A stable single-site palladium catalyst for hydrogenations. *Angew. Chem. Int. Ed.* **2015**, *54*, 11265–11269. [CrossRef]
62. Khatun, R.; Siddiqui, N.; Pal, R.S.; Bhandari, S.; Khan, T.S.; Singh, S.; Poddar, M.K.; Samanta, C.; Bal, R. Low temperature reforming of methane with CO_2 over Pt/CeO_2, Ni/CeO_2 and Pt–Ni/CeO_2 catalysts prepared by a solution-combustion method. *Catal. Sci. Technol.* **2023**, *13*, 6431–6445. [CrossRef]
63. Ma, D.; Zeng, Z.; Liu, L.; Huang, X.; Jia, Y. Computational Evaluation of Electrocatalytic Nitrogen Reduction on TM Single-, Double-, and Triple-Atom Catalysts (TM = Mn, Fe, Co, Ni) Based on Graphdiyne Monolayers. *J. Phys. Chem. C* **2019**, *123*, 19066–19076. [CrossRef]
64. Back, S.; Kim, H.; Jung, Y. Selective Heterogeneous CO_2 Electroreduction to Methanol. *ACS Catal.* **2015**, *5*, 965–971. [CrossRef]
65. Zhao, Z.; Lu, G. Cu-Based Single-Atom Catalysts Boost Electroreduction of CO_2 to CH_3OH: First-Principles Predictions. *J. Phys. Chem. C* **2019**, *123*, 4380–4387. [CrossRef]
66. Wannakao, S.; Artrith, N.; Limtrakul, J.; Kolpak, A.M. Engineering Transition-Metal-Coated Tungsten Carbides for Efficient and Selective Electrochemical Reduction of CO_2 to Methane. *ChemSusChem* **2015**, *8*, 2745–2751. [CrossRef]
67. Elhenawy, S.E.M.; Khraisheh, M.; AlMomani, F.; Walker, G. Metal-organic frameworks as a platform for CO_2 capture and chemical processes: Adsorption, membrane separation, catalytic-conversion, and electrochemical reduction of CO_2. *Catalysts* **2020**, *10*, 1293. [CrossRef]
68. Zhang, N.; Zhang, X.; Kang, Y.; Ye, C.; Jin, R.; Yan, H.; Lin, R.; Yang, J.; Xu, Q.; Wang, Y.; et al. A Supported Pd_2 Dual-Atom Site Catalyst for Efficient Electrochemical CO_2 Reduction. *Angew. Chem. Int. Ed.* **2021**, *60*, 13388–13393. [CrossRef]
69. Hao, Q.; Zhong, H.; Wang, J.; Liu, K.; Yan, J.; Ren, Z.; Zhou, N.; Zhao, X.; Zhang, H.; Liu, D.; et al. Nickel dual-atom sites for electrochemical carbon dioxide reduction. *Nat. Synth.* **2022**, *1*, 719–728. [CrossRef]
70. Huang, K.; Li, R.; Qi, H.; Yang, S.; An, S.; Lian, C.; Xu, Q.; Liu, H.; Hu, J. Regulating Adsorption of Intermediates via the Sulfur Modulating Dual-Atomic Sites for Boosting CO_2RR. *ACS Catal.* **2024**, *14*, 8889–8898. [CrossRef]
71. Chen, B.; Shi, D.; Deng, R.; Xu, X.; Liu, W.; Wei, Y.; Liu, Z.; Zhong, S.; Huang, J.; Yu, Y. Leveraging Atomic-Scale Synergy for Selective CO_2 Electrocatalysis to CO over CuNi Dual-Atom Catalysts. *ACS Catal.* **2024**, *14*, 16224–16233. [CrossRef]
72. Roy, S.; Li, Z.; Chen, Z.; Mata, A.C.; Kumar, P.; Sarma, S.C.; Teixeira, I.F.; Silva, I.F.; Gao, G.; Tarakina, N.V.; et al. Cooperative Copper Single-Atom Catalyst in 2D Carbon Nitride for Enhanced CO_2 Electrolysis to Methane. *Adv. Mater.* **2024**, *36*, e2300713. [CrossRef]
73. Xue, H.; Zhao, Z.H.; Liao, P.Q.; Chen, X.M. "Ship-in-a-Bottle" Integration of Ditin(IV) Sites into a Metal-Organic Framework for Boosting Electroreduction of CO_2 in Acidic Electrolyte. *J. Am. Chem. Soc.* **2023**, *145*, 16978–16982. [CrossRef]
74. Monteiro, M.C.O.; Dattila, F.; Hagedoorn, B.; García-Muelas, R.; López, N.; Koper, M.T.M. Absence of CO_2 electroreduction on copper, gold and silver electrodes without metal cations in solution. *Nat. Catal.* **2021**, *4*, 654–662. [CrossRef]
75. Kresse, G.; Furthmüller, J. Efficient iterative schemes for ab initio total-energy calculations using a plane-wave basis set. *Phys. Rev. B* **1996**, *54*, 11169. [CrossRef]
76. Kresse, G.; Furthmüller, J. Efficiency of ab-initio total energy calculations for metals and semiconductors using a plane-wave basis set. *Comput. Mater. Sci.* **1996**, *6*, 15–50. [CrossRef]
77. Blöchl, P.E. Projector augmented-wave method. *Phys. Rev. B* **1994**, *50*, 17953–17979. [CrossRef]
78. Kresse, G.; Joubert, D. From ultrasoft pseudopotentials to the projector augmented-wave method. *Phys. Rev. B* **1999**, *59*, 1758. [CrossRef]

79. Perdew, J.P.; Burke, K.; Ernzerhof, M. Generalized gradient approximation made simple. *Phys. Rev. Lett.* **1996**, *77*, 3865. [CrossRef]
80. Grimme, S. Semiempirical GGA-type density functional constructed with a long-range dispersion correction. *J. Comput. Chem.* **2006**, *27*, 1787–1799. [CrossRef]
81. Mao, X.; Kour, G.; Zhang, L.; He, T.; Wang, S.; Yan, C.; Zhu, Z.; Du, A. Silicon-doped graphene edges: An efficient metal-free catalyst for the reduction of CO_2 into methanol and ethanol. *Catal. Sci. Technol.* **2019**, *9*, 6800–6807. [CrossRef]
82. Cui, X.; An, W.; Liu, X.; Wang, H.; Men, Y.; Wang, J. C_2N-graphene supported single-atom catalysts for CO_2 electrochemical reduction reaction: Mechanistic insight and catalyst screening. *Nanoscale* **2018**, *10*, 15262–15272. [CrossRef] [PubMed]
83. Nørskov, J.K.; Studt, F.; Abild-Pedersen, F.; Bligaard, T. Fundamental Concepts in Heterogeneous Catalysis. *Focus Catal.* **2015**, *2015*, 7. [CrossRef]
84. Nørskov, J.K.; Rossmeisl, J.; Logadottir, A.; Lindqvist, L.; Kitchin, J.R.; Bligaard, T.; Jonsson, H. Origin of the overpotential for oxygen reduction at a fuel-cell cathode. *J. Phys. Chem. B* **2004**, *108*, 17886–17892. [CrossRef]
85. Peterson, A.A.; Abild-Pedersen, F.; Studt, F.; Rossmeisl, J.; Nørskov, J.K. How copper catalyzes the electroreduction of carbon dioxide into hydrocarbon fuels. *Energ. Environ. Sci.* **2010**, *3*, 1311–1315. [CrossRef]

inorganics

MDPI

Article

Synthesis of M-Doped $MoSe_2$ (M = Fe, Co, Ni) via Chemical Vapor Deposition for an Electrocatalytic Hydrogen Evolution Reaction

Xinya Chen [1], Xingchen Zhang [1], Jinying Zhang [2] and Zhiyong Wang [1,*]

1 Key Laboratory of Advanced Light Conversion Materials and Biophotonics, School of Chemistry and Life Resources, Renmin University of China, Beijing 100872, China
2 State Key Laboratory of Electrical Insulation and Power Equipment, Center of Nanomaterials for Renewable Energy (CNRE), School of Electrical Engineering, Xi'an Jiaotong University, Xi'an 710049, China
* Correspondence: zhiyongwang@ruc.edu.cn

Abstract: Given the high cost and limited availability of noble-metal-based catalysts in acidic media water electrolysis, developing cost-effective and high-performance non-noble metal catalysts is crucial for realizing large-scale hydrogen production. In this study, Fe-, Co-, and Ni-doped $MoSe_2$ nanomaterials were synthesized via chemical vapor deposition, and their electrocatalytic performance for the hydrogen evolution reaction (HER) was systematically evaluated. Characterization techniques including X-ray diffraction, X-ray photoelectron spectroscopy, transmission electron microscopy, scanning electron microscopy, and Raman spectroscopy were used to confirm the incorporation of doping elements and investigate their effects on the crystal structure and morphology of $MoSe_2$. Electrochemical tests, including linear sweep voltammetry and cyclic voltammetry, revealed that the doping of Fe, Co, and Ni significantly enhanced the HER catalytic activity of $MoSe_2$, with the Co-doped sample exhibiting the best performance, showing an overpotential of 0.293 V at 100 mA/cm^{-2} and a Tafel slope of 47 mV/dec. Furthermore, density functional theory calculations were employed to analyze the adsorption energy of hydrogen atoms on the catalysts, providing deeper insights into the role of doping in tuning the catalytic activity of $MoSe_2$. This study offers new theoretical support and experimental evidence for the application of transition metal-doped $MoSe_2$ in electrocatalysis.

Keywords: $MoSe_2$; chemical vapor deposition; doping; electrocatalysis; hydrogen evolution reaction

1. Introduction

With the rapid advancement of renewable energy technologies, water electrolysis has garnered considerable attention as a clean and sustainable approach to hydrogen production [1–4]. In particular, acidic water splitting offers simplicity in operations and high efficiency, making it promising for large-scale hydrogen generation. Currently, noble metals (e.g., platinum) exhibit the best catalytic performance in acidic media, but their high cost and limited availability impede widespread application [5–7]. Therefore, the development of cost-effective, high-performance non-noble metal electrocatalysts is crucial to advancing hydrogen energy technologies.

In this context, two-dimensional transition metal dichalcogenides (TMDs) have emerged as a major focus of electrocatalytic research, owing to their tunable band gaps, abundant active sites, and distinctive layered structures [8–10]. Among them, molybdenum diselenide ($MoSe_2$) has demonstrated promising theoretical catalytic activity due to its layered structure and moderate hydrogen adsorption free energy [11]. However, the low intrinsic electrical conductivity of $MoSe_2$ severely limits its practical applications in electrocatalysis.

Recent advances demonstrate that transition metal doping can effectively tailor TMD's electronic configuration through synergistic lattice strain and charge redistribution effects [12–23]. For instance, theoretical studies by Chen et al. have predicted that Fe doping enhances $MoSe_2$'s metallic character by shifting the d-band center, while cobalt incorporation has been shown to optimize hydrogen adsorption kinetics on selenium vacancies [19]. Liu et al. computationally confirmed that Co doping optimizes the hydrogen adsorption energy ($\Delta G_{H^*} \approx 0.08$ eV) of $MoSe_2$ [20]. Currently, there is a lack of experimental studies systematically comparing catalytic performance among different doped $MoSe_2$ materials.

This study aims to synthesize Fe-, Co-, and Ni-doped $MoSe_2$ using the chemical vapor deposition (CVD) method and systematically compare their electrocatalytic performance in a hydrogen evolution reaction (HER) through a combination of experimental characterization and theoretical calculations. Their structure and electronic properties were studied using techniques such as X-ray diffraction (XRD), X-ray photoelectron spectroscopy (XPS), transmission electron microscopy (TEM), and Raman spectroscopy. In addition, density functional theory calculations are used to gain insight into the effects of doping on the hydrogen adsorption behavior and catalytic reaction pathways. This study not only elucidates the mechanism of transition metal doping to improve the catalytic performance of $MoSe_2$, but also provides important theoretical and experimental guidance for the design and optimization of novel HER electrocatalysts.

2. Results and Discussion

2.1. Morphology and Structure Analysis

Scanning electron microscopy (SEM) was used to systematically characterize the surface morphology of $MoSe_2$ and its Fe-, Co-, and Ni-doped variants. As shown in Figure 1a, undoped $MoSe_2$ exhibits a petal-like lamellar structure with interlaced layers, creating numerous edges and pores. In Figure 1b, Fe-doped $MoSe_2$ reveals a denser lamellar arrangement, accompanied by additional nanometer-scale protrusions and defects. Compared to the Fe-doped sample, Co-doped $MoSe_2$ (Figure 1c) displays larger grain sizes and a granular appearance, whereas Ni-doped $MoSe_2$ (Figure 1d) exhibits a similar granular structure to the Co-doped sample, but with smoother surfaces and a more uniform particle distribution. These findings suggest that introducing Fe, Co, or Ni significantly impacts the original petal-like layers of $MoSe_2$, producing denser and more regular lamellar or granular features. The film thicknesses of the pristine and doped $MoSe_2$ sample are about 1–2 μm.

Figure 2 provides a detailed analysis of the microstructure of $MoSe_2$ and its doped variants using high-resolution TEM (HRTEM), Selected area electron diffraction (SAED), and STEM-HAADF. Figure 2a,b show the HRTEM images of undoped $MoSe_2$, displaying clear lattice fringes with a lattice spacing of approximately 0.28 nm, corresponding to the (100) plane of $MoSe_2$. The Fast Fourier Transform (FFT) images within the red box further confirm the single-crystal nature of $MoSe_2$. Figure 2c presents the SAED pattern of $MoSe_2$, displaying a set of hexagonally arranged diffraction spots, indicating the presence of a well-ordered single-crystal structure. Figure 2d shows the STEM-HAADF image and its elemental distribution map of $MoSe_2$, demonstrating that Mo and Se are evenly distributed throughout the sample, confirming the uniformity of the $MoSe_2$ structure.

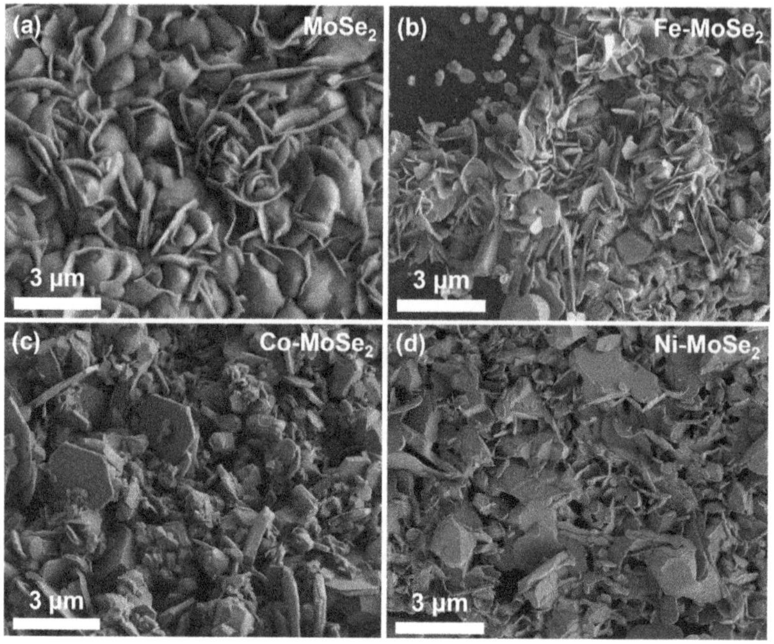

Figure 1. SEM images of (**a**) $MoSe_2$, (**b**) Fe-$MoSe_2$, (**c**) Co-$MoSe_2$, and (**d**) Ni-$MoSe_2$.

Figure 2e,f depict the HRTEM images of Fe-doped $MoSe_2$, with lattice spacings of 0.28 nm (corresponding to the $MoSe_2$ (100) plane) and 0.26 nm (corresponding to the $MoSe_2$ (102) plane). Figure 2g shows the corresponding SAED pattern, with diffraction spots arranged in a hexagonal pattern, confirming that the sample retains its single-crystal characteristics. Figure 2h presents the STEM-HAADF image and elemental distribution map of Fe-doped $MoSe_2$, where Fe is uniformly distributed within $MoSe_2$ without observable aggregation, suggesting that Fe has been successfully incorporated into $MoSe_2$ and may introduce additional defects and active sites, which could enhance its electrocatalytic performance.

Figure 2i,j show the HRTEM images of Co-doped $MoSe_2$, with lattice fringe spacings similar to those of the Fe-doped samples. The FFT images indicate that the single-crystal structure of $MoSe_2$ remains intact after Co incorporation. Figure 2k shows the SAED pattern of Co-doped $MoSe_2$, suggesting that Co doping does not significantly alter the single-crystal structure of $MoSe_2$. Figure 2l shows the STEM-HAADF image and elemental distribution map of Co-doped $MoSe_2$, where Co is evenly distributed, which may further increase the number of catalytic active sites, thereby enhancing the electrocatalytic performance.

Figure 2m,n depict the HRTEM images of Ni-doped $MoSe_2$, exhibiting lattice fringes similar to those of the Fe and Co-doped samples, indicating that the crystal structure of $MoSe_2$ remains largely unchanged after Ni doping. The FFT image is basically unchanged. Figure 2o shows the SAED pattern of Ni-doped $MoSe_2$, with regular diffraction spots, further confirming that the single-crystal structure of $MoSe_2$ is preserved after Ni doping. Figure 2p presents the STEM-HAADF image and elemental distribution map of Ni-doped $MoSe_2$, where Ni is uniformly distributed, providing additional active sites that may enhance its electrocatalytic performance.

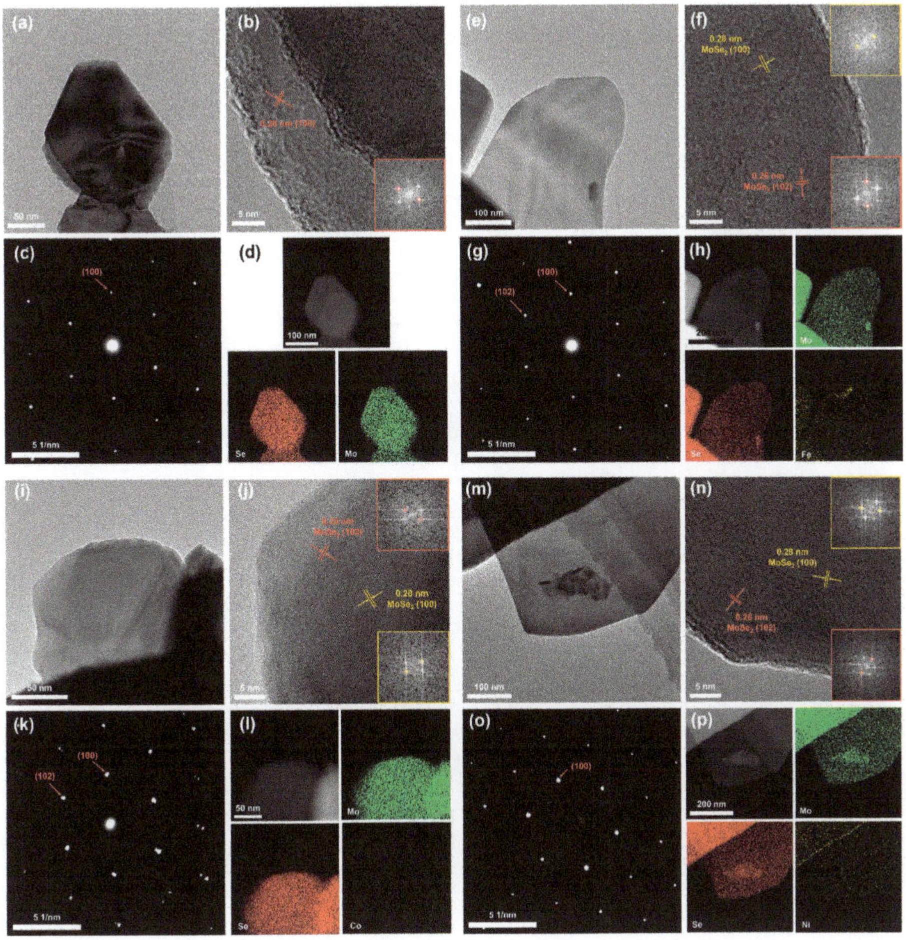

Figure 2. (**a**,**b**) HRTEM images of $MoSe_2$, with the corresponding FFT in the red box. (**c**) SAED pattern of $MoSe_2$. (**d**) STEM-HAADF image and EDS elemental mapping of Mo and Se. (**e**,**f**) HRTEM images of Fe-$MoSe_2$, with the insets showing the corresponding FFT patterns. (**g**) SAED pattern of Fe-$MoSe_2$. (**h**) STEM-HAADF image and EDS elemental mapping of Mo, Se, and Fe. (**i**,**j**) HRTEM images of Co-$MoSe_2$, with the insets showing the corresponding FFT patterns. (**k**) SAED pattern of Co-$MoSe_2$. (**l**) STEM-HAADF image and EDS elemental mapping of Mo, Se, and Co. (**m**,**n**) HRTEM images of Ni-$MoSe_2$, with the insets showing the corresponding FFT patterns. (**o**) SAED pattern of Ni-$MoSe_2$. (**p**) STEM-HAADF image and EDS elemental mapping of Mo, Se, and Ni.

2.2. Component Analysis

XPS analysis was conducted for Fe-doped $MoSe_2$. As shown in Figure 3a, the XPS analysis of Fe reveals that the Fe 2p spectrum exhibits remarkable splitting characteristics. The Fe $2p_{3/2}$ peak presents two main peaks at approximately 710.5 eV and 713.2 eV. The peaks are at approximately 723.9 eV and 727.0 eV in the Fe $2p_{1/2}$ spectrum. The XPS spectrum of Fe $2p_{3/2}$ and Fe $2p_{1/2}$ at 710.5 eV and 723.9 eV correspond to the Fe^{2+} state, and the XPS spectrum of Fe $2p_{3/2}$ and Fe $2p_{1/2}$ at 713.2 eV and 727.0 eV correspond to the Fe^{3+} state. Moreover, the shake-up satellite peaks near approximately 718 eV and 733 eV, typical features of Fe^{3+}, further confirm the presence of Fe^{3+} in the sample. Collectively, Fe in the sample primarily appears in the Fe^{2+} state, with a lesser fraction of Fe^{3+}, resulting in a mixed valence environment. This valence state distribution is attributed to the partial cleavage of Fe-Se bonds during the preparation process, leading to the oxidation of Fe atoms, which is highly likely to have a significant impact on the electrochemical and catalytic properties of the sample.

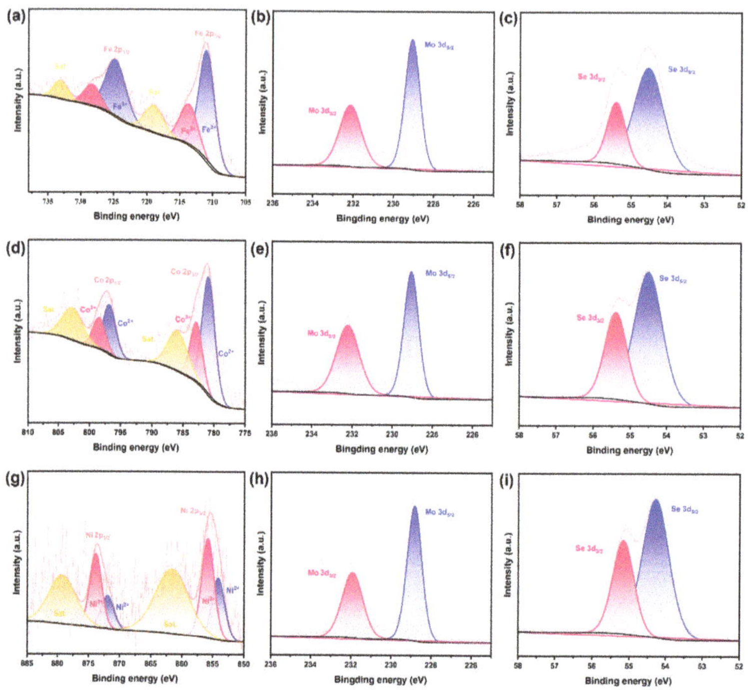

Figure 3. XPS spectra of Fe-$MoSe_2$ showing the signals of (**a**) Fe 2p, (**b**) Mo 3d, and (**c**) Se 3d. XPS spectra of Co-$MoSe_2$ showing the signals of (**d**) Co 2p, (**e**) Mo 3d, and (**f**) Se 3d. XPS spectra of Ni-$MoSe_2$ showing the signals of (**g**) Ni 2p, (**h**) Mo 3d, and (**i**) Se 3d.

As depicted in Figure 3d, analysis of the Co XPS spectrum shows that both the Co $2p_{3/2}$ and Co $2p_{1/2}$ peaks are split. The Co $2p_{3/2}$ peaks at approximately 781.0 eV and 782.7 eV, as well as the Co $2p_{1/2}$ peaks at approximately 797.0 eV and 798.5 eV, correspond to Co^{2+} and Co^{3+}, respectively. Meanwhile, the typical Co^{2+} shake-up satellite peaks near approximately 786.5 eV and 803.0 eV indicate that Co in the sample mainly exists in a mixed valence state of Co^{2+} and Co^{3+}, with Co^{2+} being dominant. This strongly demonstrates that Co has been successfully incorporated into the $MoSe_2$ lattice to form Co-Se bonds, a structure that plays a crucial role in maintaining material stability and regulating its electronic properties.

As shown in Figure 3g, regarding the XPS analysis of Ni, the splitting features of the Ni $2p_{3/2}$ and Ni $2p_{1/2}$ peaks indicate that the Ni $2p_{3/2}$ peaks at approximately 854.0 eV and 855.5 eV, along with the Ni $2p_{1/2}$ peaks at approximately 872.0 eV and 874.2 eV, correspond to Ni^{2+} and Ni^{3+}, respectively. The shake-up satellite peaks at 861.7 eV and 879.5 eV are consistent with the coordination environment of Ni^{2+}. The Ni in the sample mainly exists in a mixed valence state of Ni^{3+} and Ni^{2+}, and Ni^{3+} is the main one. This implies that the partial oxidation of Ni occurred during the preparation process, accompanied by changes in the Ni-Se structure, which may significantly influence the electrochemical and catalytic properties of the sample.

Notably, as illustrated in Figure 3b,e,h for Mo 3d and Figure 3c,f,i for Se 3d, the XPS spectra of Mo 3d and Se 3d show that, regardless of Fe, Co, or Ni doping, the Mo $3d_{5/2}$ peak remains at 229.5 eV, the Mo $3d_{3/2}$ peak at 232.5 eV, and the Se $3d_{5/2}$ and Se $3d_{3/2}$ peaks at 54.5 eV and 55.5 eV, respectively. This clearly indicates that the doping process has a minimal effect on the chemical states of Mo and Se, and that the Mo-Se skeleton structure of $MoSe_2$ remains highly stable. Additionally, quantitative analysis results show that the contents of Fe, Co, and Ni are 9.01%, 5.44%, and 2.24%, respectively, further confirming the successful implementation of the doping process.

2.3. *Optical and Crystal Structure Analysis*

Additionally, the Raman scattering spectra and XRD patterns of both doped and undoped $MoSe_2$ provide insights into the effects of doping on the material's structure and vibrational modes.

Figure 4a shows the Raman spectra of Fe/Co/Ni-$MoSe_2$ and $MoSe_2$. Undoped $MoSe_2$ exhibits two main Raman active peaks at approximately 238 cm^{-1} (E_{2g}) and 285 cm^{-1} (A_{1g}), corresponding to the interlayer vibrational mode (E_{2g}) and the in-plane vibrational mode (A_{1g}) of $MoSe_2$. These peaks indicate the well-defined layered structure of $MoSe_2$. For the Fe-, Co-, and Ni-doped $MoSe_2$ samples, the shapes and positions of the E_{2g} and A_{1g} peaks undergo slight shifts towards lower frequencies, suggesting that doping has caused minor changes in the lattice. Particularly, Co-doped $MoSe_2$ shows a slight splitting of the E_{2g} peak, possibly due to lattice stress and distortion after Co incorporation. This splitting may result from local lattice deformation or dislocation introduced by Co, leading to subtle changes in the vibrational modes. Fe- and Ni-doped samples show similar peak shifts, but no obvious splitting like the Co-doped sample, indicating that Fe and Ni doping have a relatively smaller effect on the lattice.

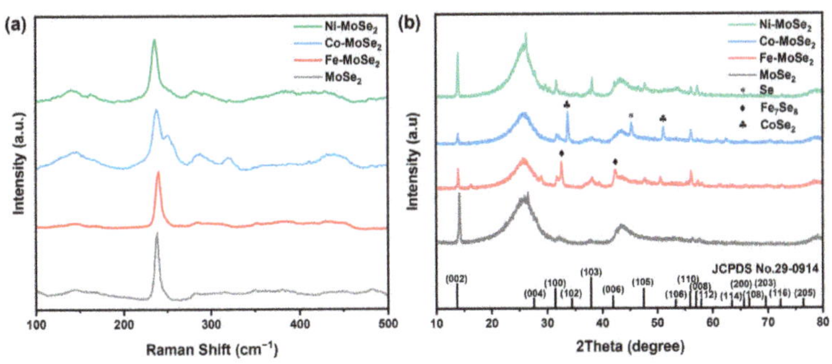

Figure 4. (**a**) Raman scattering spectra and (**b**) XRD patterns of Fe/Co/Ni-$MoSe_2$ and $MoSe_2$.

The crystal structures of pristine and M-doped $MoSe_2$ were investigated using XRD, as shown in Figure 4b. All samples exhibit dominant diffraction peaks corresponding to the hexagonal 2H-$MoSe_2$ phase (JCPDS No.29-0914), including the characteristic (002) peak at ~13.7° (interlayer stacking), (100) peak at ~31.4°, and (110) peak at ~55.9°, confirming the retention of the layered structure after doping.

The sharp (002) peak with a full width at half maximum (FWHM) of ~0.3° correlates with the vertically aligned nanosheet morphology observed in SEM (Figure 1a–d). For the Fe-doped sample, the peaks at 32.5° and 43.4° are assigned to Fe_7Se_8, while for the Co-doped sample, the peaks at 34.4° and 51.9° correspond to $CoSe_2$, and the peak at 45.7° is attributed to elemental Se. These results indicate that partial selenization of the metal dopants occurred during the reaction process. The broad peaks near 25° and 43° originate from the glassy carbon substrate, as verified by control experiments (Figure S1 of the Supplementary Materials).

2.4. Electrochemical Characterization

The electrocatalytic performance of $MoSe_2$, glassy carbon (GC), Pt/C, and Fe-/Co-/Ni-doped $MoSe_2$ catalysts was evaluated through polarization curves, Tafel slopes, and double-layer capacitance (C_{dl}) calculations. As shown in Figure 5a, in 0.5 M H_2SO_4 solution, the polarization curves clearly show the HER activity of each catalyst: When the current density reaches 100 mA/cm^2, Pt/C has the lowest overpotential (only requiring approximately −0.149 V), serving as a benchmark to exhibit excellent HER activity. Pure $MoSe_2$ requires approximately −0.5 V, while Fe-$MoSe_2$ (−0.367 V), Co-$MoSe_2$ (−0.293 V), and Ni-$MoSe_2$ (−0.401 V) show significantly reduced overpotentials compared with pure $MoSe_2$. Among them, Co-$MoSe_2$ has the most obvious reduction, indicating that doping elements effectively improve the electrocatalytic performance of $MoSe_2$. The GC electrode has extremely weak catalytic performance due to its high overpotential of −1.424 V (Figure S2 of the Supplementary Materials). The overpotential of the Co-$MoSe_2$ at 100 mA/cm^2 is smaller than that of the previously reported Ni-$MoSe_2$ catalyst [24]. Combined with the Tafel slopes in Figure 5b, Pt/C has the smallest slope (fastest charge transfer). Co-$MoSe_2$ has a Tafel slope of 47 mV/dec (close to Pt/C), Ni-$MoSe_2$ shows 110 mV/dec, and Fe-$MoSe_2$ shows 75 mV/dec—all lower than pure $MoSe_2$ (172 mV/dec), further indicating that doping optimizes the catalytic reaction kinetics. Additionally, to systematically explore the catalytic enhancement effects of Fe, Co, and Ni doping on $MoSe_2$, $FeSe_2$, $CoSe_2$, and $NiSe_2$ were synthesized in this study. Their electrocatalytic performance was evaluated via LSV. The results showed that the performance of these selenides was lower than that of pure $MoSe_2$, thus ruling out the interference of selenides themselves on the catalytic performance. More details can be found in Figure S3 of the Supplementary Materials.

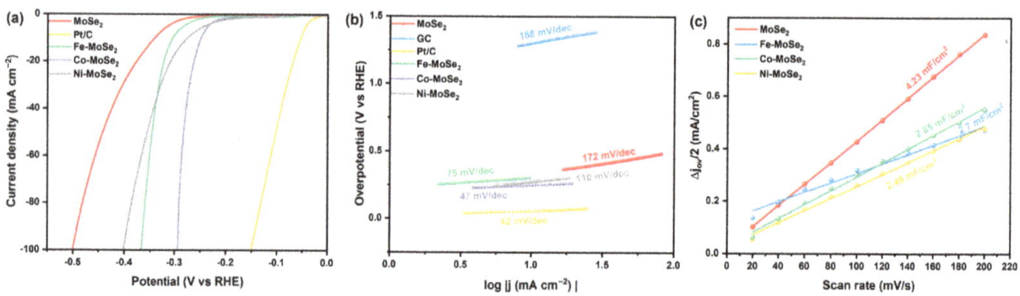

Figure 5. (**a**) Polarization curves of $MoSe_2$, Pt/C, and Fe/Co/Ni-$MoSe_2$ catalysts. (**b**) Calculated Tafel slopes. Testing was performed in 0.5 M H_2SO_4 with a scan rate of 50 mV/s. (**c**) C_{dl} of $MoSe_2$ and Fe/Co/Ni-$MoSe_2$.

To thoroughly assess the electrochemical catalytic performance of our electrocatalysts and enable meaningful comparison with previous studies, we have compiled the catalytic properties of various MoX_2-based catalysts in Table 1. Comparative analysis reveals that the Co-doped $MoSe_2$ catalyst in this work demonstrates superior electrocatalytic performance, as manifested by its reduced overpotential and lower Tafel slope relative to other catalysts. For example, the reported $Ni_{0.15}Mo_{0.85}Se_2$ catalyst requires an overpotential of 0.32 V to reach 90 mA/cm^2 [25], whereas our Co-doped $MoSe_2$ catalyst achieves a higher current density of 100 mA/cm^2 at a lower overpotential of 0.29 V. The Co-doped $MoSe_2$ catalyst also shows better performance than $MoSe_2$ decorated with Co and Ni nanoparticles [26] and Fe-, Co-, and Ni-doped MoS_2 [27]. While the Tafel slopes of Fe-, Co-, or Ni-doped MoS_2 typically range from 66 to 94 mV/dec, our Co-doped $MoSe_2$ exhibits a significantly lower Tafel slope of 47 mV/dec, indicating enhanced reaction kinetics. Compared with the graphene/$MoSe_2$ composites reported previously [28], the Co-doped $MoSe_2$ catalyst also exhibits a lower Tafel slope (80 vs. 47 mV/dec).

Table 1. The electrochemical catalytic performance of MoX_2-based catalysts.

Electrocatalyst	Electrolyte	Current Density ($mA\ cm^{-2}$)	Overpotential (mV)	Tafel Slope ($mV\ dev^{-1}$)
Co-$MoSe_2$ (this work)	0.5 M H_2SO_4	100	293	47
Fe-$MoSe_2$ (this work)	0.5 M H_2SO_4	100	367	75
Ni-$MoSe_2$ (this work)	0.5 M H_2SO_4	100	401	110
Ni-$MoSe_2$ [24]	1.0 M KOH	50	455	96
$MoSe_2$/carbon cloth [29]	0.5 M H_2SO_4	21 60	250 300	76
$Ni_{0.15}Mo_{0.85}Se_2$ [25]	0.5 M H_2SO_4	10 90	180 ~320	67
$MoSe_2$@Co [26]	1 M KOH	10	391	109
$MoSe_2$@Ni [26]	1 M KOH	10	367	227
$MoSe_2$@CoNi [26]	1 M KOH	10	378	170
graphene/$MoSe_2$ [28]	0.5 M H_2SO_4	100	~210	80
Ni-MoS_2 [27]	0.5 M H_2SO_4	10 35	302 ~360	66
Co-MoS_2 [27]	0.5 M H_2SO_4	10 35	350 ~550	69
Fe-MoS_2 [27]	0.5 M H_2SO_4	10 35	490 ~700	94

Furthermore, this study analyzed the electrochemical active surface areas (ECSAs) of the catalysts. The ECSA was evaluated by C_{dl} using cyclic voltammetry (CV) at various scan rates in the potential range from 0 to 0.2 V versus RHE (Figure S4 of the Supplementary Materials). Typically, the ECSA is directly proportional to the C_{dl} of the electrode, expressed as ECSA = C_{dl}/C_s, where C_s represents the specific capacitance of the catalysts per unit area under identical electrolyte conditions. The specific capacitance typically falls within a reported range of 0.015–0.110 mF/cm^2 in acidic solutions. The general specific capacitance of 0.035 mF/cm^2 was used to estimate the ECSA in this work [30]. In Figure 5c, the ECSA values of Fe-/Co-/Ni-doped $MoSe_2$ are 77.14, 70.86, and 81.43 cm^2 respectively, slightly lower than that of pure $MoSe_2$ (120.9 cm^2). Consequently, while the ECSA is an important parameter for understanding electrocatalytic behavior, it may not be the predominant factor governing the enhanced HER performance of Fe-/Co-/Ni-doped $MoSe_2$.

2.5. Theoretical Calculation

To gain deeper insights into the active sites of Fe-, Co-, and Ni-doped $MoSe_2$ (M-$MoSe_2$) in the HER, we performed density functional theory (DFT) calculations on Fe-$MoSe_2$, Co-$MoSe_2$, Ni-$MoSe_2$, and pristine $MoSe_2$. The catalytic activity was evaluated using the ΔG_H, which reflects the hydrogen adsorption capability on the material surface. An effective HER catalyst typically exhibits a ΔG_H value close to zero, indicating moderate hydrogen adsorption strength that facilitates efficient HER. A ΔG_H that is too positive (weak adsorption) or too negative (strong adsorption) would hinder the reaction. Thus, the ideal HER catalyst should have $\Delta G_H \approx 0$ eV.

In this study, we analyzed multiple potential hydrogen adsorption sites on M-$MoSe_2$, including basal plane Se sites (plane Se), metallic edge Mo sites (edge Mo), and non-metallic edge M/Se sites (edge M/edge Se). For comparison, we also calculated hydrogen adsorption on pristine $MoSe_2$ at basal plane Se sites (plane Se), metallic edge Mo sites (edge Mo), and non-metallic edge Se sites (edge Se). The optimized structure is shown in Figure 6, and the calculated ΔG value is shown in Figure 7.

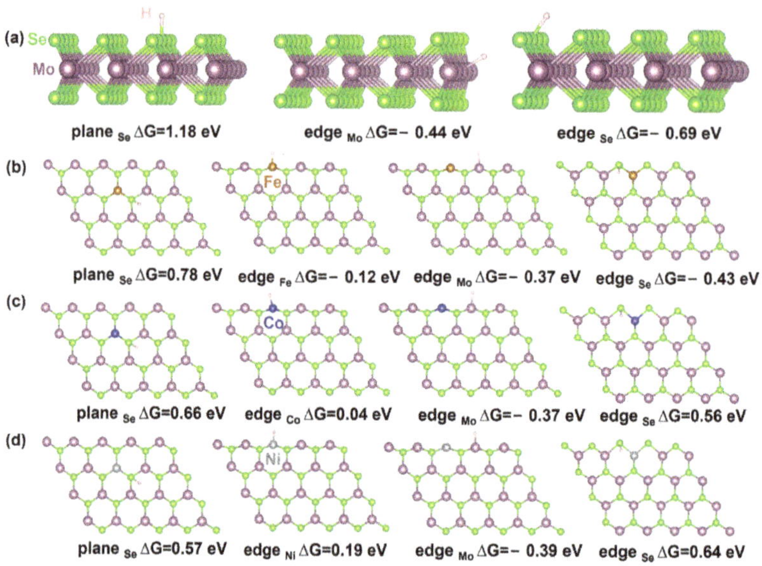

Figure 6. DFT-optimized structures showing H atom adsorption on different positions of (**a**) $MoSe_2$, (**b**) Fe-$MoSe_2$, (**c**) Co-$MoSe_2$, and (**d**) Ni-$MoSe_2$.

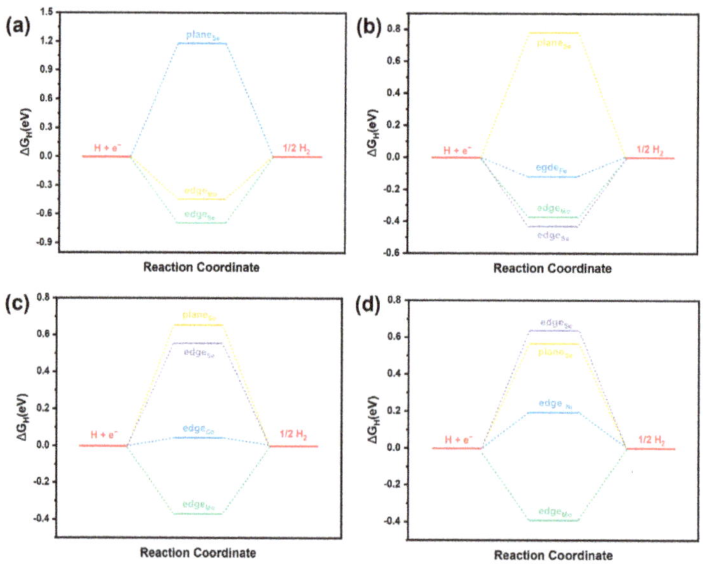

Figure 7. The calculated ΔG for H atom adsorption on different positions of (**a**) $MoSe_2$, (**b**) Fe-$MoSe_2$, (**c**) Co-$MoSe_2$, and (**d**) Ni-$MoSe_2$.

For pristine MoSe2, the ΔG values for hydrogen adsorption at basal plane Se sites, metallic edge Mo sites, and non-metallic edge Se sites are 1.18, −0.44, and −0.69 eV, respectively. Upon Fe doping, the ΔG value at basal plane Se sites decreases to 0.78 eV, while the metallic edge Mo sites show a ΔG of −0.37 eV; notably, the edge Fe site exhibits a significantly lower ΔG of −0.12 eV, and the adjacent edge Se site is −0.43 eV. With Co doping, the ΔG at basal plane Se sites is further reduced to 0.66 eV, the metallic edge Mo site remains at −0.37 eV, and the edge Co site approaches a near-ideal ΔG of 0.04 eV, with adjacent edge Se sites at 0.56 eV. In the Ni-doped sample, the ΔG values are 0.57 eV (basal plane Se), −0.39 eV (metallic edge Mo), and 0.19 eV (edge Ni), while the adjacent edge Se sites near Ni exhibit a ΔG of 0.64 eV. Overall, these findings suggest that doping MoSe2 with transition metals can optimize hydrogen adsorption and thus improve its electrocatalytic performance.

3. Experimental

3.1. Sample Preparation

Fe-, Co-, and Ni-doped $MoSe_2$ were synthesized using the chemical vapor deposition method (Figure 8a). Initially, a 1 cm × 1 cm GC was sonicated in distilled water and ethanol for 10 min each to remove surface impurities. The precursor solution was prepared as follows: $(NH_4)_6Mo_7O_{24}\cdot 4H_2O$ was dissolved in 1 mL of water to obtain a 0.0121 mol/L solution, which contained approximately 8.16 mg Mo per mL. For doping purposes, $FeCl_2$ (0.146 mol/L, ~8.16 mg Fe/mL), $CoCl_2\cdot 6H_2O$ (0.139 mol/L, ~8.16 mg Co/mL), and $NiCl_2$ (0.139 mol/L, ~8.16 mg Ni/mL) were each dissolved in 1 mL of deionized water. Approximately 4–5 drops (equivalent to ~0.200–0.250 mL) of the precursor solution were uniformly drop-cast onto the surface of a pre-cleaned 1 cm × 1 cm GC, forming a full surface coverage. The sample was then dried at 60 °C for 1 h.

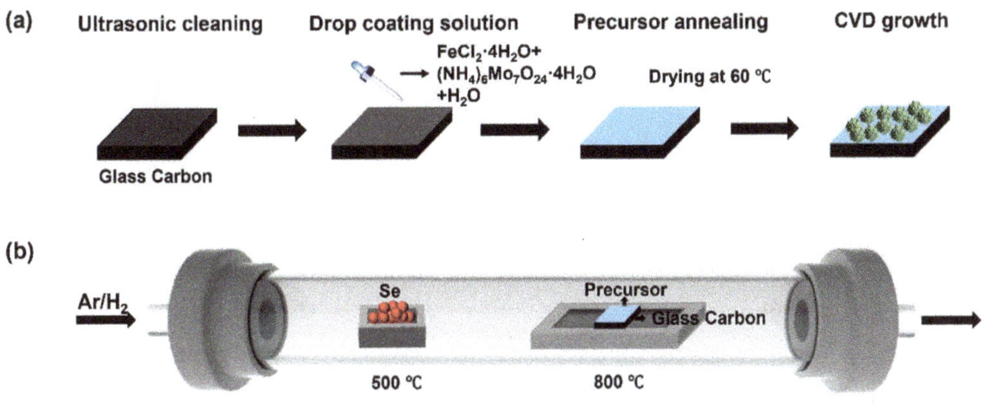

Figure 8. (**a**) Schematic diagram of the preparation process for Fe-doped $MoSe_2$. (**b**) Schematic diagram of the CVD system used to synthesize transition metal-doped $MoSe_2$ nanosheets.

The dried GC substrate was placed on the right side of a tube furnace, with 75 mg of Se powder placed in a porcelain boat on the left side as the Se source (Figure 8b). During the experiment, the temperatures on both sides were raised simultaneously, with the left side reaching 500 °C and the right side reaching 800 °C over a 40 min heating period. Ar gas was introduced at a flow rate of 100 sccm. Once the target temperatures were reached, the samples were kept at these temperatures for 15 min, while the atmosphere was adjusted to a 100 sccm Ar/H_2 gas mixture (9:1). Afterward, the heating system was turned off, and the samples were allowed to cool naturally to room temperature, yielding Fe-, Co-, and

Ni-doped $MoSe_2$ nanomaterials. To evaluate the impact of doping on the electrocatalytic performance of $MoSe_2$, undoped $MoSe_2$ was synthesized under the same conditions and compared with the doped samples.

3.2. Characterization Methods

Various techniques were employed to systematically analyze the materials. A Hitachi SU-8010 scanning electron microscope (manufactured by Hitachi High-Tech Corporation, Tokyo, Japan) was used to examine the surface morphology, while a JEOL F200 transmission electron microscope (200 kV, equipped with an Oxford X-MaxN 80T IE250 EDS; manufactured by JEOL Ltd., Tokyo, Japan) utilized DigitalMicrograph 3.5 software for image processing and data analysis, providing insights into the microstructure and lattice fringes. The crystal structure was assessed via X-ray diffraction using Cu Kα radiation (λ = 1.5406 Å) with a Shimadzu XRD-7000 X-ray diffractometer (manufactured by Shimadzu Corporation, Kyoto, Japan), and the data was processed with Origin 2022 software. The Raman spectra were acquired on a Horiba Xplora spectrometer (manufactured by Horiba Scientific, Kyoto, Japan) with a 532 nm laser source, and the spectral data was analyzed using LabSpec 6.3 software. Furthermore, X-ray photoelectron spectroscopy was performed on a Thermo Scientific K-Alpha instrument (manufactured by Thermo Fisher Scientific Inc., Waltham, MA, USA), and the elemental composition and chemical states of the samples were determined through Avantage 5.9 software.

3.3. Electrochemical Measurements

Electrochemical tests were performed using a CHI660E electrochemical work station (manufactured by Shanghai Chenhua Instrument Co., Ltd., Shanghai, China). GC electrodes loaded with Fe-, Co-, and Ni-doped $MoSe_2$ catalysts were used as the working electrodes, with Ag/AgCl and Pt electrodes serving as the reference and counter electrodes, respectively. All potentials were referenced to the reversible hydrogen electrode (RHE), calculated using the following formula: E_{RHE} = E(Ag/AgCl) + 0.059 pH + 0.197 V. LSV was carried out in a 0.5 M H_2SO_4 solution at a scan rate of 50 mV/s, with a 95% ohmic drop (IR) correction applied to all results. The Tafel slope, derived from the LSV curves, was used to evaluate the electrocatalytic reaction kinetics. Additionally, CV was performed in the potential range of 0–0.2 V vs. RHE at various scan rates (20–200 mV/s) to determine the double-layer capacitance.

3.4. Computational Details

First-principle calculations based on DFT were performed using the Vienna Ab initio Simulation Package (VASP) [31–33]. The exchange–correlation interactions were described by the Perdew–Burke–Ernzerhof (PBE) [34] functional within the generalized gradient approximation (GGA). For structural optimization, the energy change, maximum force, and maximum displacement convergence criteria were set to 2.0×10^{-5} Ha, 0.004 Ha Å^{-1}, and 0.005 Å, respectively. A $3 \times 3 \times 1$ k-point grid based on the Monkhorst–Pack scheme was used for Brillouin zone integration. The plane wave cutoff energy in VASP was set to 400 eV, and spin polarization effects were included in all calculations. Structural optimizations continued until the residual forces on each atom were reduced to less than 0.01 eV/Å.

To evaluate the catalytic activity of Fe-/Co-/Ni-doped $MoSe_2$ for the HER, we calculated the hydrogen adsorption energy (ΔE_H) at different active sites, defined as follows:

$$\Delta E_H = E_A - E_P - 0.5E_{H_2}$$

where E_A represents the energy of the catalyst with a single hydrogen atom adsorbed, E_P is the energy of the pristine catalyst, and E_{H_2} is the energy of a hydrogen molecule in the gas phase.

To further assess the hydrogen adsorption thermodynamics, we incorporated zero-point energy (ΔE_{ZPE}) and entropy (ΔS) corrections to determine the Gibbs free energy of hydrogen adsorption (ΔG_H), calculated as follows:

$$\Delta G_H = \Delta E_H + \Delta E_{ZPE} - T\Delta S$$

where T is the temperature (set to 298.15 K). These calculations provide insights into how Fe, Co, and Ni doping influence the hydrogen adsorption properties of $MoSe_2$, potentially enhancing its catalytic performance for HER.

4. Conclusions

We successfully synthesized Fe-, Co-, and Ni-doped $MoSe_2$ via CVD and systematically investigated their morphology, structure, and electrocatalytic performance. The results demonstrate that doping elements not only significantly altered the morphology of $MoSe_2$, but also enhanced its HER performance by introducing defects and optimizing the electronic structure. DFT calculations further confirmed that metal doping effectively reduces ΔG at active sites and enhances catalytic activity. Among the doped samples, Co-doped $MoSe_2$ exhibited the best catalytic activity, with an onset overpotential of 0.293 mV at 100 mA/cm^{-2} and a Tafel slope of 47 mV/dec. This study provides new insights into the application of transition metal-doped $MoSe_2$ in electrocatalysis.

Supplementary Materials: The following supporting information can be downloaded at: https://www.mdpi.com/article/10.3390/inorganics13050155/s1. Figure S1: XRD patterns of GC, Fe/Co/Ni-$MoSe_2$, and $MoSe_2$. Figure S2: Polarization curves of $MoSe_2$, GC, Pt/C, and Fe/Co/Ni-$MoSe_2$ catalysts. Figure S3: Polarization curves of (a) $MoSe_2$, Fe-$MoSe_2$, and $FeSe_2$ catalysts, (b) $MoSe_2$, Co-$MoSe_2$, and $CoSe_2$ catalysts, (c) $MoSe_2$, Ni-$MoSe_2$, and $NiSe_2$ catalysts. Figure S4: CV curves of $MoSe_2$ (a), Fe-$MoSe_2$ (b), Co-$MoSe_2$ (c), and Ni-$MoSe_2$ (d) at various scan rates (20 mV s^{-1}, 40 mV s^{-1}, 60 mV s^{-1}, 80 mV s^{-1}, 100 mV s^{-1}, 120 mV s^{-1}, 140 mV s^{-1}, 160 mV s^{-1}, 180 mV s^{-1}, 200 mV s^{-1}) in the potential range of 0~0.2 (vs. RHE).

Author Contributions: X.C.: investigation, data curation, formal analysis, writing—original draft preparation. X.Z.: data curation, software, formal analysis. J.Z.: methodology, data curation, resources. Z.W.: conceptualization, methodology, validation, writing—review and editing, visualization, supervision, project administration. All authors have read and agreed to the published version of the manuscript.

Funding: This research received no external funding.

Data Availability Statement: The original contributions presented in this study are included in the article and Supplementary Materials. Further inquiries can be directed to the corresponding author.

Acknowledgments: Resources supporting this work were provided by the Public Computing Cloud Platform of Renmin University of China. The authors thank the Office of Research Infrastructure of Renmin University of China for the laboratory facilities and technical support in sample characterization.

Conflicts of Interest: There are no conflicts of interest to declare.

References

1. Qiao, D.; Yun, S.; Sun, M.L.; Dang, J.E.; Zhang, Y.W.; Yuan, S.X.; Yang, G.P.; Yang, T.X.; Gao, Z.; Wang, Z.G. 1D/3D Trepang-Like N-Modified Carbon Confined Bimetal Carbides and Metal Cobalt: Boosting Electron Transfer via Dual Mott-Schottky Heterojunctions Triggered Built-In Electric Fields for Efficient Hydrogen Evolution and Tri-Iodide Reduction. *Appl. Catal. B Environ.* **2023**, *334*, 122830. [CrossRef]
2. Wang, L.Q.; Hao, Y.X.; Deng, L.M.; Hu, F.; Zhao, S.; Li, L.L.; Peng, S.J. Rapid Complete Reconfiguration Induced Actual Active Species for Industrial Hydrogen Evolution Reaction. *Nat. Commun.* **2022**, *13*, 5785. [CrossRef]
3. Sarwar, S.; Ali, A.; Wang, Y.F.; Ahasan, M.R.; Wang, R.G.; Adamczyk, A.J.; Zhang, X.Y. Enhancement of Hydrogen Evolution Reaction Activity Using Metal–Rich Molybdenum Sulfotelluride with Graphene Support: A Combined Experimental and Computational Study. *Nano Energy* **2021**, *90*, 106599. [CrossRef]
4. Zhang, Z.Y.; Tian, H.; Bian, L.; Liu, S.Z.; Liu, Y.; Wang, Z.L. Cu-Zn-Based Alloy/Oxide Interfaces for Enhanced Electroreduction of CO_2 to C_2^+ Products. *J. Energy Chem.* **2023**, *83*, 90–97. [CrossRef]
5. Wan, C.Z.; Zhang, Z.S.; Dong, J.C.; Xu, M.J.; Pu, H.T.; Baumann, D.; Lin, Z.Y.; Wang, S.B.; Huang, J.; Shah, A.H.; et al. Amorphous Nickel Hydroxide Shell Tailors Local Chemical Environment on Platinum Surface for Alkaline Hydrogen Evolution Reaction. *Nat. Mater.* **2023**, *22*, 1022–1029. [CrossRef]
6. Zhang, J.S.; Zhang, X.L.; Shi, C.; Xia, G.Q.; Li, H.; Wang, P.; Di, L.B. Plasma Synthesis of Defect-Rich Flexible Carbon Cloth Decorated with PtRu Alloyed Nanoclusters for Highly Efficient pH-Universal Electrocatalytic Hydrogen Evolution. *Nanoscale* **2022**, *14*, 15942–15949. [CrossRef]
7. Wang, Y.H.; Yun, S.N.; Shi, J.; Zhang, Y.W.; Dang, J.E.; Dang, C.W.; Liu, Z.L.; Deng, Y.Y.; Yang, T.X. Defect Engineering Tuning Electron Structure of Biphasic Tungsten-Based Chalcogenide Heterostructure Improves Its Catalytic Activity for Hydrogen Evolution and Triiodide Reduction. *J. Colloid Interface Sci.* **2022**, *625*, 800–816. [CrossRef]
8. Da, P.F.; Zheng, Y.; Hu, Y.; Wu, Z.L.; Zhao, H.Y.; Wei, Y.C.; Guo, L.C.; Wang, J.J.; Wei, Y.P.; Xi, S.B.; et al. Synthesis of Bandgap-Tunable Transition Metal Sulfides through Gas-Phase Cation Exchange-Induced Topological Transformation. *Angew. Chem. Int. Ed.* **2023**, *62*, e202301802. [CrossRef]
9. Chen, X.Y.; Han, Z.; Zhang, B.; Sun, B.J.; Wang, Y.; Du, Y.C.; Han, X.J.; Xu, P. Construction of Plasmonic 1T-WS_2/2H-WS_2/CdS Heterostructures for Enhanced Solar-Driven Hydrogen Evolution. *J. Mater. Chem. A* **2022**, *10*, 24030–24040. [CrossRef]
10. Chen, S.-F.; Wu, T.-S.; Soo, Y.-L. Highly Defective Graphene Quantum Dots-Doped 1T/2H-MoS_2 as an Efficient Composite Catalyst for the Hydrogen Evolution Reaction. *Sci. Rep.* **2023**, *13*, 15184. [CrossRef]
11. Eftekhari, A. Electrocatalytic hydrogen evolution reaction: A comprehensive review. *Int. J. Hydrogen Energy* **2017**, *42*, 11053–11077. [CrossRef]
12. Dai, X.P.; Du, K.L.; Li, Z.Z.; Liu, M.Z.; Ma, Y.D.; Sun, H.; Zhang, X.; Yang, Y. Co-Doped MoS_2 Nanosheets with the Dominant CoMoS Phase Coated on Carbon as an Excellent Electrocatalyst for Hydrogen Evolution. *ACS Appl. Mater. Interfaces* **2015**, *7*, 27242–27253. [CrossRef] [PubMed]
13. Lu, Z.; Li, S.; Ning, L.; Tang, K.; Guo, Y.; You, L.; Chen, C.; Wang, G. Nickel–Iron-Layered Double Hydroxide Electrocatalyst with Nanosheets Array for High Performance of Water Splitting. *Molecules* **2024**, *29*, 2092. [CrossRef]
14. Feng, Q.L.; Mao, N.N.; Wu, J.X.; Xu, H.; Wang, C.M.; Zhang, J.; Xie, L.M. Growth of $MoS_{2(1-x)}Se_{2x}$ (x = 0.41–1.00) Monolayer Alloys with Controlled Morphology by Physical Vapor Deposition. *ACS Nano* **2015**, *9*, 7450–7455. [CrossRef]
15. Chhowalla, M.; Shin, H.S.; Eda, G.; Li, L.J.; Loh, K.P.; Zhang, H. The Chemistry of Two-Dimensional Layered Transition Metal Dichalcogenide Nanosheets. *Nat. Chem.* **2013**, *5*, 263–275. [CrossRef]
16. Shen, D.Y.; Zhao, B.; Zhang, Z.C.; Zhang, H.M.; Yang, X.D.; Huang, Z.W.; Li, B.L.; Song, R.; Jin, Y.J.; Wu, R.X.; et al. Synthesis of Group VIII Magnetic Transition Metal-Doped Monolayer $MoSe_2$. *ACS Nano* **2022**, *16*, 10623–10631. [CrossRef]
17. Zimron, O.; Zilberman, T.; Kadam, S.R.; Ghosh, S.; Kolatker, S.-L.; Neyman, A.; Bar-Ziv, R.; Bar-Sadan, M. Co-Doped $MoSe_2$ Nanoflowers as Efficient Catalysts for Electrochemical Hydrogen Evolution Reaction (HER) in Acidic and Alkaline Media. *Isr. J. Chem.* **2020**, *60*, 624–629. [CrossRef]
18. Yang, Y.Q.; Zhao, X.; Mao, H.; Ning, R.; Zheng, X.H.; Sui, J.H.; Cai, W. Nickel-doped $MoSe_2$ nanosheets with Ni-Se bond for alkaline electrocatalytic hydrogen evolution. *J. Mater. Chem. A* **2023**, *11*, 12345–12352. [CrossRef]
19. Chen, Y.P.; Huang, Y.C.; Xiong, K.Y.; Ye, X.Y.; Liu, Q.Q.; Zhuang, J.R.; Zhang, Y.F.; Zhu, J. Cooperative 3d-Metal-Doped $MoSe_2$ Catalysts for Enhanced Electrocatalytic Hydrogen Evolution Reaction. *Mol. Catal.* **2024**, *566*, 114401. [CrossRef]
20. Liu, T.; Xu, H.; Zhou, W.; Zhang, K.; Sun, Y. Computational Design of Single-Atom Catalysts: Cobalt-Doped $MoSe_2$ as a High-Performance Electrocatalyst for Hydrogen Evolution. *ACS Catal.* **2020**, *10*, 12345–12356.
21. Xu, C.; Peng, S.J.; Tan, C.L.; Ang, H.X.; Tan, H.T.; Zhang, H.; Yan, Q.Y. Ultrathin S-Doped $MoSe_2$ Nanosheets for Efficient Hydrogen Evolution. *J. Mater. Chem. A* **2014**, *2*, 5597–5601. [CrossRef]
22. Xiong, Q.Z.; Zhang, X.; Wang, H.J.; Liu, G.Q.; Wang, G.Z.; Zhang, H.M.; Zhao, H.J. One-Step Synthesis of Cobalt-Doped MoS_2 Nanosheets as Bifunctional Electrocatalysts for Overall Water Splitting under Both Acidic and Alkaline Conditions. *Chem. Commun.* **2018**, *54*, 3859–3862. [CrossRef]

23. Sun, X.; Dai, J.; Guo, Y.Q.; Wu, C.Z.; Hu, F.T.; Zhao, J.Y.; Zeng, X.C.; Xie, Y. Semimetallic Molybdenum Disulfide Ultrathin Nanosheets as an Efficient Electrocatalyst for Hydrogen Evolution. *Nanoscale* **2014**, *6*, 8359–8367. [CrossRef]
24. Zong, H.; Yu, K.; Zhu, Z. Heterostructure nanohybrids of Ni-doped $MoSe_2$ coupled with Ti_2NT_x toward efficient overall water splitting. *Electrochim. Acta* **2020**, *353*, 136598. [CrossRef]
25. Zhao, G.Q.; Wang, X.Y.; Wang, S.L.; Rui, K.; Chen, Y.P.; Yu, H.B.; Ma, J.; Dou, S.X.; Sun, W.P. Evolution Kinetics for Alkaline Water Splitting. *Chem. Asian J.* **2019**, *14*, 301–306. [CrossRef]
26. Dymerska, A.; Kukułka, W.; Wenelska, K.; Mijowska, E. Two-Dimensional Molybdenum Diselenide Tuned by Bimetal Co/Ni Nanoparticles for Oxygen Evolution Reaction. *ACS Omega* **2020**, *5*, 28730–28737. [CrossRef]
27. Venkatesh, P.S.; Kannan, N.; Babu, M.G.; Paulraj, G.; Jeganathan, K. Transition Metal Doped MoS_2 Nanosheets for Electrocatalytic Hydrogen Evolution Reaction. *Mater. Today Proc.* **2022**, *47*, 37256–37263.
28. Najafi, L.; Bellani, S.; Oropesa-Nuñez, R.; Ansaldo, A.; Prato, M.; Del Rio Castillo, A.E.; Bonaccorso, F. Engineered $MoSe_2$-based heterostructures for efficient electrochemical hydrogen evolution reaction. *Adv. Energy Mater.* **2018**, *8*, 1703212. [CrossRef]
29. Liu, Y.D.; Ren, L.; Zhang, Z.; Qi, X.; Li, H.X.; Zhong, J.X. 3D Binder-free $MoSe_2$ Nanosheets/Carbon Cloth Electrodes for Efficient and Stable Hydrogen Evolution Prepared by Simple Electrophoresis Deposition Strategy. *Sci. Rep.* **2016**, *6*, 22516. [CrossRef]
30. McCrory, C.C.L.; Jung, S.; Peters, J.C.; Jaramillo, T.F. Benchmarking Heterogeneous Electrocatalysts for the Oxygen Evolution Reaction. *J. Am. Chem. Soc.* **2013**, *135*, 16977–16987. [CrossRef]
31. Kresse, G.; Hafner, J. Ab Initio Molecular Dynamics for Liquid Metals. *Phys. Rev. B* **1993**, *47*, 558(R). [CrossRef] [PubMed]
32. Kresse, G.; Furthmüller, J.; Hafner, J. Theory of the Crystal Structures of Selenium and Tellurium: The Effect of Generalized-Gradient Corrections to the Local-Density Approximation. *Phys. Rev. B* **1994**, *50*, 13181–13190. [CrossRef] [PubMed]
33. Kresse, G.; Furthmüller, J. Efficient Iterative Schemes for Ab Initio Total-Energy Calculations Using a Plane-Wave Basis Set. *Phys. Rev. B* **1996**, *54*, 11169–11186. [CrossRef] [PubMed]
34. Perdew, J.P.; Burke, K.; Ernzerhof, M. Generalized Gradient Approximation Made Simple. *Phys. Rev. Lett.* **1996**, *77*, 3865–3868. [CrossRef]

inorganics

MDPI

Article

Decorating Ti_3C_2 MXene Nanosheets with Fe-N_x-C Nanoparticles for Efficient Oxygen Reduction Reaction

Han Zheng, Fagang Wang * and Weimeng Si *

School of Materials Science and Engineering, Shandong University of Technology, Xincunxi Road 266th, Zibo 255000, China
* Correspondence: a_gang@sdut.edu.cn (F.W.); siweimeng@foxmail.com (W.S.); Tel./Fax: +86-533-2781317 (F.W.)

Abstract: Finding alternatives to platinum that exhibit high activity, stability, and abundant reserves as oxygen reduction electrocatalysts is crucial for the advancement of fuel cells. In this study, we first mixed $FeCl_2 \cdot 4H_2O$, 1,10-phenanthroline, and Vulcan XC-72, followed by pyrolysis in a nitrogen atmosphere, to obtain FeNC. Subsequently, we combined FeNC with MXene produce FeNC/MXene composites. The FeNC/MXene catalyst achieved a half-wave potential of 0.857 V in an alkaline medium, exhibiting better oxygen reduction reaction (ORR) activity and durability than commercial Pt/C catalysts. The layered structure of MXene endows the material with a high specific surface area and facilitates efficient electron transfer pathways, thereby promoting rapid charge transfer and material diffusion. The cleavage of Ti-C bonds in Ti_3C_2 at elevated temperatures results in the transformation of MXene into TiO_2, where the coexistence of anatase and rutile phases generates a synergistic effect that enhances both the mass transfer rate and the electrical conductivity of the catalytic layer. Additionally, the unique electronic structure of the FeN_x sites simultaneously optimizes electrocatalytic activity and stability. Leveraging these structural advantages, the FeNC/MXene composite catalysts demonstrate exceptional catalytic activity and long-term stability in oxygen reduction reactions.

Keywords: Fe-N-C catalysts; MXene; Non-Pt catalysts; oxygen reduction reaction

1. Introduction

As the global community increasingly prioritizes energy and environmental issues, hydrogen fuel cells have garnered significant attention [1,2]. In these cells, hydrogen is oxidized through the hydrogen oxidation process, while oxygen undergoes reduction, enabling the conversion of stored energy into electrical energy via the oxygen reduction mechanism [3–5]. However, the ORR is characterized by slow kinetics and necessitates a high overpotential for effective operation. Currently, precious metal-based electrocatalysts represent the most advanced ORR catalysts [6]; however, they possess several limitations, including high costs, a low abundance of the constituent metals, single functionality, and poor stability [7]. Therefore, cost-effective and durable non-precious metal ORR catalysts are vital for improving fuel cell performance and promoting sustainable energy development.

Nitrogen–carbon (M-N-C) materials doped with transition metals are regarded as promising substitutes for commercial Pt/C catalysts [8–11]. The incorporation of transition metals benefits the ORR by weakening the bond strength of key intermediates (OOH*, O*, OH*), as these metals can effectively accept external electrons [12–14]. This phenomenon arises from the three-dimensional electronic orbitals of transition metals, which have the ability to accommodate external electrons. Additionally, the introduction of nitrogen through doping can facilitate the formation of additional active sites,

https://doi.org/10.3390/inorganics13060188

thereby enhancing the electrochemical efficiency of the catalyst and promoting the ORR. Notably, carbon materials exhibit electrical conductivity, and nitrogen doping can further improve this property, increasing the rate of electron transfer within the catalyst [15,16]. The synergy between transition metals and nitrogen-doped carbon enhances both durability and catalytic activity, leading to improved ORR performance. Among these materials, iron–nitrogen–carbon (Fe-N-C) has attracted considerable attention for its promise as a high-performance alternative to noble metal catalysts like Pt/C [17,18]. Fe-N-C materials offer several advantages, including moderate cost, high conductivity, stability, and high catalytic efficiency [19–22]. However, the pyrolysis process unavoidably induces significant aggregation and the formation of thick catalyst layers, thereby impeding mass transport and ionic conductivity, which ultimately results in suboptimal ORR performance [23]. To overcome this problem, researchers have designed various Fe-N-C catalysts supported on diverse substrates, including covalent organic frameworks (COFs) [24], imidazole zeolite frameworks (ZIFs) [25], and MXene [26,27]. In particular, MXene materials have garnered significant interest for heterogeneous catalysis, attributed to their layered architectures and rich functional groups [28].

MXene ($Mn_{(1-x)}X_{(x+1)}T_x$, where M is an early transition metal, X represents C and/or N, and T denotes surface terminations) is a novel two-dimensional transition metal carbide or nitride. It is widely applied in energy storage and conversion due to its excellent electrical conductivity, hydrophilicity [29,30], and unique surface properties [28,31–33]. However, MXene features a readily stackable sheet structure, which is held together by van der Waals forces and hydrogen bonds [34]. Variations in MXene's lattice structure and surface terminations can significantly alter its electronic properties, allowing the fine-tuning of its surface environment and catalyst interactions [35]. Consequently, MXene could play a pivotal role in enhancing the catalytic activity of Fe-N-C.

In the present study, a composite material was designed, consisting of Fe-N_x-C and Ti_3C_2-type MXene. A series of FeNC/MXene electrocatalysts was prepared, and their electrochemical performance as ORR catalysts in an alkaline medium was investigated. Furthermore, we optimized the pyrolysis temperature of Fe-N_x-C and MXene to achieve the most effective ORR activity. The FeNC/MXene catalyst demonstrates a high half-wave potential of 0.857 V and supports an efficient four-electron ORR pathway. It also exhibits excellent stability and strong resistance to methanol crossover. These FeNC/MXene catalysts meet the necessary criteria for both purity and high performance, and are expected to serve as a cost-effective, efficient, and sustainable alternative to commercial Pt/C catalysts.

2. Experimental Section

2.1. Materials

MAX phase (Ti_3AlC_2) was bought from Jilin 11 Technology Co., Ltd. (Changchun, China). Hydrofluoric acid (40%), methanol (99.5%), and propan-2-ol (99.7%) were purchased from Sinopharm Chemical Reagent Co., Ltd. (Shanghai, China). Iron(II) chloride tetrahydrate and 1,10-phenanthroline were purchased from Macklin chemical reagent. (Shanghai, China). Vulcan XC-72 carbon black was purchased from Cabot Corporation (Boston, MA, USA), and Nafion was obtained from DuPont (Wilmington, DE, USA).

2.2. Synthesis of MXene

1 g of Ti_3AlC_2 was gradually introduced into 20 mL of a 40% HF solution and stirred at 45 °C for 24 h [36]. Following the stirring process, the mixture was subjected to repeated centrifugation with deionized water until the pH reached 6. The precipitate obtained from centrifugation was subsequently collected and vacuum-freeze-dried for 12 h to yield multilayered $Ti_3C_2T_x$.

2.3. Synthesis of eNC/MXene

A total of 120 mg of carbon black (Vulcan XC72) was combined with 8 mL of methanol. In a separate preparation, 22 mg of Iron(II) chloride tetrahydrate was added to 1 mL of methanol, and 60 mg of 1,10-phenanthroline was dissolved in 1 mL of methanol. The three resulting solutions were then combined and stirred thoroughly. The mixture was subsequently placed in an open beaker, where it was continuously stirred at a temperature of 60 °C until complete evaporation of the solvent occurred. Following this, the material was heated in an oven at 80 °C for two hours. The carbon black-containing $[Fephen_3]^{2+}$ was then transferred to a tubular heating furnace and subjected to pyrolysis at 800 °C for three hours, utilizing a heating rate of 5 °C min^{-1} in a nitrogen atmosphere. The final product was designated as FeNC.

To prepare the FeNC/MXene composite, add 30 mg of MXene and 30 mg of FeNC to 20 mL of methanol. Stir the mixture at 60 °C until the methanol has completely evaporated. Subsequently, transfer the resulting mixture to a tube furnace and subject it to pyrolysis in a nitrogen atmosphere at 800 °C for a duration of 2 h.

2.4. Electrochemical Characterization

In this experiment, all electrochemical measurements were performed using a CS2350 electrochemical workstation with a standard three-electrode configuration. A saturated Hg/HgO electrode was used as the reference electrode, a graphite rod as the counter electrode, and a 5 mm diameter glassy carbon rotating disk electrode (RDE) as the working electrode. The electrolyte was 0.1 M KOH, which was saturated with oxygen prior to testing to ensure reliable evaluation of the catalyst's ORR performance. To standardize the results, all potentials were converted to the reversible hydrogen electrode (RHE) scale using the following equation:

$$\mathrm{RHE} = \mathrm{Hg/HgO} + 0.059\mathrm{pH} + 0.098$$

Before testing, the catalyst was prepared into an ink by dispersing 5 mg of catalyst in 200 μL of isopropanol, 780 μL of ultrapure water, and 20 μL of 5% Nafion solution. The mixture was sonicated for 1 h to ensure uniform dispersion. The working electrode was polished with 50 nm Al_2O_3 powder to remove surface impurities. Then, 20 μL of the catalyst ink was drop-cast onto the RDE surface and allowed to dry. Electrochemical activation was performed using cyclic voltammetry (CV) in O_2-saturated 0.1 M KOH at a scan rate of 50 mV s^{-1} over a potential range of 0.1–1.1 V (vs. RHE). ORR activity was evaluated via linear sweep voltammetry (LSV) under O_2-saturated conditions at a scan rate of 10 mV s^{-1} and a rotation speed of 1600 rpm, within the potential range of 0–1.1 V (vs. RHE). To analyze the reaction kinetics, LSV curves were also recorded at various rotation rates (625–2025 rpm). The Koutecky–Levich (K-L) equation was used to calculate the number of electrons transferred (n), expressed as

$$j^{-1} = j_K^{-1} + j_L^{-1} = j_K^{-1} + (B\omega^{0.5})^{-1}$$

where B is defined by the following equation:

$$B = 0.2nFC_0D_0^{2/3}v^{-1/6}$$

Here, j represents the measured limiting current density; j_K and j_L denote the kinetic and limiting diffusion current densities, respectively; ω indicates the angular velocity of the rotating disk electrode (RDE); n is the number of electron transfers; and F is Faraday's constant (F = 96,485 C mol^{-1}). Additionally, C_0 is the volume concentration of O_2 in

0.1 M KOH (C_0 = 1.2 × 10^{-6} mol cm^{-3}), D_0 is the diffusion coefficient of O_2, which is 1.9 × 10^{-5} cm^2 s^{-1}, and the kinematic viscosity of the electrolyte (γ) is 0.01 cm^2 s^{-1} in 0.1 M KOH. The rotation speed (rpm) is expressed using the constant 0.2.

3. Results and Discussion

3.1. Characterization of FeNC/MXene

Figure 1a presents the X-ray diffraction (XRD) image of the MAX phase and MXene. A distinct diffraction peak at 2θ = 38.4°, indexed as (104), is observed in the MAX phase, attributed to aluminum (Al), which must be fully removed to obtain the MXene [37] layer. The disappearance of the 38.4° peak after 24 h HF etching confirms the complete removal of aluminum and the formation of high-purity, layered MXene. Additionally, new diffraction peaks corresponding to the crystal planes (002), (004), and (006) appear at 2θ values of 8.6°, 17.9°, and 27.1°, respectively [36]. The diffraction pattern provides additional evidence for the successful conversion of Ti_3AlC_2 to Ti_3C_2. Notably, the (002) peak shifts to a lower angle in the MXene phase (2θ = 8.6°), likely due to the increased lattice spacing caused by the expansion of the Ti_3C_2 layers along the (002) plane, thus validating the characteristic accordion-like structure of MXene. As seen in Figure 1b, the XRD image of FeNC exhibits characteristic diffraction peaks corresponding to Vulcan XC72 at angles of 25° and 43° [38]. The XRD plots of FeNC/MXene-1 exhibit characteristic peaks at 27.3°, 35.9°, 41.1°, 43.9°, 54.1°, and 56.5°, which correspond to the (110), (101), (111), (210), (211), and (220) crystal facets of rutile TiO_2 [39]. Additionally, a weak peak at 25.1° corresponds to the (110) crystal facet. The absence of peaks corresponding to Ti_3C_2 in the XRD pattern of FeNC/MXene-1 may be attributed to the dominance of TiO_2 peaks, which hinder the detection of Ti_3C_2. In addition, no diffraction peaks corresponding to iron-based species were observed in the XRD patterns of either FeNC or FeNC/MXene-1, indicating that the iron species were atomically dispersed within the carbon matrix [40]. These Fe-N-C catalysts may interact with TiO_2 nanoparticles, resulting in a synergistic effect [41].

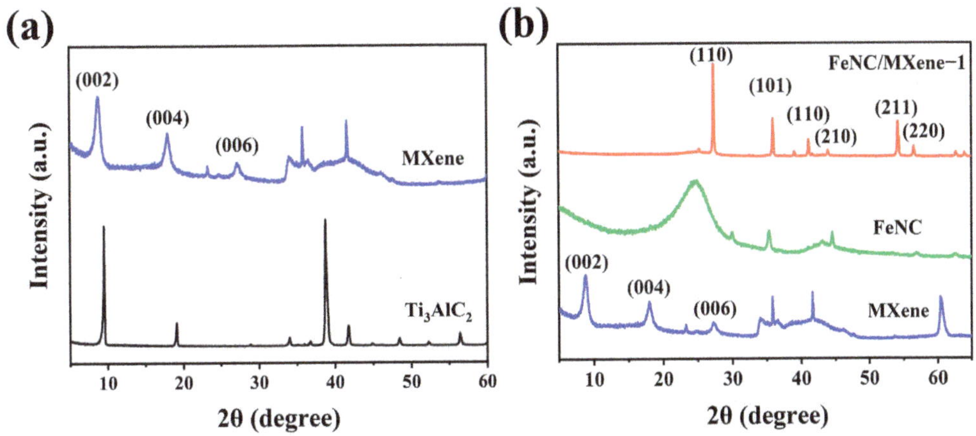

Figure 1. (**a**) XRD pattern of MXene and Ti_3AlC_2; (**b**) XRD pattern of FeNC/MXene-1, FeNC, and MXene.

Figure 2a illustrates the XRD plots of FeNC/MXene at various pyrolysis temperatures. At a pyrolysis temperature of 700 °C, MXene is converted to anatase TiO_2 due to the facile cleavage of the Ti-C bond in Ti_3C_2 at elevated temperatures, leading to the formation of TiO_2. Anatase TiO_2 exhibits a more open crystal structure, a relatively high specific surface area, an abundance of surface defects, and a rich presence of oxygen vacancies, all of which contribute to its high catalytic activity. When the pyrolysis temperature is elevated to 800 °C, anatase TiO_2 transforms into rutile TiO_2, with the peak intensity of rutile TiO_2 increasing further with higher pyrolysis temperatures, indicating the generation of additional TiO_2. Rutile TiO_2 possesses relatively good electrical conductivity, particularly at high temperatures, which enhances its stability in electrocatalytic reactions characterized by high voltage and current density. The anatase and rutile phases in mixed-phase TiO_2 exhibit a beneficial synergistic effect [42,43]. An optimal ratio of anatase to rutile enhances the mass transfer rate and conductivity of the catalytic layer [44]. As shown in Figure 2b, Raman spectroscopy was conducted to investigate FeNC/MXene at different pyrolysis temperatures. The Raman spectra (Figure 2b) show the D (1340 cm^{-1}) and G (1580 cm^{-1}) bands of carbon crystals, which are associated with defects in the carbon lattice and the vibrations of sp^2-hybridized carbon atoms, respectively. The intensity ratio (I_D/I_G) reflects the surface defects in the carbon film, with values of 1.03 for FeNC/MXene-700, 1.08 for FeNC/MXene-800, and 1.08 for FeNC/MXene-900, indicating that the carbon layer in FeNC/MXene-800 has a higher defect density [45,46].

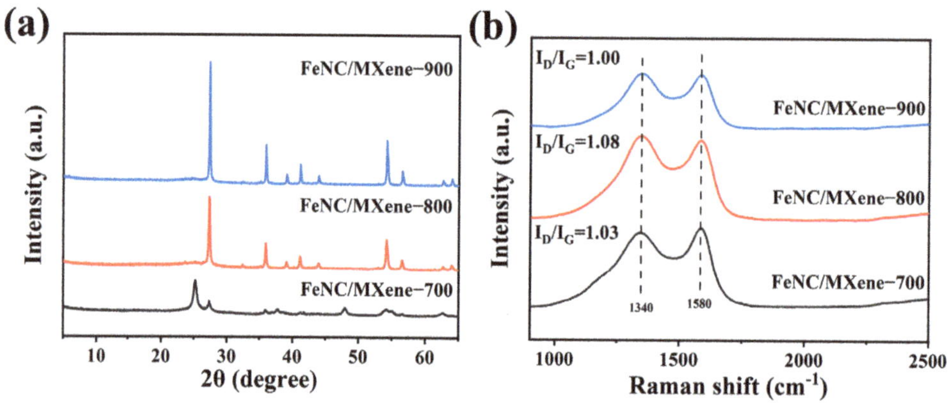

Figure 2. (**a**) XRD pattern and (**b**) Raman spectra of FeNC/MXene-700, FeNC/Mene-800, and FeNC/Mene-900.

The surface morphology of the MAX phase (MXene), FeNC, and FeNC/MXene was observed via scanning electron microscopy (SEM). As illustrated in Figure S1, the MAX phase exhibited closely stacked layers, whereas the MAX phase underwent a successful transformation into an accordion-like (see Figure 3b) MXene layer following 24 h of etching. Figure 3c depicts the FeNC structure without MXene, which exhibits agglomerated particles. Following the doping of MXene into FeNC at a mass ratio of 1:1, the surface of MXene was modified by FeNC particles, as illustrated in Figure 3d. This process retained the lamellar structure of MXene while simultaneously facilitating the dispersion of FeNC clusters, thereby improving the porosity of the material. The dispersion of FeNC clusters over time led to an improvement in porosity. Furthermore, the distribution of FeNC particles in the interlayer of MXene hindered lamellar stacking and enhanced mass transport, greatly improving ORR performance [47].

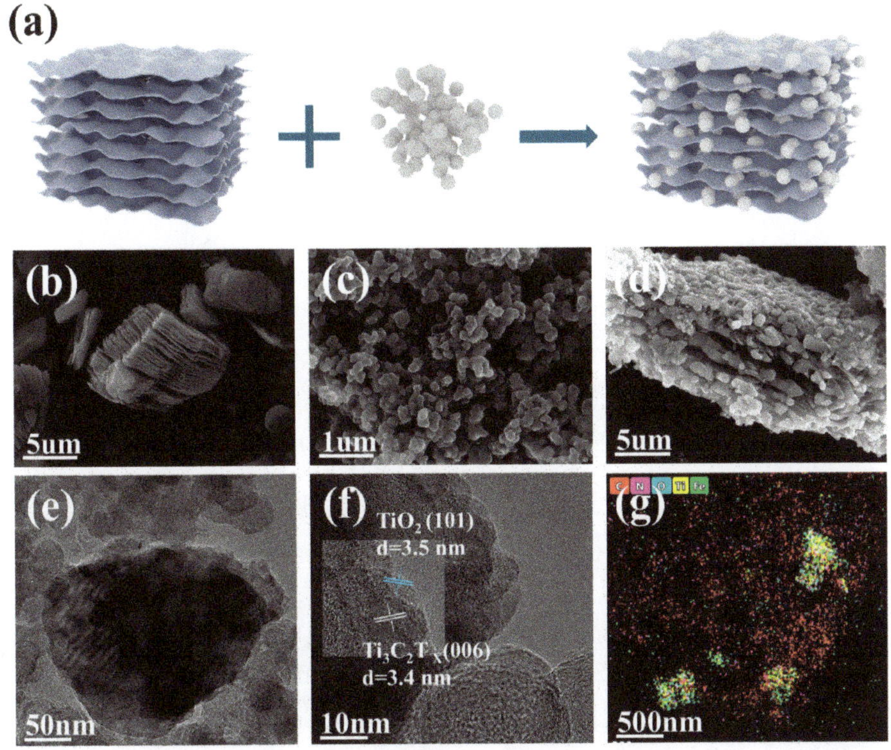

Figure 3. (**a**) Schematic illustration of FeNC/MXene hybridization; (**b**) SEM images of MXene, (**c**) FeNC, and (**d**) FeNC/MXene-1; (**e**,**f**) TEM images of FeNC/MXene-1; (**g**) overlay of the elemental mapping of the energy-scattered X-ray spectra of FeNC/MXene-1.

The figure shows a TEM image of a highly porous FeNC/MXene electrocatalyst (see Figure 3e,f). The TEM image further verifies that the MXene layer covers the FeNC, in agreement with the SEM image. It shows that the edges of FeNC/MXene appear more worn or rough, indicating an increase in porosity. The lattice stripe spacing of TiO_2 has been measured to be 0.35 nm, which corresponds to the (101) crystal plane, and the lattice stripe spacing of $Ti_3C_2T_x$ MXene has been measured to be 0.34 nm, which corresponds to the (006) crystal plane. Additionally, the lack of carbon lattice fringes confirms the amorphous nature of FeNC, which contributes to enhanced resistance to volume change during electrochemical reactions [48]. Furthermore, elemental mapping revealed the presence of C, N, O, Ti, and Fe (see Figure 3g), indicating that MXene and FeNC effectively hybridized to form a novel FeNC/MXene with a unique microstructure, in which the aggregation of metallic Fe atoms can be efficiently avoided during annealing due to the discrete nature of the Fe–phenanthroline complex [49], which is conducive to the formation of high-density and uniformly distributed Fe-N_X sites.

It is hypothesized that the enhanced activity may result from the interaction at the FeNC/MXene interface. The electronic structure of FeN_4 groups, before and after coupling with MXene, was examined using multi-spectral techniques. The X-ray photoelectron spectroscopy (XPS) measurement produced a spectrum confirming the presence of Ti, Fe, N, C, O, and F (see Figure S2). It can be inferred that the signals corresponding to O and F may originate from the hydroxyl and fluorine ends of MXene, respectively. In the pure FeNC spectrum, the $Fe2p_{3/2}$ and $Fe2p_{1/2}$ peaks in the XPS narrow scan are observed at

710.1 eV and 722.9 eV, respectively. Following coupling with MXene, both peaks shift to 711.8 eV and 725.4 eV, as illustrated in Figure 4a,b. Although the XPS signal of Fe2p is relatively weak, the noticeable shift in the Fe2p peak to higher binding energies suggests a strong interaction between MXene and the FeN_4 component, which significantly diminishes the local electronic density at the Fe center [35].

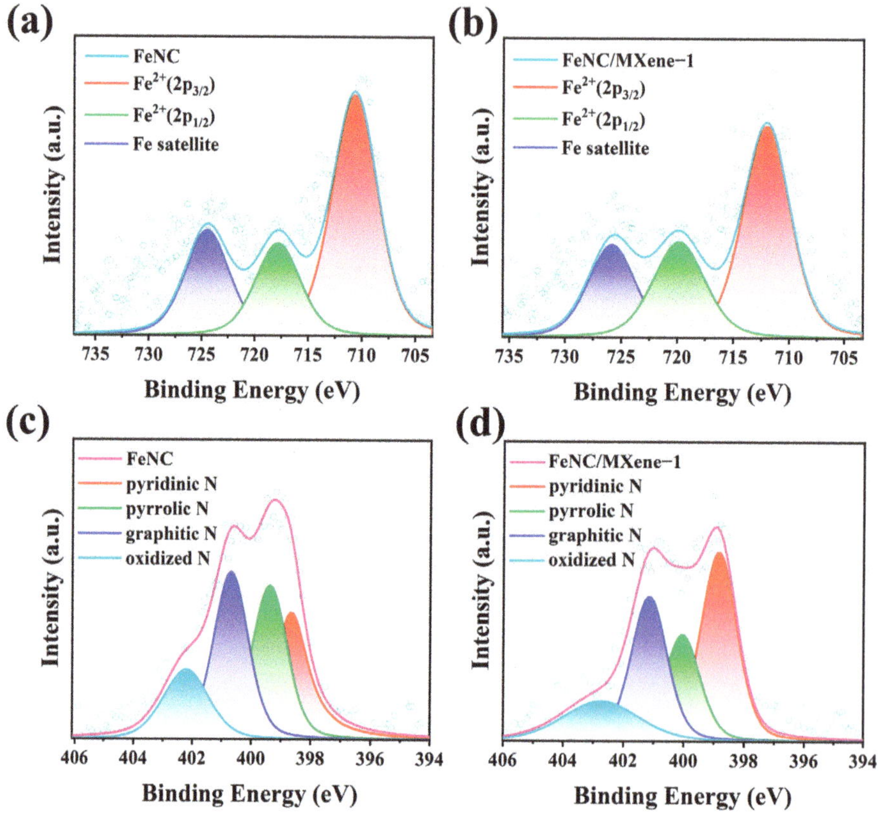

Figure 4. (**a**–**d**) XPS survey spectra and (**a**) high-resolution Fe2p scans of FeNC. (**b**) High-resolution Fe2p scan of FeNC/MXene-1. (**c**) High-resolution N1s scans of FeNC. (**d**) High-resolution N1s scans of FeNC/MXene-1.

Additionally, as depicted in Figure 4c, the N1s spectrum was fitted into four main N-related peaks along with supplementary MXene-N bonds, corresponding to pyridinic N, pyrrolic N, graphitic N, and oxidized N, which are observed at 400.5 eV, 399.4 eV, 398.3 eV, and 402.7 eV, respectively. As can be observed in Table 1, the proportions of pyridinic N, pyrrolic N, and graphitic N in FeNC/MXene are 31.95%, 25.15%, and 21.94%, respectively. The pyridinic N is connected to two carbon atoms and contributes two p-electrons to the aromatic system, while the graphitic N is bonded to three carbon atoms. The pyrrolic N-substituted nitrogen atom is bonded to two carbon atoms at the edge of the carbon plane and donates one p-electron, whereas the oxidized N-substituted nitrogen atom is bonded to two carbon atoms and one oxygen atom [50]. The four types of nitrogen species exhibit different effects on the ORR activity of the electrocatalyst, with pyridinic N showing the strongest contribution. It enhances electron-donating capability and facilitates O_2 adsorption, thereby improving catalytic performance. Notably, the enhanced ORR activity is primarily attributed to this effect [51]. Graphitic N provides a substantial number of electrocatalytic active centers for the ORR, thereby enhancing the limiting current density. Additionally, pyrrolic N sites contribute to improved ORR performance, although their impact on overall catalyst performance is minimal [52]. As shown in Table 1, the increase in the pyridinic N content of the FeNC/MXene catalysts indicates that the catalysts have improved limiting current densities and onset potentials (E_{onset}) toward ORR.

Table 1. Deconvoluted N1s atomic percentage of FeNC and FeNC/MXene-1.

Catalysts	Pyridinic N	Pyrrolic N	Graphite N	Oxygen N
FeNC	27.12%	27.15	29.56%	16.17%
FeNC/MXene-1	36.08%	20.29%	27.66%	15.97%

3.2. *Electrochemical Properties of FeNC/MXene*

The ORR performance of the synthesized catalyst FeNC/MXene was analyzed in an alkaline electrolyte (0.1 M KOH) with the rotating disk electrode (RDE) method. Initially, the electrocatalytic behavior of the catalyst was assessed in an electrolyte saturated with O_2 and N_2 through cyclic voltammetry (CV) cycling at a scan rate of 10 mV s^{-1}. As illustrated in Figure 5a, the absence of a redox peak in the N_2-saturated electrolyte supports the assertion that the electrocatalyst exhibits capacitive behavior [53]. Conversely, the O_2-saturated electrolyte exhibited a cathodic redox peak at a potential of 0.85 V for FeNC/MXene. A series of catalysts, designated as FeNC/MXene-1, FeNC/MXene-2, FeNC/MXene-3, and FeNC/MXene-4, were synthesized by adjusting the ratios of FeNC to MXene to 1:1, 1:2, 1:3, and 2:1, respectively. Subsequently, the ORR catalytic activity of these catalysts was compared with that of commercial Pt/C using linear sweep voltammetry (LSV). As shown in Figure 5b, single FeNC exhibits discernible ORR activity. The electrochemical activity of FeNC is altered when loaded on MXene, particularly at a 1:1 ratio, which demonstrates a significant enhancement in ORR activity. As shown in Figure 5c, a detailed comparison of the onset potential, half-wave potential, and limiting current density was conducted for the different catalysts. The FeNC/MXene-1 catalyst exhibits an onset potential of 0.987 V, comparable to that of commercial Pt/C (0.997 V), and a half-wave potential of 0.857 V, which even surpasses that of Pt/C (0.845 V). Moreover, compared to FeNC, FeNC/MXene-1 shows a significantly enhanced limiting current density, reaching 5.5 mA cm^{-2}. This is attributed to the synergistic combination of conductive Vulcan XC-72, the layered structure of MXene with high electronic conductivity, and the graphitized carbon surrounding Fe-N_x sites, which together provide efficient electron transfer pathways. These characteristics facilitate rapid charge transfer and mass diffusion. As shown in Figure S3, the ORR performance of the MXene substrate and metal-free N-C was evaluated using LSV. The MXene substrate exhibited minimal ORR activity due to the absence of active sites. Metal-free N-C showed moderate activity, attributed to the presence of pyridinic nitrogen. However, both materials displayed significantly lower ORR performance compared to the FeNC/MXene catalyst, highlighting the critical role of Fe-N_x sites in enhancing catalytic activity. The ORR kinetics of all samples were assessed using the Tafel slope, as shown in Figure 5d. The Tafel slope of the FeNC/MXene-1, measured at 67 mV dec^{-1}, is smaller than that of Pt/C (90 mV dec^{-1}) and FeNC (72 mV dec^{-1}), suggesting more favorable -oxygen reduction kinetics. These results indicate that the FeNC/MXene-1 catalyst exhibits superior ORR activity, likely due to its high concentration of FeN_x sites and its layered structure [54]. The ORR performance of FeNC/MXene was compared with other reported Fe-based catalysts (see Table S1 [55–64]). The results indicate that FeNC/MXene shows ORR performance similar to other Fe-based catalysts. To further examine the ORR kinetics, the catalysts were tested in a 0.1 M KOH O_2-saturated electrolyte using an LSV system at a scan rate of 10 mV s^{-1} and rotation speeds between 400 and 1600 rpm. As seen in Figure 5e, the limiting current density increases with the rotation speed [65]. The corresponding K-L plot, illustrated in Figure 5f, is nearly linear in three dimensions within the 0.1 mol L^{-1} KOH solution, confirming first-order reaction kinetics. Through the K-L equation, the value of n for the FeNC/MXene-1 catalyst is calculated to be around 4.0 within the potential range of 0.2 V–0.6 V, suggesting a direct four-electron transfer pathway for the ORR [66].

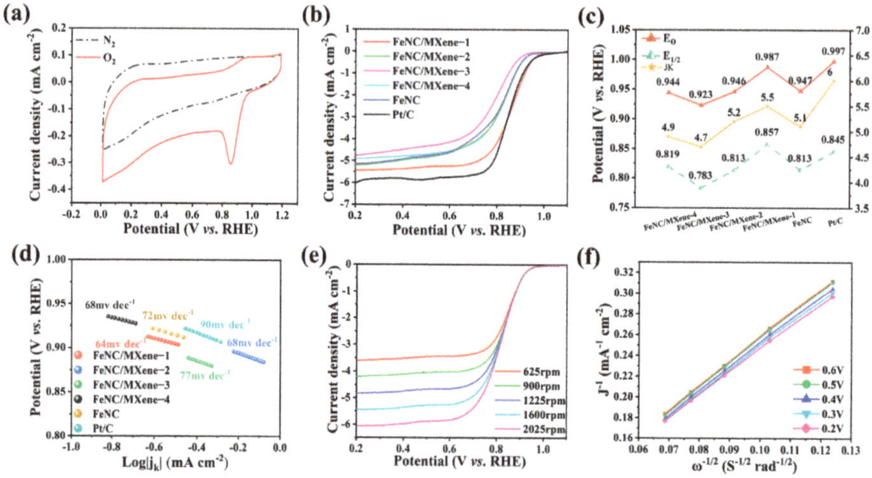

Figure 5. (**a**) CV curves of FeNC/MXene-1, (**b**) LSV curves of FeNC/MXene-1, FeNC/MXene-2, FeNC/MXene-3, FeNC/MXene-4, and Pt/C at 1600 rpm, (**c**) the corresponding E_{onset}, $E_{1/2}$, and J_k, (**d**) Tafel slope of the above catalyst, (**e**) LSV curves of FeNC/MXene-1 at various rotation speeds of the RDE, and (**f**) Koutecky-Levich (K-L) curves of FeNC/MXene-1 at various potentials.

The LSV test reveals that the FeNC/MXene composite at a 1:1 ratio exhibits the highest ORR activity in alkaline electrolytes, thereby establishing this ratio as optimal. Catalysts synthesized at this ratio, differentiated by distinct pyrolysis temperatures, are designated as FeNC/MXene-700, FeNC/MXene-800 (FeNC/MXene-1), and FeNC/MXene-900. The ORR performance of these catalysts was assessed through cyclic voltammetry (CV) curves under saturated nitrogen (N_2) and oxygen (O_2) conditions in a 0.1 mol L^{-1} KOH electrolyte. In the alkaline solution saturated with N_2, none of the catalysts displayed a cathodic ORR peak. In contrast, under O_2 saturation, a cathodic oxygen reduction peak was identified at a potential of 0.82 V for FeNC/MXene-700 (Figure S4a), while FeNC/MXene-900 demonstrated a peak at a slightly lower potential of 0.80 V (Figure S4b). The findings indicate that the FeNC/MXene-800 composite achieves the most positive ORR peak potential, suggesting an enhanced likelihood of O_2 reduction to H_2O [67].

Additionally, the ORR performance of the catalysts in the O_2-saturated 0.1 M KOH electrolyte at various pyrolysis temperatures was recorded via LSV at a scan rate of 10 mV s^{-1} and a rotation speed of 1600 rpm. As shown in Figure 6a, the onset potential (E_{onset}) and half-wave potential ($E_{1/2}$) for FeNC/MXene-700 are 0.933 V and 0.812 V, respectively, whereas those for FeNC/MXene-900 are 0.905 V and 0.778 V. These results are comparatively lower than those of FeNC/MXene-1. This indicates that maintaining a 1:1 mass ratio and performing pyrolysis at 800 °C effectively enhance the loading of FeNC onto the MXene substrate. Furthermore, the ORR kinetics of all samples were assessed using the Tafel slope, as shown in Figure 6b. The Tafel slope of FeNC/MXene-800 is lower than that of other pyrolysis temperatures, indicating better reaction kinetics. This enhancement arises from the conversion of MXene at pyrolysis temperatures of 800 °C, which results in a beneficial synergistic effect due to the coexistence of rutile and anatase TiO_2. This effect significantly improves the mass transfer rate and electrical conductivity of the catalytic layer [42,43].

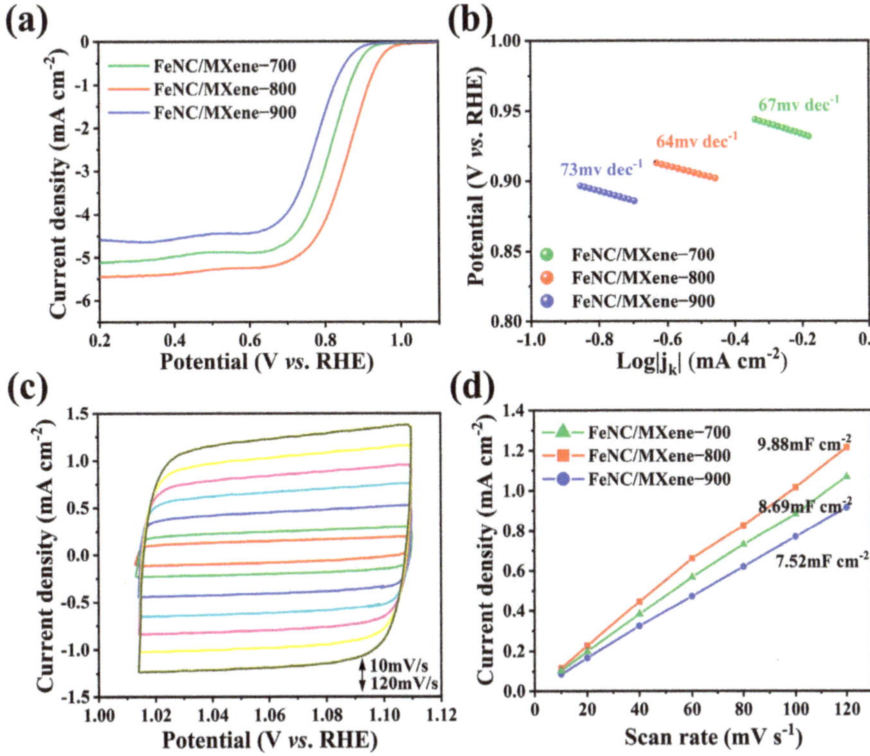

Figure 6. (**a**,**b**) The LSV curves and Tafel slope of FeNC/MXene-1 at various temperatures, (**c**) the ECSA curves of FeNC/MXene-1 at CV scan rates ranging from 10 mV s^{-1} to 120 mV s^{-1} (The different colors represent the corresponding scan rates), (**d**) the direct correlation between the recorded capacitive current and different scanning speeds.

The electrochemical surface area (ECSA) plays an important role in influencing electrocatalytic performance for the ORR. The ECSA was determined from the electrochemical double-layer capacitance (C_{dl}) measured during the CV cycling of the obtained electrocatalysts, namely, FeNC/MXene-700, FeNC/MXene-800, and FeNC/MXene-900, at different scan rates between 10 mV s^{-1} and 120 mV s^{-1} (see Figures 6c and S4c,d). The C_{dl} values were determined using the following equation:

$$\Delta j = j_a - j_c$$

$$C_{dl} = d(\Delta j)/2dv$$

$$ECSA = C_{dl}/C_s$$

here, j_a and j_c denote the currents measured at the anode and cathode during cyclic voltammetry in mA cm^{-2}, Δj refers to the current difference between the anode and cathode, and C_s represents the specific capacitance of the planar electrode, with units of μF cm^{-2}. The C_s value for non-precious metals typically ranges from 20 to 40 μF cm^{-2}. In this study, an average value of 40 μF cm^{-2} was used. The capacitance per unit area (C_{dl}) of FeNC/MXene-700, FeNC/MXene-800, and FeNC/MXene-900, as shown in Figure 6d, is

8.69 μF cm^{-2}, 9.88 μF cm^{-2}, and 7.52 μF cm^{-2}, respectively. The highest calculated ECSA was 247 m^2 g^{-1} for FeNC/MXene-800, followed by FeNC/MXene-700 (217 m^2 g^{-1}) and FeNC/MXene-900 (188 m^2 g^{-1}). These ECSA results align with the LSV measurements of the synthetic catalysts. As the pyrolysis temperature increases, Vulcan XC72 gradually eliminates the oxygen-containing functional groups on its surface, which enhances the electrical conductivity and hydrophobicity of the carbon material. At a pyrolysis temperature of 800 °C, the microporous structure of Vulcan XC72 significantly enhances the utilization efficiency of the MXene layer. When the pyrolysis temperature reaches 900 °C, the arrangement of the carbon layers begins to graphitize; however, at elevated temperatures, the micropores collapse, resulting in a decrease in specific surface area [68]. These ECSA results correlate well with the LSV measurements of the synthesized catalysts.

3.3. Stability and Durability of FeNC/MXene

In the context of actual operating conditions for a hydrogen fuel cell, a number of factors contribute to catalyst degradation. These include carbon corrosion, metal peeling, the protonation of active centers, and intermediate formation [69,70]. The durability of FeNC/MXene-1 and commercial Pt/C was evaluated through chronoamperometry, as depicted in Figure 7a. The FeNC/MXene-1 catalyst exhibited excellent stability, maintaining a current loss of only 9% after 20,000 s of operation at 0.6 V. This is a notable improvement over the current loss rate of 25% observed for the Pt/C catalyst. Apart from its stable performance, the FeNC/MXene-1 catalyst demonstrated remarkable methanol tolerance, retaining 91% of its initial current density after 1500 s of testing in the presence of 3M methanol (CH_3OH) for 300 s (see Figure 7b). In contrast, the commercial Pt/C catalyst exhibited a 29% reduction in current density. As conventional metals are inherently less tolerant of methanol interference, the good resistance to toxicity of FeNC/MXene can be attributed to the rich microporous carbon structure of Vulcan XC72, which enhances the smooth diffusion rate of the electrolyte. The experimental results demonstrate that FeNC/MXene-1 exhibits excellent methanol poisoning resistance and long-term cycling stability. The FeNC/MXene-1 catalyst has the potential to be a low-cost ORR electrocatalyst for the circular economy and sustainable hydrogen-fueled fuel cells and direct methanol fuel cells.

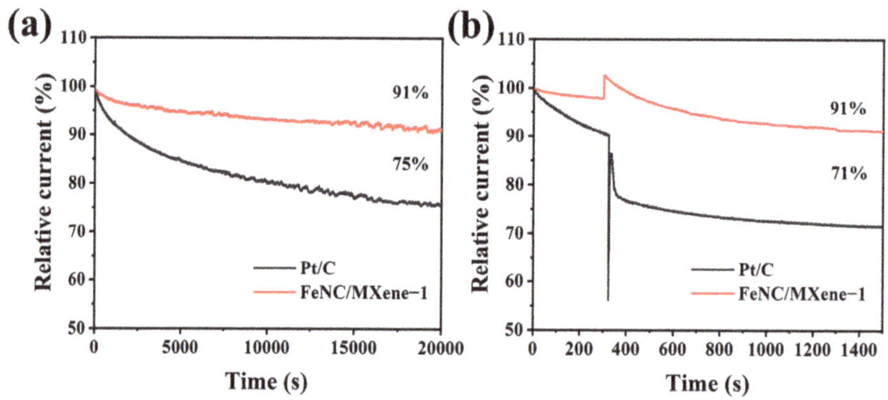

Figure 7. (**a**) Chronoamperometry curves of FeNC/MXene-1 and Pt/C were recorded at a constant potential of 0.6 V (vs. RHE) over a duration of 20,000 s. (**b**) Chronoamperometry curves of FeNC/MXene-1 and Pt/C at a constant voltage of 0.6 V (vs. RHE) for 1500 s after the addition of methanol.

4. Conclusions

In conclusion, oxygen reduction reaction activity was enhanced by anchoring iron–nitrogen–carbon (Fe-N_x-C) complexes on a Ti_3C_2-based MXene substrate. SEM and TEM analyses show that FeNC particles are uniformly dispersed across both the surface and interlayer of the MXene. XPS results reveal an increase in pyridine nitrogen concentration, which improves the catalyst's performance. In a 0.1 M KOH electrolyte, the FeNC/MXene catalyst exhibits a half-wave potential of 0.857 V, surpassing the 0.845 V of Pt/C catalysts. Additionally, the FeNC/MXene materials demonstrate excellent durability and methanol resistance compared to commercial Pt/C catalysts. The high catalytic activity and stability of FeNC/MXene provide a valuable reference for developing non-precious metal ORR catalysts and expanding the use of Pt/C in alternative fuel cells.

Supplementary Materials: The following supporting information can be downloaded at https://www.mdpi.com/article/10.3390/inorganics13060188/s1, Figure S1: SEM of MAX; Figure S2: XPS survey spectra of FeNC/MXene-1; Figure S3: LSV curves of MXene and metal-free N-C at 1600 rpm; Figure S4: (a,b) CV curves of FeNC/Mene-700 and FeNC/Mene-900, (c,d) ECSA curves at different CV scan rate of 10 mV/s to 120 mV/s of FeNC/Mene-700 and FeNC/Mene-900. Table S1: Performance comparison of ORR activities for recently reported Fe-based catalysts. References [55–64] are cited in the supplementary materials.

Author Contributions: Conceptualization, H.Z.; Methodology, H.Z.; Software, H.Z.; Validation, H.Z.; Formal analysis, F.W. and W.S.; Investigation, H.Z.; Data curation, F.W.; Writing—original draft, H.Z.; Writing—review & editing, F.W. and W.S.; Supervision, F.W.; Project administration, W.S.; Funding acquisition, F.W. All authors have read and agreed to the published version of the manuscript.

Funding: This work was supported by the Natural Science Foundation of Shandong Province (ZR2020ME041, ZR2022QB173, ZR2023MB054), Joint Zibo-SDUT Fund (2019ZBXC358), and the Foundation of State Key Laboratory of Biobased Material and Green Papermaking, Qilu University of Technology (Shandong Academy of Sciences) (KF2019-06).

Institutional Review Board Statement: Not applicable.

Informed Consent Statement: Not applicable.

Data Availability Statement: All data generated or analyzed during this study are included in this published article and its ESI.

Conflicts of Interest: The authors declare that they have no known competing financial interests or personal relationships that could have appeared to influence the work reported in this paper.

References

1. Li, Z.; He, H.; Cao, H.; Sun, S.; Diao, W.; Gao, D.; Lu, P.; Zhang, S.; Guo, Z.; Li, M.; et al. Atomic Co/Ni dual sites and Co/Ni alloy nanoparticles in N-doped porous Janus-like carbon frameworks for bifunctional oxygen electrocatalysis. *Appl. Catal. B Environ.* **2019**, *240*, 112–121. [CrossRef]
2. Eftekhari, A.; Fang, B. Electrochemical hydrogen storage: Opportunities for fuel storage, batteries, fuel cells, and supercapacitors. *Int. J. Hydrogen Energy* **2017**, *42*, 25143–25165. [CrossRef]
3. Wang, X.; Yu, M.; Feng, X. Electronic structure regulation of noble metal-free materials toward alkaline oxygen electrocatalysis. *eScience* **2023**, *3*, 100141. [CrossRef]
4. Yang, X.; Lin, L.; Guo, X.; Zhang, S. Design of Multifunctional Electrocatalysts for ORR/OER/HER/HOR: Janus Makes Difference. *Small* **2024**, *20*, e2404000. [CrossRef]
5. Yang, Y.; Li, P.; Zheng, X.; Sun, W.; Dou, S.X.; Ma, T.; Pan, H. Anion-exchange membrane water electrolyzers and fuel cells. *Chem. Soc. Rev.* **2022**, *51*, 9620–9693. [CrossRef]
6. Zeng, Q.; Chang, S.; Wang, M.; Li, M.; Deng, Q.; Xiong, Z.; Zhou, B.; Liu, Y. Highly-active, metal-free, carbon-based ORR cathode for efficient organics removal and electricity generation in a PFC system. *Chin. Chem. Lett.* **2021**, *32*, 2212–2216. [CrossRef]
7. Bhoyate, S.D.; Kim, J.; de Souza, F.M.; Lin, J.; Lee, E.; Kumar, A.; Gupta, R.K. Science and engineering for non-noble-metal-based electrocatalysts to boost their ORR performance: A critical review. *Coord. Chem. Rev.* **2023**, *474*, 214854. [CrossRef]

8. Chen, D.; Yuan, J.; Zhang, J.; Li, G.; Tian, L.; Wang, Z.; Hu, J.; Zhang, Y.; Li, C.; Yu, J.; et al. Construction of atom-scale Co/Fe/Ni M-N-C active sites within nitrogen-doped carbon spheres for high-performance bifunctional oxygen catalysts in rechargeable zinc-air batteries. *J. Energy Storage* **2024**, *101*, 113772. [CrossRef]
9. Ye, C.; Zhang, L.; Shen, Y. Activity Origin and Catalytic Mechanism of the M–N–C Catalysts for the Oxygen Reduction Reaction. *ACS Mater. Lett.* **2024**, *6*, 2858–2887. [CrossRef]
10. Liu, J.; Wan, X.; Liu, S.; Liu, X.; Zheng, L.; Yu, R.; Shui, J. Hydrogen Passivation of M–N–C (M = Fe, Co) Catalysts for Storage Stability and ORR Activity Improvements. *Adv. Mater.* **2021**, *33*, e2103600. [CrossRef]
11. Li, X.; Wang, D.; Zha, S.; Chu, Y.; Pan, L.; Wu, M.; Liu, C.; Wang, W.; Mitsuzaki, N.; Chen, Z. Active sites identification and engineering of M-N-C electrocatalysts toward oxygen reduction reaction. *Int. J. Hydrogen Energy* **2024**, *51*, 1110–1127. [CrossRef]
12. Liu, H.; Zhou, Y.; Zhu, S.; Zheng, L.; Cui, Z.; Li, Z.; Wu, S.; Ma, L.; Liang, Y. Highly durable Cu–N–C active sites towards efficient oxygen reduction for zinc-air battery: Carbon matrix effect, reaction mechanism and pathways. *J. Alloys Compd.* **2021**, *857*, 158321. [CrossRef]
13. Zhao, X.; Liu, X.; Huang, B.; Wang, P.; Pei, Y. Hydroxyl group modification improves the electrocatalytic ORR and OER activity of graphene supported single and bi-metal atomic catalysts (Ni, Co, and Fe). *J. Mater. Chem. A* **2019**, *7*, 24583–24593. [CrossRef]
14. Wan, X.; Liu, Q.; Liu, J.; Liu, S.; Liu, X.; Zheng, L.; Shang, J.; Yu, R.; Shui, J. Iron atom–cluster interactions increase activity and improve durability in Fe–N–C fuel cells. *Nat. Commun.* **2022**, *13*, 2963. [CrossRef]
15. Dong, X.; Wang, J.; Wang, X.; Xu, J.; Yang, J.; Zeng, W.; Zhang, Y.; Huang, G.; Wang, J.; Pan, F. Prussian Blue Analogue-Derived Bimetallic CoFe@NC as Effective and Extremely Stable Oxygen Reduction Electrocatalysts for Mg-Air Battery. *ACS Appl. Energy Mater.* **2022**, *5*, 12272–12282. [CrossRef]
16. Guo, S.; Lian, J.; Zhao, J.; Zhang, Y.; Wang, X. Se Atom Doping Enhances the Catalytic Activity of Co@NC for Oxygen Reduction Reaction. *Energy Technol.* **2023**, *11*, 2300205. [CrossRef]
17. Wang, H.; Cao, L.; Feng, Y.; Chen, J.; Feng, W.; Luo, T.; Hu, Y.; Yuan, C.; Zhao, Y.; Zhao, Y.; et al. Facile synthesis of defect-rich Fe-N-C hybrid from fullerene/ferrotetraphenylporphyrin as efficient oxygen reduction electrocatalyst for Zn-air battery. *Chin. Chem. Lett.* **2023**, *34*, 107601. [CrossRef]
18. Zhang, F.; Liu, X.; Chen, Y.; Tian, M.; Yang, T.; Zhang, J.; Gao, S. Ordered mesoporous carbon fiber bundles with high-density and accessible Fe-NX active sites as efficient ORR catalysts for Zn-air batteries. *Chin. Chem. Lett.* **2023**, *34*, 108142. [CrossRef]
19. Liu, K.; Fu, J.; Luo, T.; Ni, G.; Li, H.; Zhu, L.; Wang, Y.; Lin, Z.; Sun, Y.; Cortés, E.; et al. Potential-Dependent Active Moiety of Fe–N–C Catalysts for the Oxygen Reduction Reaction. *J. Phys. Chem. Lett.* **2023**, *14*, 3749–3756. [CrossRef] [PubMed]
20. Qiu, X.; Wang, Y.; Shao, M. High-performance and durable Fe–N–C fuel cell catalysts. *Joule* **2024**, *8*, 881–882. [CrossRef]
21. Zhou, Y.; Lu, R.; Tao, X.; Qiu, Z.; Chen, G.; Yang, J.; Zhao, Y.; Feng, X.; Müllen, K. Boosting Oxygen Electrocatalytic Activity of Fe–N–C Catalysts by Phosphorus Incorporation. *J. Am. Chem. Soc.* **2023**, *145*, 3647–3655. [CrossRef]
22. Ma, Q.; Jin, H.; Zhu, J.; Li, Z.; Xu, H.; Liu, B.; Zhang, Z.; Ma, J.; Mu, S. Stabilizing Fe–N–C Catalysts as Model for Oxygen Reduction Reaction. *Adv. Sci.* **2021**, *8*, e2102209. [CrossRef]
23. Qin, G.; Sun, S.; Zhang, X.; Han, Z.; Li, Y.; Han, G.; Li, Y.; Zhu, S. Millisecond activity modulation of atomically-dispersed Fe–N–C catalysts. *Energy Storage Mater.* **2024**, *69*, 103421. [CrossRef]
24. Peng, P.; Shi, L.; Huo, F.; Mi, C.; Wu, X.; Zhang, S.; Xiang, Z. A pyrolysis-free path toward superiorly catalytic nitrogen-coordinated single atom. *Sci. Adv.* **2019**, *5*, eaaw2322. [CrossRef]
25. Wang, T.; Sun, C.; Yan, Y.; Li, F. Understanding the active sites of Fe–N–C materials and their properties in the ORR catalysis system. *RSC Adv.* **2022**, *12*, 9543–9549. [CrossRef]
26. Ahmad Junaidi, N.H.; Wong, W.Y.; Loh, K.S.; Rahman, S.; Choo, T.F.; Wu, B. Enhanced oxygen reduction reaction catalyst stability and durability of MXene-supported Fe-N-C catalyst for proton exchange membrane fuel cell application. *J. Alloys Compd.* **2023**, *968*, 171898. [CrossRef]
27. Gu, W.; Wu, M.; Xu, J.; Zhao, T. MXene boosted metal-organic framework-derived Fe–N–C as an efficient electrocatalyst for oxygen reduction reactions. *Int. J. Hydrogen Energy* **2022**, *47*, 17224–17232. [CrossRef]
28. Lisheshar, I.W.; Rouhi, S.; Ay, F.; Kosku Perkgöz, N. High-performance supercapacitors based on nonfunctionalized MXenes. *J. Power Sources* **2025**, *628*, 235894. [CrossRef]
29. Wu, M.; Fan, X.; Zhang, W.; Chen, B.; Ye, T.; Zhang, Q.; Fang, Y.; Wang, Y.; Tang, Y. Highly dispersed Ru nanospecies on N-doped carbon/MXene composite for highly efficient alkaline hydrogen evolution. *Chin. Chem. Lett.* **2024**, *35*, 109258. [CrossRef]
30. Cheng, H.; Liu, Y.; Cheng, Z.; Wang, X.; Huang, N.; Zhang, H. Ti3C2Tx MXene wrapped, carbon-coated porous Si sheets for improved lithium storage performance. *Chin. Chem. Lett.* **2024**, *35*, 108923. [CrossRef]
31. Tyagi, N.; Bhardwaj, V.; Moka, S.; Singh, M.K.; Khanuja, M.; Sharma, G. An overview on synthesis of MXene and MXene based nanocomposites for supercapacitors. *Mater. Today Commun.* **2024**, *41*, 110223. [CrossRef]
32. Zhao, L.; Dong, B.; Li, S.; Zhou, L.; Lai, L.; Wang, Z.; Zhao, S.; Han, M.; Gao, K.; Lu, M.; et al. Interdiffusion Reaction-Assisted Hybridization of Two-Dimensional Metal–Organic Frameworks and Ti3C2Tx Nanosheets for Electrocatalytic Oxygen Evolution. *ACS Nano* **2017**, *11*, 5800–5807. [CrossRef]

33. Ran, J.; Gao, G.; Li, F.-T.; Ma, T.-Y.; Du, A.; Qiao, S.-Z. Ti3C2 MXene co-catalyst on metal sulfide photo-absorbers for enhanced visible-light photocatalytic hydrogen production. *Nat. Commun.* **2017**, *8*, 13907. [CrossRef]
34. Peng, X.; Zhao, S.; Mi, Y.; Han, L.; Liu, X.; Qi, D.; Sun, J.; Liu, Y.; Bao, H.; Zhuo, L.; et al. Trifunctional Single-Atomic Ru Sites Enable Efficient Overall Water Splitting and Oxygen Reduction in Acidic Media. *Small* **2020**, *16*, e2002888. [CrossRef]
35. Li, Z.; Zhuang, Z.; Lv, F.; Zhu, H.; Zhou, L.; Luo, M.; Zhu, J.; Lang, Z.; Feng, S.; Chen, W.; et al. The Marriage of the FeN4 Moiety and MXene Boosts Oxygen Reduction Catalysis: Fe 3d Electron Delocalization Matters. *Adv. Mater.* **2018**, *30*, e1803220. [CrossRef]
36. Alhabeb, M.; Maleski, K.; Anasori, B.; Lelyukh, P.; Clark, L.; Sin, S.; Gogotsi, Y. Guidelines for Synthesis and Processing of Two-Dimensional Titanium Carbide ($Ti_3C_2T_x$ MXene). *Chem. Mater.* **2017**, *29*, 7633–7644. [CrossRef]
37. Mahmood, M.; Rasheed, A.; Ayman, I.; Rasheed, T.; Munir, S.; Ajmal, S.; Agboola, P.O.; Warsi, M.F.; Shahid, M. Synthesis of Ultrathin MnO_2 Nanowire-Intercalated 2D-MXenes for High-Performance Hybrid Supercapacitors. *Energy Fuels* **2021**, *35*, 3469–3478. [CrossRef]
38. Pinheiro, V.S.; Paz, E.C.; Aveiro, L.R.; Parreira, L.S.; Souza, F.M.; Camargo, P.H.C.; Santos, M.C. Ceria high aspect ratio nanostructures supported on carbon for hydrogen peroxide electrogeneration. *Electrochim. Acta* **2018**, *259*, 865–872. [CrossRef]
39. Sabry, R.S.; Al-Haidarie, Y.K.; Kudhier, M.A. Synthesis and photocatalytic activity of TiO2 nanoparticles prepared by sol–gel method. *J. Sol-Gel Sci. Technol.* **2016**, *78*, 299–306. [CrossRef]
40. Muhyuddin, M.; Berretti, E.; Mirshokraee, S.A.; Orsilli, J.; Lorenzi, R.; Capozzoli, L.; D'Acapito, F.; Murphy, E.; Guo, S.; Atanassov, P.; et al. Formation of the active site structures during pyrolysis transformation of Fe-phthalocyanine into Fe-Nx-C electrocatalysts for the oxygen reduction reaction. *Appl. Catal. B Environ.* **2024**, *343*, 123515. [CrossRef]
41. Zhang, J.; Zhang, X.; Yue, W. Boosting electrocatalytic oxygen reduction performance of CoNC catalysts on Ti_3C_2 MXene by the synergistic effect with oxygen vacancy-rich TiO_2. *Chem. Eng. J.* **2023**, *456*, 141101. [CrossRef]
42. Li, J.; Zhou, H.; Zhuo, H.; Wei, Z.; Zhuang, G.; Zhong, X.; Deng, S.; Li, X.; Wang, J. Oxygen vacancies on TiO_2 promoted the activity and stability of supported Pd nanoparticles for the oxygen reduction reaction. *J. Mater. Chem. A* **2018**, *6*, 2264–2272. [CrossRef]
43. Pei, D.-N.; Gong, L.; Zhang, A.-Y.; Zhang, X.; Chen, J.-J.; Mu, Y.; Yu, H.-Q. Defective titanium dioxide single crystals exposed by high-energy {001} facets for efficient oxygen reduction. *Nat. Commun.* **2015**, *6*, 8696. [CrossRef]
44. Shi, F.; Baker, L.R.; Hervier, A.; Somorjai, G.A.; Komvopoulos, K. Tuning the Electronic Structure of Titanium Oxide Support to Enhance the Electrochemical Activity of Platinum Nanoparticles. *Nano Lett.* **2013**, *13*, 4469–4474. [CrossRef]
45. Xie, X.; Zhao, M.-Q.; Anasori, B.; Maleski, K.; Ren, C.E.; Li, J.; Byles, B.W.; Pomerantseva, E.; Wang, G.; Gogotsi, Y. Porous heterostructured MXene/carbon nanotube composite paper with high volumetric capacity for sodium-based energy storage devices. *Nano Energy* **2016**, *26*, 513–523. [CrossRef]
46. Yu, L.; Hu, L.; Anasori, B.; Liu, Y.-T.; Zhu, Q.; Zhang, P.; Gogotsi, Y.; Xu, B. MXene-Bonded Activated Carbon as a Flexible Electrode for High-Performance Supercapacitors. *ACS Energy Lett.* **2018**, *3*, 1597–1603. [CrossRef]
47. Li, X.; Chen, Y.; Zhao, Z.; Huang, Y.; Zhu, K.; Feng, S.; Luo, S.; Wang, L.; Yan, K. A single-atom Fe–N_2 embedded in nitrogen-doped porous carbon as a bifunctional photocatalyst for efficient removal of marine petroleum pollutants. *Environ. Funct. Mater.* **2023**, *2*, 66–75. [CrossRef]
48. Chang, K.; Chen, W. Single-layer MoS_2/graphene dispersed in amorphous carbon: Towards high electrochemical performances in rechargeable lithium ion batteries. *J. Mater. Chem.* **2011**, *21*, 17175–17184. [CrossRef]
49. Zhao, L.; Zhang, Y.; Huang, L.-B.; Liu, X.-Z.; Zhang, Q.-H.; He, C.; Wu, Z.-Y.; Zhang, L.-J.; Wu, J.; Yang, W.; et al. Cascade anchoring strategy for general mass production of high-loading single-atomic metal-nitrogen catalysts. *Nat. Commun.* **2019**, *10*, 1278. [CrossRef] [PubMed]
50. Kaur, P.; Verma, G.; Sekhon, S.S. Biomass derived hierarchical porous carbon materials as oxygen reduction reaction electrocatalysts in fuel cells. *Prog. Mater. Sci.* **2019**, *102*, 1–71. [CrossRef]
51. Nie, Y.; Xie, X.; Chen, S.; Ding, W.; Qi, X.; Wang, Y.; Wang, J.; Li, W.; Wei, Z.; Shao, M. Towards Effective Utilization of Nitrogen-Containing Active Sites: Nitrogen-doped Carbon Layers Wrapped CNTs Electrocatalysts for Superior Oxygen Reduction. *Electrochim. Acta* **2016**, *187*, 153–160. [CrossRef]
52. Bao, X.; Nie, X.; von Deak, D.; Biddinger, E.J.; Luo, W.; Asthagiri, A.; Ozkan, U.S.; Hadad, C.M. A First-Principles Study of the Role of Quaternary-N Doping on the Oxygen Reduction Reaction Activity and Selectivity of Graphene Edge Sites. *Top. Catal.* **2013**, *56*, 1623–1633. [CrossRef]
53. Bera, B.; Chakraborty, A.; Kar, T.; Leuaa, P.; Neergat, M. Density of States, Carrier Concentration, and Flat Band Potential Derived from Electrochemical Impedance Measurements of N-Doped Carbon and Their Influence on Electrocatalysis of Oxygen Reduction Reaction. *J. Phys. Chem. C* **2017**, *121*, 20850–20856. [CrossRef]
54. Cao, L.; Zhang, M.; Su, K.; Gu, F.; Li, Z. Fe–Nx–C sites decorated porous carbon nanorods with huge specific surface area boost oxygen reduction catalysis for Zn-air batteries. *J. Alloys Compd.* **2021**, *868*, 159015. [CrossRef]

55. Chen, L.; Wang, X.; Lu, T.; Pang, H.; Zhang, S.; Xu, L.; Yang, G.; Zhou, Q.; Tang, Y. Maneuver of the d-Orbital Spin Polarization of Fe Single Atomic Sites via a Near-Range Axial Ligand Effect for Boosting Oxygen Reduction Reaction. *Appl. Catal. B* **2025**, *366*, 125007. [CrossRef]
56. Wu, D.; Zhuo, Z.; Song, Y.; Rao, P.; Luo, J.; Li, J.; Deng, P.; Yang, J.; Wu, X.; Tian, X. Synergistic Spin-Valence Catalysis Mechanism in Oxygen Reduction Reactions on Fe-N-C Single-Atom Catalysts. *J. Mater. Chem. A* **2023**, *11*, 13502–13509. [CrossRef]
57. Xie, H.; Du, B.; Huang, X.; Zeng, D.; Meng, H.; Lin, H.; Li, W.; Asefa, T.; Meng, Y. High Density Single Fe Atoms on Mesoporous N-Doped Carbons: Noble Metal-Free Electrocatalysts for Oxygen Reduction Reaction in Acidic and Alkaline Media. *Small* **2023**, *19*, 2303214. [CrossRef]
58. He, Y.; Yang, X.; Li, Y.; Liu, L.; Guo, S.; Shu, C.; Liu, F.; Liu, Y.; Tan, Q.; Wu, G. Atomically Dispersed Fe-Co Dual Metal Sites as Bifunctional Oxygen Electrocatalysts for Rechargeable and Flexible Zn–Air Batteries. *ACS Catal.* **2022**, *12*, 1216–1227. [CrossRef]
59. Karimi-Maleh, H.; Karaman, C.; Karaman, O.; Karimi, F.; Vasseghian, Y.; Fu, L.; Baghayeri, M.; Rouhi, J.; Kumar, P.S.; Show, P.-L.; et al. Nanochemistry Approach for the Fabrication of Fe and N Co-Decorated Biomass-Derived Activated Carbon Frameworks: A Promising Oxygen Reduction Reaction Electrocatalyst in Neutral Media. *J. Nanostruct. Chem.* **2022**, *12*, 429–439. [CrossRef]
60. Li, H.; Wen, Y.; Jiang, M.; Yao, Y.; Zhou, H.; Huang, Z.; Li, J.; Jiao, S.; Kuang, Y.; Luo, S. Understanding of Neighboring Fe-N4-C and Co–N4–C Dual Active Centers for Oxygen Reduction Reaction. *Adv. Funct. Mater.* **2021**, *31*, 2011289. [CrossRef]
61. Li, Y.; Ding, Y.; Zhang, B.; Huang, Y.; Qi, H.; Das, P.; Zhang, L.; Wang, X.; Wu, Z.-S.; Bao, X. N,O Symmetric Double Coordination of an Unsaturated Fe Single-Atom Confined within a Graphene Framework for Extraordinarily Boosting Oxygen Reduction in Zn–Air Batteries. *Energy Environ. Sci.* **2023**, *16*, 2629–2636. [CrossRef]
62. Sheng, J.; Zhu, S.; Jia, G.; Liu, X.; Li, Y. Carbon Nanotube Supported Bifunctional Electrocatalysts Containing Iron–Nitrogen–Carbon Active Sites for Zinc–Air Batteries. *Nano Res.* **2021**, *14*, 4541–4547. [CrossRef]
63. Yang, Q.; Liu, R.; Pan, Y.; Cao, Z.; Zuo, J.; Qiu, F.; Yu, J.; Song, H.; Ye, Z.; Zhang, S. Ultrahigh-Loaded Fe Single Atoms and Fe3C Nanoparticle Catalysts as Air Cathodes for High-Performance Zn–Air Batteries. *ACS Appl. Mater. Interfaces* **2023**, *15*, 5720–5731. [CrossRef]
64. Wu, Y.; Liu, J.; Sun, Q.; Chen, J.; Zhu, X.; Abazari, R.; Qian, J. Molecular Catalyst of Fe Phthalocyanine Loaded into In-Based MOF-Derived Defective Carbon Nanoflowers for Oxygen Reduction. *Chem. Eng. J.* **2024**, *483*, 149243. [CrossRef]
65. Li, T.; Lu, Y.; Zhao, S.; Gao, Z.-D.; Song, Y.-Y. Co_3O_4-doped Co/CoFe nanoparticles encapsulated in carbon shells as bifunctional electrocatalysts for rechargeable Zn–Air batteries. *J. Mater. Chem. A* **2018**, *6*, 3730–3737. [CrossRef]
66. Liang, Y.; Wang, H.; Diao, P.; Chang, W.; Hong, G.; Li, Y.; Gong, M.; Xie, L.; Zhou, J.; Wang, J.; et al. Oxygen Reduction Electrocatalyst Based on Strongly Coupled Cobalt Oxide Nanocrystals and Carbon Nanotubes. *J. Am. Chem. Soc.* **2012**, *134*, 15849–15857. [CrossRef]
67. Han, X.; Yu, C.; Yang, J.; Zhao, C.; Huang, H.; Liu, Z.; Ajayan, P.M.; Qiu, J. Mass and Charge Transfer Coenhanced Oxygen Evolution Behaviors in CoFe-Layered Double Hydroxide Assembled on Graphene. *Adv. Mater. Interfaces* **2016**, *3*, 1500782. [CrossRef]
68. Öztürk, A.; Bayrakçeken Yurtcan, A. Preparation and characterization of melamine-led nitrogen-doped carbon blacks at different pyrolysis temperatures. *J. Solid State Chem.* **2021**, *296*, 121972. [CrossRef]
69. He, C.; Xia, C.; Li, F.-M.; Zhang, J.; Guo, W.; Xia, B.Y. Rational Design of Oxygen Species Adsorption on Nonnoble Metal Catalysts for Two-Electron Oxygen Reduction. *Adv. Energy Mater.* **2024**, *14*, 2303233. [CrossRef]
70. Song, W.; Ren, Z.; Chen, S.-Y.; Meng, Y.; Biswas, S.; Nandi, P.; Elsen, H.A.; Gao, P.-X.; Suib, S.L. Ni- and Mn-Promoted Mesoporous Co_3O_4: A Stable Bifunctional Catalyst with Surface-Structure-Dependent Activity for Oxygen Reduction Reaction and Oxygen Evolution Reaction. *ACS Appl. Mater. Interfaces* **2016**, *8*, 20802–20813. [CrossRef] [PubMed]

inorganics

MDPI

Article

First-Principles Calculations for the H Adsorption of Monolayer $MoTe_2$ for Hydrogen Evolution Reaction

Xujing Gao and Jianling Meng *

College of Mathematics and Physics, Beijing University of Chemical Technology, Beijing 100029, China; 2022050313@buct.edu.cn
* Correspondence: mengjianling@buct.edu.cn

Abstract: Hydrogen from water splitting is seen as a promising future energy source. Pt electrochemical catalysts with an ideal hydrogen evolution reaction (HER) performance face problems relating to their cost and scarcity. Research into transition metal dichalcogenides (TMDs) as alternative catalysts is in demand. In our work, H adsorption on monolayer $MoTe_2$ is investigated at different sites and rates. Through structure and charge distribution analysis, it is found that uniform charge distribution facilitates H adsorption. In addition, the enhanced electronic density of states and reduced band gap calculated by the electronic energy band structure are advantageous for H adsorption. And the Mo edge of $MoTe_2$ is sensitive to the H adsorption rate. Finally, the H adsorbed on the sites is stable at 600 K, as shown in molecular dynamics (MD) calculations. Our work provides a further mechanism for H adsorption on $MoTe_2$.

Keywords: DFT calculation; HER; 2D materials

1. Introduction

Hydrogen as an energy carrier is seen as an alternative renewable energy due to its environmentally friendly and sustainable advantages over fossil fuels [1]. Electrical-energy driven water splitting has been used to produce hydrogen [2]. Pt is well-known as the most efficient electrocatalyst for HER with a near-zero overpotential, a remarkable exchange current density, and a small Tafel slope [3]. Its high cost and inadequate reserves make Pt less competitive in practical applications. Over the past few years, great efforts have been devoted to developing an alternative electrocatalyst [4]. $MoTe_2$ is a promising catalyst for the hydrogen evolution reaction (HER) due to its excellent mobility (~4000 cm^2/V s) and conductivity (0.2/Ω cm) [5]. Monolayer $MoTe_2$ can improve the charge transfer for HER due to its atomic thickness. Moreover, the prominent chemical stability and robustness of $MoTe_2$ during electrochemical reactions are advantages for HER electrocatalysts [6].

The edge sites of $MoTe_2$ are active catalytic sites while the basal planes of $MoTe_2$ are inert for catalytic activity. So, great effort has been put into activating the catalytic sites. Experimentally, the HER catalytic capability of 2H-$MoTe_2$ has been improved by Te vacancies engineering [7], nanowire structure engineering [8], hybrid composites [9], phase engineering [10,11] and so on. Theoretically, the better HER activity of the $MoTe_2$ plane via metal/non-metal atom substituted doping [6,12] or Te vacancy [13] engineering has been predicted. Among them, the HER activity of $MoTe_2$ as a function of hydrogen adsorption coverage is investigated. For example, the equilibrium hydrogen adsorption coverage at vacancy sites of $MoTe_2$ is calculated as bigger than 0.8 [14]. G-CN/$MoTe_2$ heterojunctions achieve higher HER performance at 4 H coverage [15]. The effective HER performance of $MoTe_2$ with a 2H-1T phase boundary is achieved below 60% hydrogen adsorption coverage, according to a density functional theory (DFT) calculation [16].

As far as we know, systematic work on monolayer $MoTe_2$ HER activity in terms of hydrogen adsorption sites and coverage is scarce. Hydrogen adsorption with different active sites and coverage rates on similar TMDS $MoSe_2$ materials has been reported [17]. These authors found that the catalytic capacity of HER for the Se edge is more stable than that of the Mo edge, with less sensibility to the absorption hydrogen coverage rates. Our work follows up on previous work, except for the $MoTe_2$ materials used.

Here, DFT is used to calculate the hydrogen adsorption Gibbs free energy of $MoTe_2$ as a function of the adsorption sites and adsorption rates. The exchange current density (j_0) as a function of the Gibbs free energy (ΔG_H) is calculated to analyze the catalytic ability. In our $2 \times 2 \times 1$ $MoTe_2$ models, the Mo edge is found to be sensitive to the adsorption rate. To further explore the mechanism, the geometry structure and the differential charge density distribution are investigated, and it is found that the uniform charge distribution explains the better HER performance. In addition, the relationship between the Gibbs free energy and energy bandgap is established by calculating the electronic energy band structure of $MoTe_2$ models. And it is found that the enhanced electronic density of states and the reduced band gap is advantageous for H atom adsorption. The adsorbed hydrogen is stable at 600 K kinetically according to MD calculations.

2. Results

2.1. Structure

Figure 1a,b show the top and side views of a $MoTe_2$ unit cell. The Mo atom (purple color) is sandwiched between the two nearest S atoms (yellow color). The lattice constant, Mo-Te bond length and Te-Mo-Te angle are 3.56 Å, 2.738 Å and ~83°, respectively, which is consistent with other reports [18]. The pristine $MoTe_2$ has a bandgap of ~0.97 eV, which is consistent with previous DFT calculations [6]. The orbital projected band structure and partial density of states (PDOS) of the $MoTe_2$ primitive cell are shown in Figure S1. Figure 1c shows the top view of the $2 \times 2 \times 1$ supercell $MoTe_2$. Three sites were chosen to identify the optimal position for H adsorption: hollow sites (#0), top Te site (#1) in the Te zigzag edge, and top Mo site (#2) in the Mo zigzag edge; these are shown in Figure 1c.

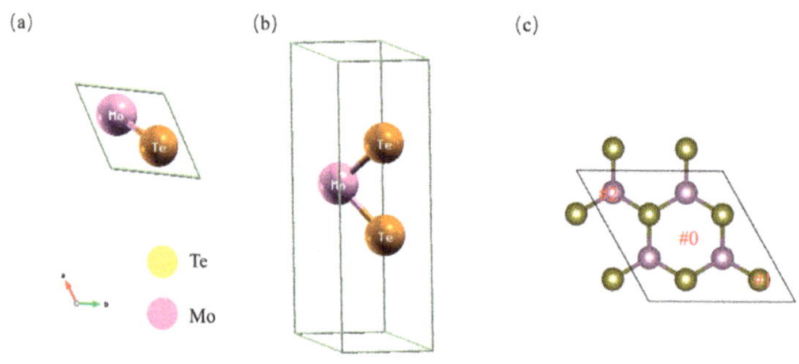

Figure 1. (**a**,**b**) The top and side view of the optimized structure of pristine $MoTe_2$. The Te and Mo atoms are represented by yellow and purple colors, respectively. (**c**) Supercell structure and hydrogen adsorption sites (#0 hollow site of $MoTe_2$; #1 site on top of Te; #2 site on top of Mo).

2.2. Gibbs Free Energy (ΔG_H) of H Adsorption Sites and Coverage

ΔG_H is a good descriptor of HER activity [19]. A value of ΔG_H close to 0 indicates the optimal catalyst, while a positive value means that the protons find it difficult to bond to the catalyst surface and a negative value meaning that the adsorbed hydrogen is hard to leave for H_2 formation. It is found that the hollow sites are weak for the adsorption of protons based on the calculated ΔG_H of adsorption sites #0–2 (shown in Table 1). So, the upper Te (#1) and Mo sites (#2) are chosen for further exploration due to the relative strength of the adsorption protons. The difference in the coverage of H at the #1 and #2 site is investigated. The coverage of H adsorption is defined as the number of atoms adsorbed by the H atom divided by the number of edge atoms. The Te (#1) edge that adsorbs one or two H atoms has a coverage of 50% and 100%, and is thus denoted as #1_1/2 and #1_2/2 for simplicity. Similarity, the Mo (#2) edge that adsorbs one and two H atoms is denoted as #2_1/2 and #2_2/2. The calculated total energies of the relaxed $MoTe_2$ and H-adsorbed $MoTe_2$ configurations (#0, #1_1/2, #1_2/2, #2_1/2 and #2_2/2) are shown in Table S1. The Gibbs free energy calculations are detailed in Section 4. The calculated ΔG_H as a function of adsorption sites and coverage is shown in Table 1. It is found that the ΔG_H of Te sites (1) increases from 1.945 to 2.024 eV. The ΔG_H of Mo sites decreases from 1.601 to 1.326 eV. The optimal H adsorption site and coverage is #2_2/2 as it had the lowest ΔG_H energy.

Table 1. ΔG_H at #0, #1 and #2 adsorption sites with a 50% and 100% hydrogen coverage rate.

Adsorption Sites and Coverage	ΔG_H(eV)
# 0	3.501
#1_1/2	1.945
#1_2/2	2.024
#2_1/2	1.601
#2_2/2	1.326

In order to further analyze the HER activity, the volcano relationship between the exchange current density (j_0) and the Gibbs free energy under standard conditions (pH = 0, T = 300 K) is shown in Figure 2 [20]. The calculation formula is as follows:

$$j_0 = -ek_0 \times \frac{1}{1+\exp\left(\frac{-\Delta G_H}{kT}\right)} \times \exp\left(\frac{-G_H}{kT}\right), \ (\Delta G_H > 0) \quad (1)$$

where e is the unit charge and k_0 is the proton transfer rate constant, k is the Boltzmann constant, and T is the temperature. The exchange current density of the #0 adsorption sites is at the lower end of the triangle. The j_0 of #2_2/2 is closer to the top of the triangle while the j_0 of #1_2/2 is away from the top of the triangle, indicating the faster reaction rate of #2_2/2 and the slower reaction rate of #1_2/2.

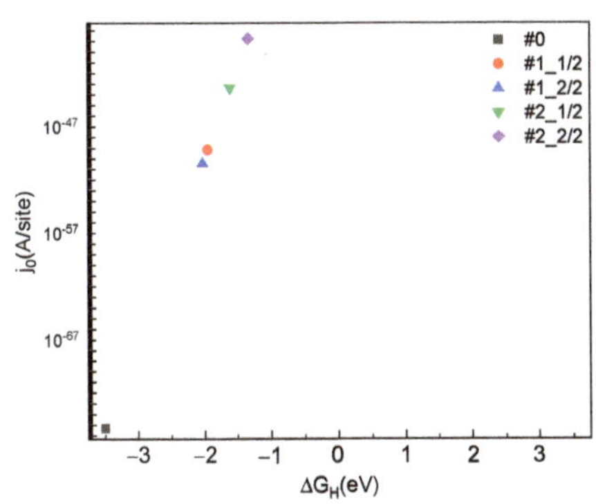

Figure 2. Exchange current density (j_0) at #0, #1 and #2 adsorption sites with a 50% and 100% hydrogen coverage rate.

3. Discussion

To further explore the mechanism from the perspective of structure and energy, the structure and charge density differences in the four situations are calculated. Figure 3 shows the geometry relaxation of #1_1/2, #1_2/2, #2_1/2, #2_2/2. The length of the Mo-H bond is 1.712, 1.795, 1.710 and 1.695 Å. The longest length is observed in structure #1_2/2, and an obvious lattice deformation is observed. Meanwhile, the ΔG_H is the largest, at 2.024 eV.

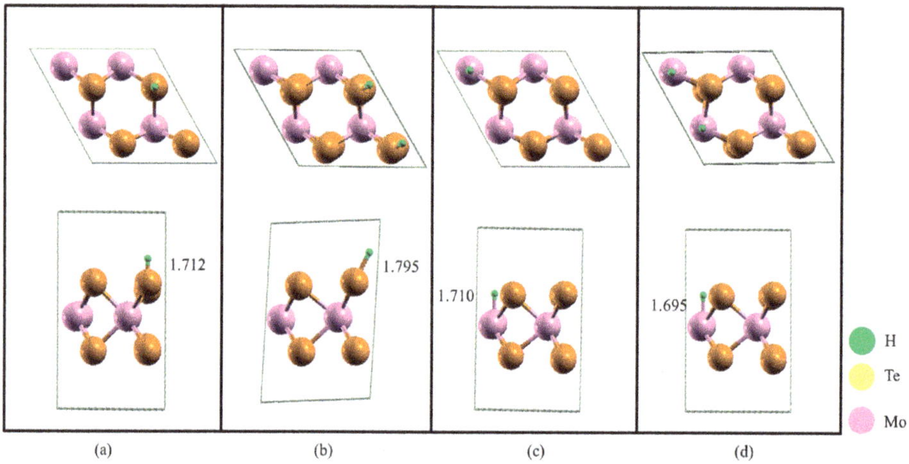

Figure 3. The top-view and side-view of the optimized structure of (**a**) #1_1/2; (**b**) #1_2/2; (**c**) #2_1/2; (**d**) #2_2/2. The green, yellow and purple spheres indicate H, Te and Mo atoms, respectively.

In addition, the charge transfer was analyzed by calculating the differenctial charge density distribution, as shown in Figure 4a–d. The isosurface is 0.0025 e * a_0^{-3}. Green regions near the atom represent electron deletion and yellow regions near the atom represent electron accumulation [21]. For structure #1_1/2, as shown in Figure 4a, it is found that the adsorbed H atom gain electrons while the adsorption site Te (#1) loses electrons. In addition, there is no transfer of electrons in the unadsorbed Te atoms and the redistribution of electrons occurs in three Mo atoms. In particular, the electron transfer occurs between the adsorbed Te atom and the two nearest Mo atoms. For structure #1_2/2, large-scale electron transfer occurs near the adsorbed Te and nearest Mo atoms. The nonlocalized electron distribution is related to the deformed structure. Electrons are distributed more locally near the adsorbed Mo atoms in structures #2_1/2 and #2_2/2 compared with the zigzag Te atoms of structures #1_1/2 and #1_2/2, especially in structure #2_2/2. Combined with the ΔG_H, it is deduced that a large ΔG_H value 2.024 eV is related to nonlocalized electron distribution and that a small ΔG_H value 1.326 eV is related to the localized electron distribution, which is consistent with another report [17]. The electronegativity of H, Mo and Te is 2.20, 2.16 and 2.10, respectively. So, electrons are enriched near the H atom and depleted near Mo and Te atoms. The zigzag Mo atom site is sensitive to H compared with the zigzag Te atom, which is consistent with the result showing that the zigzag Mo atom is more sensitive to zigzag Se atoms [17].

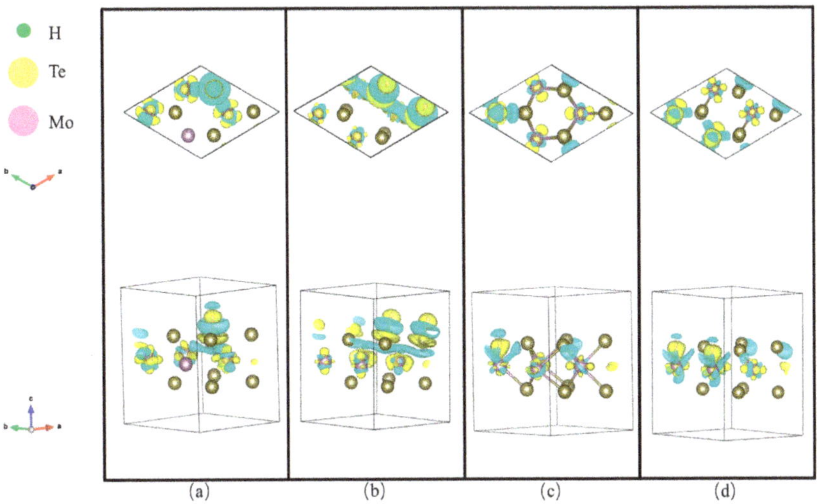

Figure 4. The top and side view of the differential charge density distribution for (**a**) #1_1/2; (**b**) #1_2/2; (**c**) #2_1/2; (**d**) #2_2/2. The blue and yellow color indicate the charge deletion and accumulation region. The green, yellow and purple spheres indicate H, Te and Mo atoms, respectively.

We also studied the energy band structure along the K′-M-K-G-K′ path in the Brillouin zone of the H-adsorbed $MoTe_2$ with four configurations (#1_1/2, #1_2/2, #2_1/2 and #2_2/2), as shown in Figure 5a–d. For the #1_1/2 structure, the conduction band minimum (CBM) and valence band maximum (VBM) are located at the K high-symmetry point, exhibiting a direct band gap with a band gap of 0.43 eV. Similarly, for the #2_1/2 and #2_2/2 structure, the direct band gap decreased to 0.40 eV and 0.17 eV, respectively. It is noted that for the #1_2/2 structure, the energy band transitions to an indirect semiconductor, exhibiting an energy bandgap of 0.49 eV. It is found that the bandgap of the four structures is related to the Gibbs free energy. The bandgap of structure #1_2/2 is the largest (0.49 eV) and the corresponding ΔG_H is also the largest (2.024 eV). Conversely, the smallest band gap (0.17 eV) of structure #2_2/2 and the smallest ΔG_H (1.326 eV) are observed. This is because the enhanced density of the conduction band and the reduced band gap are beneficial for the charge transfer from the adsorbed H to the adsorption sites of $MoTe_2$ [6]. It is noted that for structure #1_2/2, the VBM is located at the K point while the CBM is not located at the K point. To study the transition from direct band gap to indirect band gap, the orbital projected band structure and projected density of the states (PDOS) of structure #1_2/2 are shown in Figure S2 of the Supplementary Materials. Compared to the projected band structure of the pristine $MoTe_2$ shown in Figure S1b, the absorbed H 1s orbital of structure #1_2/2 contributes to the CBM located along the K to Gamma point. The PDOS of Figure S2b further demonstrates that the Mo 4d, Te 5s, 5p and H 1s contribute to the density of the states at CBM.

In order to analyze the thermal stability of adsorbed H atoms, we carried out first-principles molecular dynamics (MD) simulations. The total energy of the #2_1/2 structure as a function of time simulation at a finite temperature (600 K) is presented in Figure 6. The inset shows the snapshots for the #2_1/2 structure after the 96 fs and 10 ps MD run. It is noted that the #2_1/2 structure other than #2_2/2 is chosen due to the limited calculation resources. It can be seen that the total energy fluctuates around −10,713.6 eV and that the structure is slightly deformed after 10 ps, with no atoms leaving the substrate [22], indicating that the structure remains dynamically stable. At 700 K, the H atom oscillated violently and detached away from the adsorption sites (shown in Figure S3a–e).

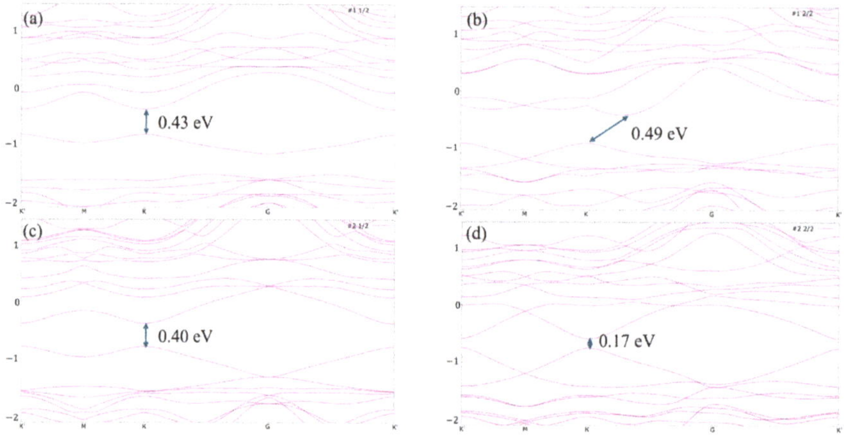

Figure 5. Energy band structure for (**a**) #1_1/2; (**b**) #1_2/2; (**c**) #2_1/2; (**d**) #2_2/2. K′, M, K, G are the points of high symmetry in the Brillouin zone with K′, K in the corners, M between two corners and G in the center.

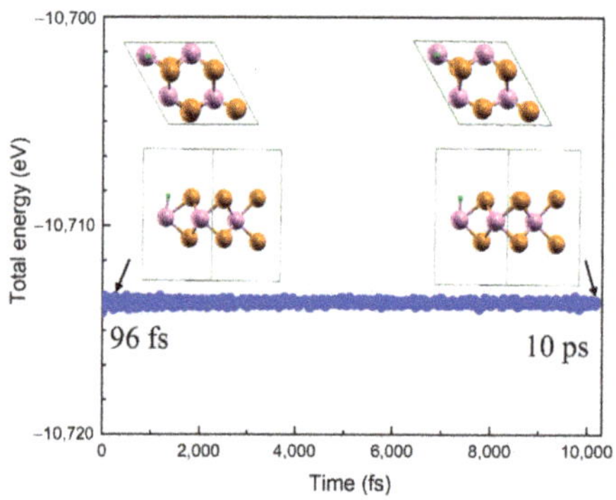

Figure 6. The total energy of the #2_1/2 structure as a function of the time simulation at finite temperature (600 K). The inset shows the top and side views for the H-$MoTe_2$ #2_1/2 structure after 96 fs and 10 ps first-principles MD simulations. The green, yellow and purple spheres indicate H, Te and Mo atoms, respectively.

4. Materials and Methods

All DFT calculations were performed by using the Quantum open-source Package for Research in Electronic Structure, Simulation, and Optimization (ESPRESSO, version 6.7) [23,24]. Electron–electron exchange correlations are treated by the Perdew–Burke–Ernzerhof (PBE) functional within the generalized gradient approximation (GGA) and electron–ion interactions are treated by the projector augmented wave (PAW) method. The crystal structures are relaxed with a cutoff energy of 49 Ry for the plane–wave expansion of the electronic wave function and the force on each atom is less than 0.04 eV/Å. A vacuum space of 15 Å is adopted to prevent interlayer interactions. The Monkhorst-Pack k-point sampling is set as $1 \times 1 \times 1$, $3 \times 3 \times 1$ for primitive and supercell structures.

In addition, the change in the Gibbs free energy of hydrogen adsorbed on $MoTe_2$ (ΔG_H) is calculated at zero potential and PH = 0 as follows:

$$\Delta G_H = \Delta E_H + \Delta E_{ZPE} - T\Delta S_H, \quad (2)$$

where ΔE_H is the hydrogen adsorption energy; ΔE_{ZPE} is the difference in the zero-point energy. ΔE_H and ΔE_{ZPE} are computed as follows:

$$\Delta E_H = \Delta E_{nH} - \Delta E_{(n-1)H} - 1/2\Delta E_{H_2}, \quad (3)$$

$$\Delta E_{ZPE} = 1/2\sum\nolimits_{nH} \hbar\omega_i - 1/2\sum\nolimits_{(n-1)H} \hbar\omega_i - 1/2\sum\nolimits_{H_2} \hbar\omega_i, \quad (4)$$

where nH and $(n-1)H$ represent the adsorption of $MoTe_2$ by n and $n-1$ hydrogen atoms, respectively. ω_i is the vibrational frequency. T is the temperature (300 K) and ΔS is the difference in entropy between the H that is adsorbed and in the gas phase at 1.01×10^5 pa. $T\Delta S_H$ is simplified as 0.20 eV, as reported elsewhere [25]. The calculated ΔE_{ZPE} is about 0.01 eV. The detailed zero-point energy is shown in Table S2 of the Supplementary Materials. We set ΔE_{ZPE}-$T\Delta S_H$ as 0.21 eV for simplicity.

The thermodynamic stability of hydrogen adsorption on $MoTe_2$ is evaluated by molecular dynamics simulations at 600 K, 700 K and 800 K under the NVT (number of simulated particles, simulation cell volume, and temperature). The total simulation time is 10 ps and the time step is 1 fs [22].

5. Conclusions

In summary, $2 \times 2 \times 1$ monolayer $MoTe_2$ shows a Gibbs free energy of 1.326 eV with 2 H adsorption on the Te edge but 2.024 eV with 2 H adsorption on the Mo edge. The exchange current density (j_0) as a function of the Gibbs free energy (ΔG_H) is calculated to analyze the HER activity. Combining the structure and charge distribution analysis, the electron distribution associated with the deformed structure is the reason for the big Gibbs free value. Local electron distribution is the reason for the small Gibbs free energy value. In addition, the relationship between the Gibbs free energy and energy band gap is established by calculating the electronic energy band structure of $MoTe_2$ models. It is found that the enhanced electronic density of states and the reduced band gap is advantageous for H adsorption. It was found that the H-adsorbed $MoTe_2$ of #1_2/2 became an indirect-gap semiconductor, which could be the reason for the large Gibbs free energy of 2.024 eV. According to the use of Gibbs free energy as a function of hydrogen adsorption rates, zigzag Mo is sensitive to H adsorption. Finally, based on MD data, the #2_1/2 structure shows thermal stability below 600 K. These findings provide valuable references for the application of two-dimensional $MoTe_2$ in the field of electrocatalytic hydrogen production.

Supplementary Materials: The following supporting information can be downloaded at: https://www.mdpi.com/article/10.3390/inorganics13060197/s1, Figure S1: (a) The band structure of monolayer 2H-$MoTe_2$; (b) The orbital projected band structure of monolayer 2H-MoTe2; (c) The partial density of states (PDOS) of monolayer 2H-MoTe2. Table S1: The calculated energy of relaxed $MoTe_2$ and the H-adsorbed $MoTe_2$ configurations (#0, #1_1/2, #1_2/2, #2_1/2 and #2_2/2). Figure S2: (a) The orbital projected band structure of #2_2/2; (b) The partial density of states (PDOS) of #2_2/2. Table S2: The calculated zero-point energy of relaxed $MoTe_2$ and the H-adsorbed $MoTe_2$ configurations (#0, #1_1/2, #1_2/2, #2_1/2 and #2_2/2). Figure S3: The snapshot of typical of H-$MoTe_2$ #2_1/2 structure at 700K at (a) 100 fs; (b) 4.02 ps; (c) 6.04 ps; (d) 7.36 ps; (e) 8.48 ps.

Author Contributions: Conceptualization, J.M.; writing—original draft preparation, X.G.; writing—review and editing, J.M.; supervision, J.M.; funding acquisition, X.G. and J.M. All authors have read and agreed to the published version of the manuscript.

Funding: This research was funded by National College Students Innovation and Entrepreneurship Training Program, grant number X202510010441.

Institutional Review Board Statement: Not applicable.

Informed Consent Statement: Not applicable.

Data Availability Statement: The original contributions presented in this study are included in the article/Supplementary Materials. Further inquiries can be directed to the corresponding author.

Acknowledgments: The authors acknowledge Beijing PARATERA Tech Co., Ltd. for providing HPC resources that have contributed to the research results reported within this paper.

Conflicts of Interest: Beijng PARATERA Tech Co., Ltd. provided HPC resources for this research. The authors declare that the research was conducted in the absence of any commercial or financial relationships that could be construed as a potential conflict of interest.

References

1. Midilli, A.; Ay, M.; Dincer, I.; Rosen, M.A. On hydrogen and hydrogen energy strategies: I: Current status and needs. *Renew. Sustain. Energy Rev.* **2005**, *9*, 255–271. [CrossRef]
2. Tee, S.Y.; Win, K.Y.; Teo, W.S.; Koh, L.-D.; Liu, S.; Teng, C.P.; Han, M.-Y. Recent Progress in Energy-Driven Water Splitting. *Adv. Sci.* **2017**, *4*, 1600337. [CrossRef] [PubMed]
3. Guo, F.; Macdonald, T.J.; Sobrido, A.J.; Liu, L.; Feng, J.; He, G. Recent Advances in Ultralow-Pt-Loading Electrocatalysts for the Efficient Hydrogen Evolution. *Adv. Sci.* **2023**, *10*, 2301098. [CrossRef] [PubMed]
4. Saji, V.S.; Pillai, V.K.; Sotiropoulos, S. Recent advances in electrochemical water splitting. *Electrochim. Acta* **2024**, *475*, 143645. [CrossRef]
5. Mao, J.; Ta, Q.T.H.; Tri, N.N.; Shou, L.; Seo, S.; Xu, W. 2D $MoTe_2$ nanomesh with a large surface area and uniform pores for highly active hydrogen evolution catalysis. *Appl. Mater. Today* **2023**, *35*, 101939. [CrossRef]
6. Xiao, Z.; Gan, X.; Zhu, T.; Lei, D.; Zhao, H.; Wang, P. Activating the Basal Planes in 2H-$MoTe_2$ Monolayers by Incorporating Single-Atom Dispersed N or P for Enhanced Electrocatalytic Overall Water Splitting. *Adv. Sustain. Syst.* **2022**, *6*, 2100515. [CrossRef]
7. Yang, H.; Zhao, Y.; Wen, Q.; Mi, Y.; Liu, Y.; Li, H.; Zhai, T. Single $MoTe_2$ sheet electrocatalytic microdevice for in situ revealing the activated basal plane sites by vacancies engineering. *Nano Res.* **2021**, *14*, 4814–4821. [CrossRef]
8. Mao, J.; Zhou, L.; Li, Y.; Tao, Y.; Chai, K.; Shi, Y.; Xu, W. Synthesis of $MoTe_2$ nanowire as an efficient hydrogen evolution reaction material. *Mater. Lett.* **2021**, *290*, 129471. [CrossRef]
9. Shinde, P.V.; Mane, P.; Late, D.J.; Chakraborty, B.; Rout, C.S. Promising 2D/2D $MoTe_2/Ti_3C_2T_x$ Hybrid Materials for Boosted Hydrogen Evolution Reaction. *ACS Appl. Energy Mater.* **2021**, *4*, 11886–11897. [CrossRef]
10. Lu, D.; Ren, X.; Ren, L.; Xue, W.; Liu, S.; Liu, Y.; Chen, Q.; Qi, X.; Zhong, J. Direct Vapor Deposition Growth of 1T′ $MoTe_2$ on Carbon Cloth for Electrocatalytic Hydrogen Evolution. *ACS Appl. Energy Mater.* **2020**, *3*, 3212–3219. [CrossRef]
11. Lei, Y.; Xiao, X.; Ma, T.; Li, W.; Zhang, H.; Ma, C. Facile hydrothermal synthesis of layered 1T′ $MoTe_2$ nanotubes as robust hydrogen evolution electrocatalysts. *Front. Chem.* **2022**, *10*, 1005782. [CrossRef] [PubMed]
12. Huang, Y.; Chen, Y.; Zhang, Y.; Zhang, H.; Xiong, K.; Ye, X.; Liu, Q.; Zhu, J. Transition-metal atoms embedded $MoTe_2$ single-atom catalyst for efficient electrocatalytic hydrogen evolution reaction. *Appl. Surf. Sci.* **2025**, *680*, 161335. [CrossRef]
13. Pu, M.; Guo, Y.; Guo, W. Wrinkle facilitated hydrogen evolution reaction of vacancy-defected transition metal dichalcogenide monolayers. *Nanoscale* **2021**, *13*, 20576–20582. [CrossRef]
14. Lee, J.; Kang, S.; Yim, K.; Kim, K.Y.; Jang, H.W.; Kang, Y.; Han, S. Hydrogen Evolution Reaction at Anion Vacancy of Two-Dimensional Transition-Metal Dichalcogenides: Ab Initio Computational Screening. *J. Phys. Chem. Lett.* **2018**, *9*, 2049–2055. [CrossRef]
15. Shen, Y.; Yuan, P.; Yuan, Z.; Cui, Z.; Ma, D.; Cheng, F.; Qin, K.; Wang, H.; Li, E. Modulation of electronic and optical properties of g-CN/XTe_2(X=W, Mo) heterojunctions by biaxial strain. *Int. J. Hydrogen Energy* **2024**, *80*, 289–297. [CrossRef]
16. Chen, Y.; Ou, P.; Bie, X.; Song, J. Basal plane activation in monolayer MoTe2 for the hydrogen evolution reaction via phase boundaries. *J. Mater. Chem. A* **2020**, *8*, 19522–19532. [CrossRef]
17. Xu, Z.-W.; Shi, C.-S.; Zhao, G.-H.; Wang, M.-Y.; Liu, G.-W.; Qiao, G.-J. Hydrogen adsorption mechanism on single-layer $MoSe_2$ for hydrogen evolution reaction: First-principles study. *Acta Phys. Sin.* **2018**, *67*, 217102. [CrossRef]
18. Lamba, A.; Kumar, R.; Srivastava, S. Band gap engineering of 2H-MX_2 (M = Mo; X = S, Se, Te) monolayers using strain effect. *Mater. Today Proc.* **2021**, *54*, 677–681. [CrossRef]

19. Tang, Q.; Jiang, D.-E. Mechanism of Hydrogen Evolution Reaction on 1T-MoS_2 from First Principles. *ACS Catal.* **2016**, *6*, 4953–4961. [CrossRef]
20. Yang, T.T.; Patil, R.B.; McKone, J.R.; Saidi, W. Revisiting trends in the exchange for hydrogen evolution. *Catal. Sci. Technol.* **2021**, *11*, 6832–6838. [CrossRef]
21. Zhang, X.; Song, Y.; Zhang, F.; Fan, Q.; Jin, H.; Chen, S.; Jin, Y.; Gao, S.; Xiao, Y.; Mwankemwa, N.; et al. The electronic properties of hydrogenated Janus MoSSe monolayer: A first principles investigation. *Mater. Res. Express* **2019**, *6*, 105055. [CrossRef]
22. Li, X.D.; Qian, Y.Y.; Guo, T.S.; Fu, L.J. DFT study on adsorption properties of TM(Ni, Co)Nx-doped graphene for high-temperature sensing of SF6 decomposed gases. *Mater. Today Chem.* **2024**, *35*, 101904. [CrossRef]
23. Giannozzi, P.; Andreussi, O.; Brumme, T.; Bunau, O.; Buongiorno Nardelli, M.; Calandra, M.; Car, R.; Cavazzoni, C.; Ceresoli, D.; Cococcioni, M.; et al. Advanced capabilities for materials modelling with Quantum ESPRESSO. *J. Phys. Condens. Matter* **2017**, *29*, 465901. [CrossRef] [PubMed]
24. Giannozzi, P.; Baroni, S.; Bonini, N.; Calandra, M.; Car, R.; Cavazzoni, C.; Ceresoli, D.; Chiarotti, G.L.; Cococcioni, M.; Dabo, I.; et al. QUANTUM ESPRESSO: A modular and open-source software project for quantum simulations of materials. *J. Phys. Condens. Matter* **2009**, *21*, 395502. [CrossRef]
25. Wang, J.S.; Liu, J.; Zhang, B.; Ji, X.; Xu, K.; Chen, C.; Miao, L.; Jiang, J.J. The mechanism of hydrogen adsorption on transition metal dichalcogenides as hydrogen evolution reaction catalyst. *Phys. Chem. Chem. Phys.* **2017**, *19*, 10125–10132. [CrossRef]

inorganics

MDPI

Article

Bimetallic Ir-Sn Non-Carbon Supported Anode Catalysts for PEM Water Electrolysis

Iveta Boshnakova *, Elefteria Lefterova, Galin Borisov, Denis Paskalev and Evelina Slavcheva *

Institute of Electrochemistry and Energy System "Acad. Evgeni Budevski", Bulgarian Academy of Sciences, Acad. G. Bonchev Str., bl. 10, 1113 Sofia, Bulgaria; e.lefterova@iees.bas.bg (E.L.); gal.rusev@iees.bas.bg (G.B.); denis.paskalev@iees.bas.bg (D.P.)

* Correspondence: boshnakova.iveta@iees.bas.bg (I.B.); eslavcheva@iees.bas.bg (E.S.)

Abstract: Nanostructured bimetallic IrSn composites deposited on the natural aluminosilicate montmorillonite were synthesized and evaluated as anode electrocatalysts for polymer electrolyte membrane electrolysis cells (PEMECs). The test series prepared via the sol–gel method consisted of samples with 30 wt. % total metal content and varying Ir:Sn ratio. The performed X-ray diffraction analysis and high-resolution transmission electron icroscopy registered very fine nanostructure of the composites with metal particles size of 2–3 nm homogeneously dispersed on the support surface and also intercalated in the basal space of its layered structure. The electrochemical behavior was investigated by cyclic voltammetry and steady-state polarization techniques. The initial screening was performed in 0.5 M H_2SO_4. Then, the catalysts were integrated as anodes in membrane electrode assemblies (MEAs) and tested in a custom-made PEMEC. The electrochemical tests revealed that the catalysts with Ir:Sn ratio 15:15 and 18:12 wt. % demonstrated high efficiency toward the oxygen evolution reaction during repetitive potential cycling and sustainable performance with current density in the range 140–120 mA cm^{-2} at 1.6 V vs. RHE during long-term stability tests. The results obtained give credence to the studied IrSn/MMT nanocomposites to be considered promising, cost-efficient catalysts for the oxygen evolution reaction (OER).

Keywords: PEM water electrolysis; oxygen evolution reaction; non-carbon catalytic supports; montmorillonite; iridium; tin; low noble-metal loading

1. Introduction

The electrolysis of water in cells with a polymer electrolyte membrane (PEMEC) is one of the most advanced technologies for producing pure hydrogen. When powered by electricity generated from renewable sources, it becomes a green, zero-emission, environmentally friendly process. However, a major challenge preventing the large-scale industrial adoption of PEMEC technology is the still high cost due to the need for noble metal catalysts from the platinum group (PGM) and the sluggish kinetics of the oxygen evolution reaction (OER)—one of the most widely studied reactions in the field of electrocatalysis [1,2]. In this context, heterointerface engineering has emerged as a promising strategy for addressing anodic intrinsic activity limitations, with a recent study reporting a $RuIrO_x$–CeO_2 ternary heterostructured catalyst featuring a strong-coupled interface that achieved impressive activity and long-term stability at high current densities [3]. On the cathode side, where the hydrogen evolution reaction (HER) occurs, a strongly electronically coupled Co@V–WO_x material with an amorphous nanosheet morphology is presented. It is synthesized via a one-step electrodeposition method and demonstrates a Faradaic efficiency exceeding 92.8%, indicating enhanced catalytic activity compared to pure Pt-based catalysts [4]. The

OER takes place at high potentials, and when combined with intensive gas evolution, high humidity, and elevated temperatures, it creates harsh operating conditions. These conditions lead to the gradual degradation of the classical carbon-based catalytic supports due to oxidation and corrosion of carbon. To address these challenges, extensive research over the past decades has been focused on developing novel, corrosion-resistant, and cost-efficient OER catalysts.

It is well established that IrO_2 is the most effective catalyst for OER in acidic media, offering high activity and stability under extreme operating conditions. However, as precious metal oxide, IrO_2 is prohibitively expensive for large-scale industrial PEMEC applications, limiting its widespread use [5].

Various efforts have been made, and new strategies have been developed to reduce catalyst costs. One approach involves enhancing the catalytic activity by synthesizing nanosized noble metal oxides with a highly developed surface area [6,7], thus increasing the number of active sites for OER and reducing the catalyst loading without compromising the process efficiency.

Another strategy is the use of high-surface-area catalytic supports, which help to minimize nanoparticle agglomeration, thereby improving catalyst utilization [8]. More recent research has focused on reducing noble-metal loading by developing mesoporous metal oxide films [9,10] and multi-component catalysts. Some of these catalysts exhibited higher electrocatalytic activity, enhanced selectivity, and stability through the realization of synergetic effects [11]. The low-cost, multi-component OER electrocatalysts are typically synthesized by combining IrO_2 with one or more less expensive metal oxides, such as SnO_2, Sb_2O_5-SnO_2, Ta_2O_5, and TiO_2 [12–17]. Among these, IrO_2–SnO_2 mixtures have been widely studied for their performance in oxygen evolution reactions under acidic conditions, in part due to the structural similarities between the two oxides. Tin, in the form of SnO_2, is commonly used as a second component due to its chemical stability under acidic media, favorable electronic properties, and compatibility with the IrO_2 structure [1,14,18,19].

In our previous study [20], we reported montmorillonite (MMT) as an alternative to conventional carbon materials for application as an anode catalytic support in hydrogen generation via electrochemical water splitting. MMT is a natural non-carbon phyllosilicate mineral—aluminosilicate from the smectite group with general formula $(Na,Ca)_{0.33}(Al,Mg)_2(Si_4O_{10})\ (OH)_2{\bullet}nH_2O$ and a complex plate-shaped layered structure with 2:1 type sheets. It has demonstrated high surface reactivity, cation exchange capacity, and capability to intercalate metal ions, such as Fe, Cu, Ir, etc., in the interlayer basal space. The material is also characterized by large surface area, chemical stability, commercial availability, and low-cost—advantages, making it the most widely used layered silicate in nanocomposite fabrication [21,22].

Figure 1 presents a visualization of the MMT structure, obtained through the density functional theory (DFT) calculations using the projector augmented wave (PAW) method with the Perdew-Burke-Ernzerhof (PBE) exchange-correlation functional.

The recent study focuses on the preparation of novel bimetallic Ir-Sn catalysts deposited on MMT and the investigation of their structure, electrochemical performance, and stability as oxygen evolution electrocatalysts for PEMECs.

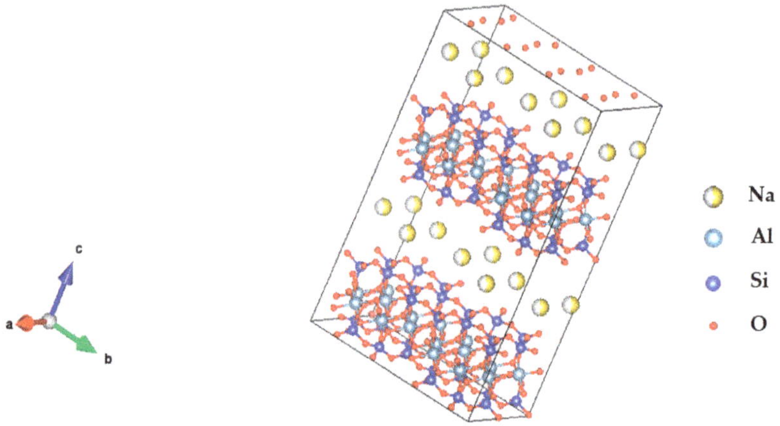

Figure 1. DFT visualization Montmorillonite structure.

2. Results and Discussion

2.1. Structural Characterization

The sol–gel process used to prepare the IrSn/MMT catalysts is a widely used wet chemical technique for synthesizing materials, particularly metal oxide nanoparticles. This method is cost-effective and operates at a low reaction temperature, allowing precise control over the chemical composition of the final products [23].

X-ray diffraction patterns (XRD) obtained to assess the phase composition and crystallinity of the synthesized composite materials are shown in Figure 2. In the pattern of the sample with 10 wt. % Ir (Ir10Sn20/MMT), only the support was identified, meaning that Ir, Sn (and/or their oxides) are in an amorphous state. For the catalysts with a higher content of Ir, besides MMT, a non-destructed Ir-acac was also registered. The presence of organic residues in the catalysts can be minimized by conducting the synthesis at higher temperatures (300 °C). However, our previous studies have demonstrated that such elevated temperatures promote particle agglomeration and growth of Ir nanoparticles [20,24]. Larger particles and agglomeration result in a significant reduction in the electrochemically active surface area, which in turn leads to lower current densities and impaired electrocatalytic efficiency. Therefore, despite the incomplete removal of ligands, a temperature of 240 °C was chosen as optimal to ensure the maintenance of a very fine nanostructure with metal particles sized 2–3 nm, homogeneously dispersed on the support surface and also intercalated in the basal space of its layered structure.

In all tested catalysts, the (001) diffraction peak of MMT is shifted to smaller 2θ angles in regard to the pure MMT from 8.03° to 7.00° (Figure 2b). Such a displacement corresponds to an increase in the interlayer spacing of MMT from 1.1 nm to 1.26 nm. The observed effect has been reported previously [20] and is related to the intercalation of foreign cations, in this case, Ir, Sn, and/or their complexes in the MMT basal space [25,26].

The broad peak registered at 40.283° could be prescribed to Ir (111), although (as seen in Figure 3) it is shifted to lower 2Θ angles compared to pure Ir iridium (2Θ = 40.662°, PDF#06-0598). Alternatively, this shift might be due to the formation of an $IrSn_2$ alloy having the strongest diffraction line (220) positioned at 2Θ = 40.217° (PDF#65-2987). The presence of oxides in an amorphous state is also not excluded. More accurate phase identification is hampered by the wide overlapping peaks of the different phases. The size of the crystallites determined by Scherrer equation [27] was very small, 2 nm.

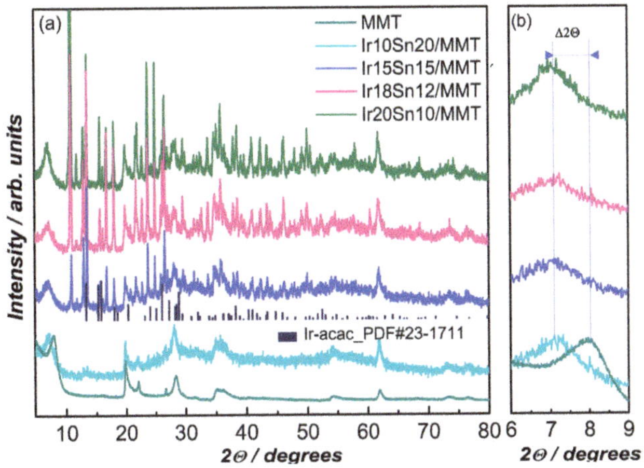

Figure 2. (**a**) XRD patternsand, (**b**) the enlarged XRD patterns from 6° to 9° of IrSn/MMT catalysts, pure MMT, Ir-acac and bar-gram of Ir-acac precursor according Inorganic Crystal Structure Database (ICSD, PDF#23-1711).

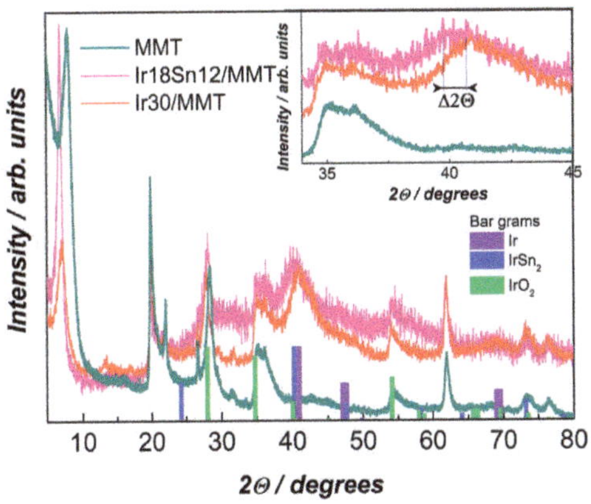

Figure 3. Comparison of XRD patterns of Ir18Sn12/MMT, Ir30/MMT, and MMT, inset graph show the zoom-in (34°–45°) XRD patterns.

A more detailed examination of catalyst morphologies was performed using High-Resolution Transmission Electron Microscopy (HRTEM). The results from the HRTEM analysis of the sample Ir18Sn12/MMT are presented in Figure 4. They reveal the presence of very small nanoparticles of both metals homogeneously distributed into the structure of the supporting montmorillonite.

As shown in Figure 4a, the pristine MMT exhibits a smooth layered surface, characteristic of its nanoplatelet structure. From the HRTEM images of the Ir18Sn12/MMT sample displayed in Figure 4b,c, small black dots scattered across the MMT surface and between the layers are seen, proving the presence of evenly distributed particles.

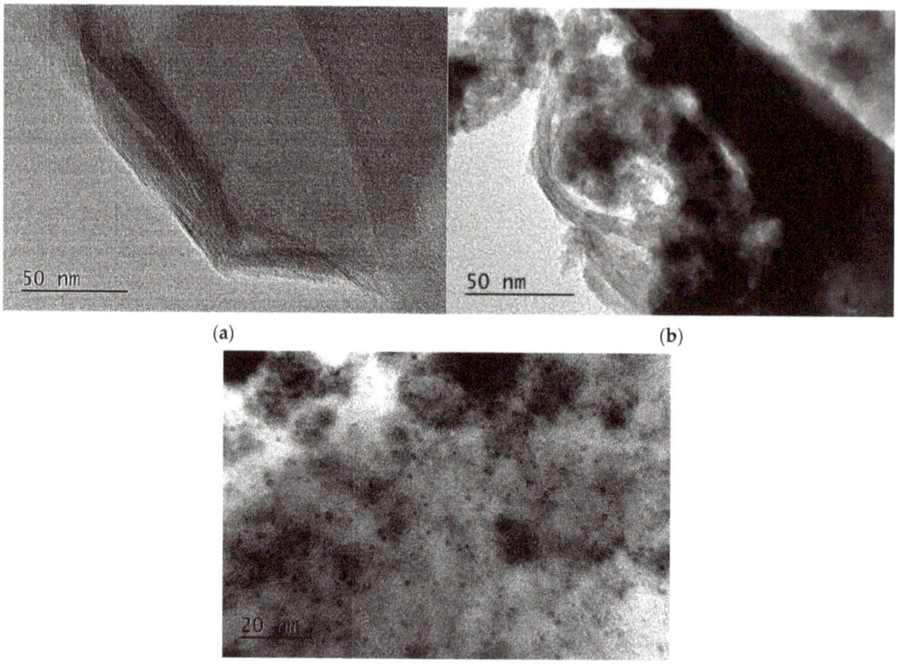

Figure 4. HRTEM images of (**a**) pure MMT and (**b**,**c**) Ir18Sn12/MMT70 catalyst.

According to these figures, the metal nanoparticles have sizes in the range of 1–2 nm, which is in full accordance with the XRD data. The sheet structure of montmorillonite is clearly observed in Figure 4b, indicating that the nanoparticle deposition did not disrupt the basic structure of the MMT support. The HRTEM results are very well in line with the information obtained from the elemental EDX mapping presented in Figure 5.

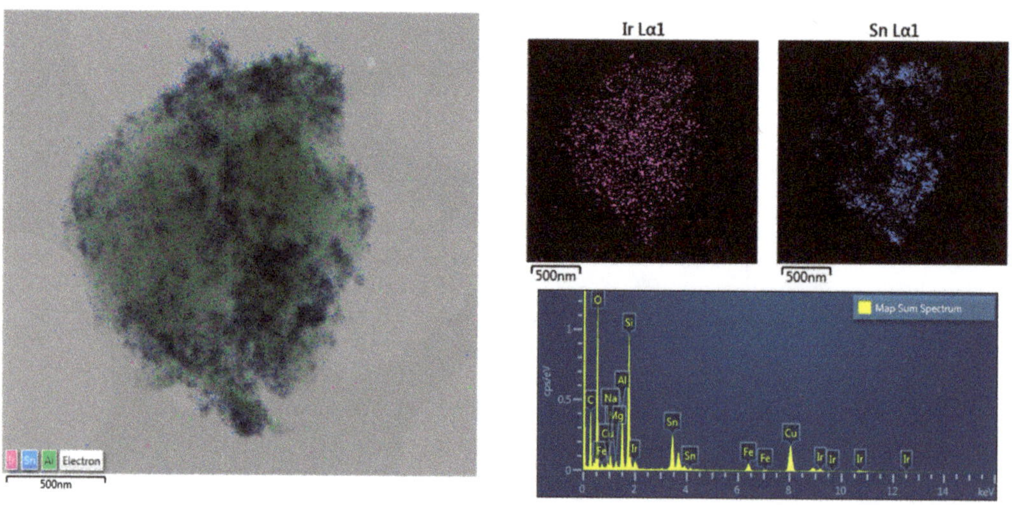

Figure 5. Elemental mapping images and EDX data of Ir18Sn12/MMT70.

2.2. Electrochemical Characterization

Anodes containing the synthesized bimetallic IrSn/MMT catalysts were first investigated in 0.5 M H_2SO_4. Anodic polarization curves were recorded for each catalyst to evaluate their electrochemical behavior during OER. In Figura 6a, the measured currents are normalized to the geometric surface area of the working electrode. The OER starts at approximately 1.3 V vs. Ag/AgCl for the Ir18Sn12/MMT and Ir15Sn15/MMT catalysts, whereas for Ir10Sn20MMT and Ir20Sn10/MMT, the OER begins at more positive potentials of around 1.4 V vs. Ag/AgCl.

Potentiostatic polarization tests at a constant potential of 1.8 V were performed to investigate the stability of each catalyst (Figure 6b). For all tested samples, the current density decreases rapidly at the beginning of the test and then gradually reaches stable values. It is seen that the OER is more intensive on the catalysts with a balanced Ir:Sn ratio (15:15, 18:12). This could be attributed to the beneficial synergetic effect between Sn and Ir at these ratios, combined with better dispersion of the metal nanoparticles on the support and as a result, higher number of active sites at the electrolyte/catalyst surface and increased efficiency toward the OER. In contrast, the catalysts with Ir content of 10 wt. % and 20 wt. % Ir exhibits a decreased reaction rate, which is due to less active sites on the catalyst's surface. For Ir10Sn20/MMT, this is due to a lower content of Ir—the more active catalyst component, while for Ir20Sn10/MMT, most probably agglomeration of metal nanoparticles commences.

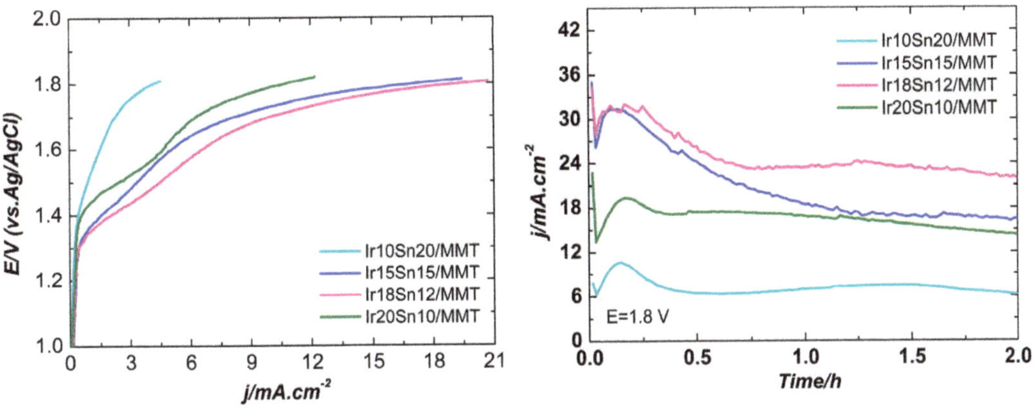

(a) (b)

Figure 6. Electrochemical performance of the prepared catalysts in 0.5 M H_2SO_4 at room temperature: (**a**) Linear sweep voltammetry with a scan rate of 1 mV s^{-1}; (**b**) Potentiostatic polarization curves at 1.8 V vs. Ag/AgCl.

To investigate the catalyst performance in a PEM electrolysis cell with Nafion 117 proton exchange membrane, MEAs were prepared using the synthesized samples as anode catalysts. Initially, the electrochemical performance was evaluated using cyclic voltammetry (CV), a well-known, highly informative, and surface-sensitive electrochemical method for fast screening in that each electrode material responds with a unique CV spectrum.

Figure 7 presents the cyclic voltammograms of IrSn/MMT catalysts integrated into MEAs and tested in PEMEC at room temperature.

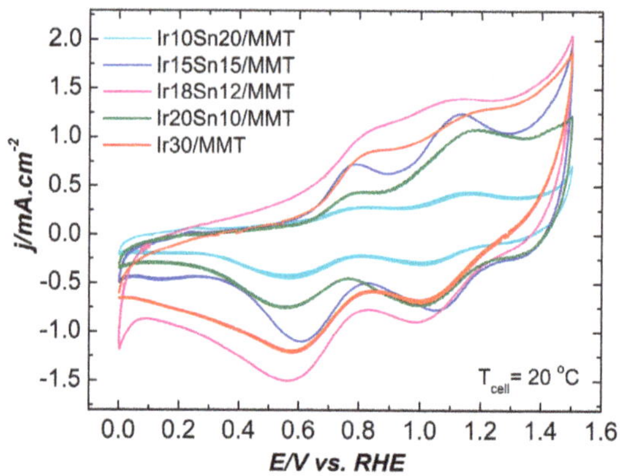

Figure 7. Cyclic voltammograms of the catalysts under study in PEM cell at room temperature, potential scan rate 100 mV s^{-1}, metal loading 0.5 $mg_{IrSn}cm^{-2}$; 200th cycle.

A typical feature of Ir is its ability to be activated when subjected to prolonged potential cycling in the "water window" potential range [24]. Therefore, for better comparison, the curves shown in Figure 7 present the 200th cycle for each sample. All CV curves exhibit broad waves resulting from redox reactions of the active surface species being oxidized or reduced—processes often accompanied by proton exchange between the metal oxide and the electrolyte [28,29]. The shape of the CVs and the potential range where the redox waves appear suggest that the proceeding electrochemical reactions are associated with transformations between different Ir oxidation states (Ir^{3+}/Ir^{4+} and/or Ir^{4+}/Ir^{5+}) [7] overlapped with changes in the valent state of Sn. Although the shape and area of CVs are similar for all tested samples, it can be seen that the redox peaks have the most distinct couture on the Ir15Sn15/MMT sample, suggesting better homogeneity of the catalyst structure, while the area under the CV is broader for the Ir18Sn12/MMT sample, implying higher charge transfer capacity.

All IrSn/MMT catalysts were evaluated at elevated temperatures in the range of 20 to 80 °C. Figure 8 presents the results for the Ir18Sn12/MMT catalyst, which showed the highest electrochemically active surface during the CV tests at room temperature.

The increase in the operating temperature results in a lower open-circuit voltage (OCV) and enhanced reaction rate, resp. decreased kinetic losses attributed to the lower activation energy of the OER [30]. As the measurement temperature increases, a significant expansion of the area under the cyclic voltammetry curves is observed (Figure 8a), which is induced by electrochemical and thermal activation of the catalysts, exposure of additional electroactive sites either from the catalyst bulk and/or by increasing the interparticle porosity. Additionally, a shift in the onset potential for oxygen evolution to lower values was noted with the rising temperature, indicating further catalyst activation (Figure 8b).

To compare the OER efficiency of all prepared catalysts, low-sweep voltammetry tests were conducted. Figure 9 presents the polarization curves for Ir15Sn15/MMT, Ir18Sn12/MMT, and Ir20Sn10/MMT alongside the previously reported Ir30/MMT [20] at 80 °C. The curve of commercial unsupported IrOx was also recorded and shown for comparison.

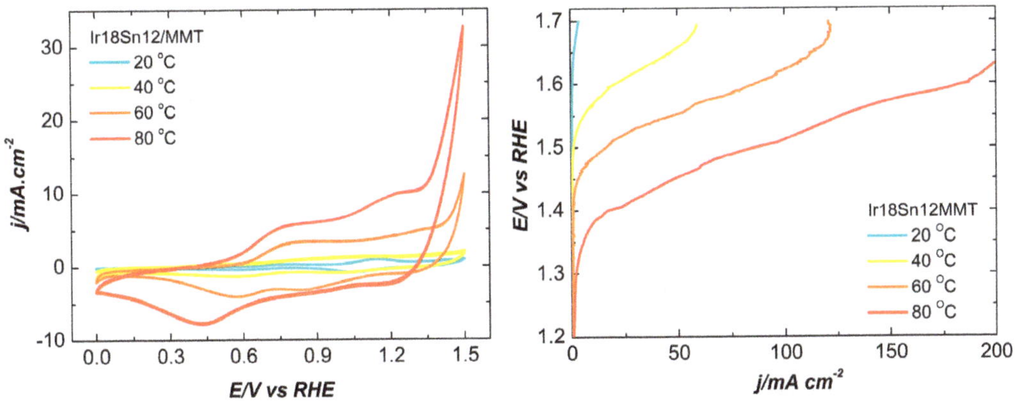

(a) (b)

Figure 8. Electrochemical performance of the catalyst Ir18Sn12MMT in PEMEC at 20 °C, 40 °C and 80 °C: (**a**) Cyclic voltammetry at a potential scan rate of 100 mV s^{-1}; (**b**) Anodic polarization curves at a potential scan rate of 1 mV s^{-1}.

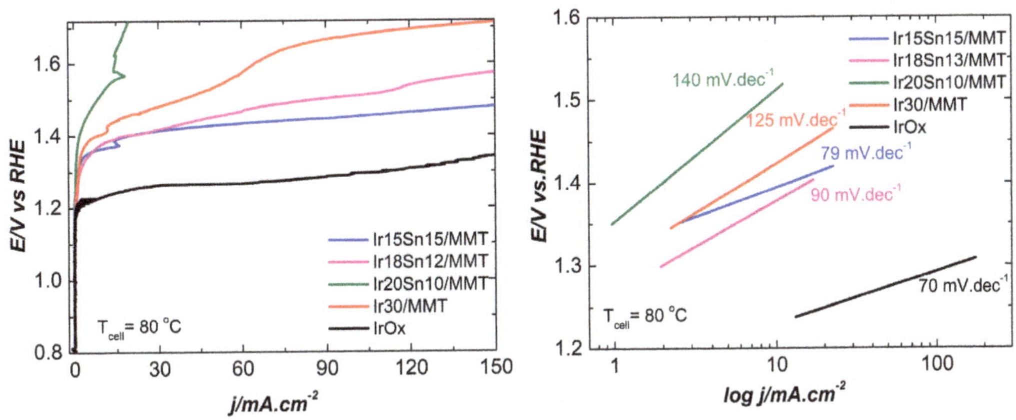

(a) (b)

Figure 9. Comparison of the bimetallic catalysts IrSn/MM, Ir30/MMT, and commercial IrOx catalysts performance in PEM single cell at 80 °C. (**a**) Linear sweep voltammetry (LSV) curves at a scan rate of 1 mV s^{-1}; (**b**) Tafel slopes corresponding to (**a**).

The data obtained from the tests at 20 °C and 80 °C are summarized in Table 1. They confirm the improved efficiency of all catalysts at elevated temperatures. The overpotentials at 20 °C are nearly twice as high as those at 80 °C. Among the tested samples, Ir15Sn12/MMT and Ir18Sn12/MMT outperformed the others, particularly in terms of overpotential at a given current density. For instance, at 80 °C and 10 mA·cm^{-2}, the Ir18Sn12/MMT electrode exhibits the lowest OER overpotential of 148 mV, while the overpotentials for Ir15Sn15/MMT, Ir20Sn10/MMT, and Ir30/MMT under identical conditions are 164 mV, 290 mV, and 195 mV, respectively.

Table 1. Summary of oxygen evolution reaction performance of the tested samples in PEM electrolysis cell at 20 °C and 80 °C.

Temperature			**20 °C**				**80 °C**	
Samples	**η/mV at j = 1 mA.cm^{-2}**	**$E_{(OER)}$ V at j = 1 mA.cm^{-2}**	**Tafel Slope mV.dec^{-1}**	**j/mA at 1.8 V**	**η/mV at j = 10 mA.cm^{-2}**	**$E_{(OER)}$ V at j = 10 mA.cm^{-2}**	**Tafel Slope mV.dec^{-1}**	**j/mA at 1.6 V**
Ir15Sn15/MMT	407	1.63	120	12.3	164	1.39	79	240
Ir18Sn12/MMT	344	1.57	180	12.8	148	1.37	90	220
Ir20Sn10/MMT	390	1.62	170	8.1	290	1.52	140	14
Ir30/MMT	---	---	240	6.9	194	1.42	125	65
IrOx 250 nm					0	1.23	70	-

The calculated Tafel slopes (b) for Ir15Sn15/MMT (79 mV·dec^{-1}) and Ir18Sn12/MMT (90 mV·dec^{-1}) are comparable to that of commercial unsupported IrO_x (70 mV·dec^{-1}), indicating that these catalysts possess more favorable kinetics for the oxygen evolution reaction (OER). The observed improvement in OER efficiency is attributed to the formation of small, well-dispersed metal nanoparticles, as clearly observed by HRTEM analysis (Figures 4 and 5). This results in an increased electrochemically active surface area available for the reaction.

Finally, to evaluate the durability of the catalytic performance, a membrane electrode assembly (MEA) with an Ir18Sn12/MMT anode catalyst was subjected to a 2 h chronoamperometry test at a constant potential of 1.6 V vs. RHE. The initially recorded current density of 140 mA·cm^{-2} during the first hour gradually decreased, then stabilized at approximately 120 mA·cm^{-2}, fluctuating slightly around this value until the end of the experiment (Figure 10). Further in-depth studies are required to thoroughly evaluate the long-term stability of the MEA incorporating the synthesized catalysts, which will help provide a more comprehensive understanding of their durability under extended operating conditions.

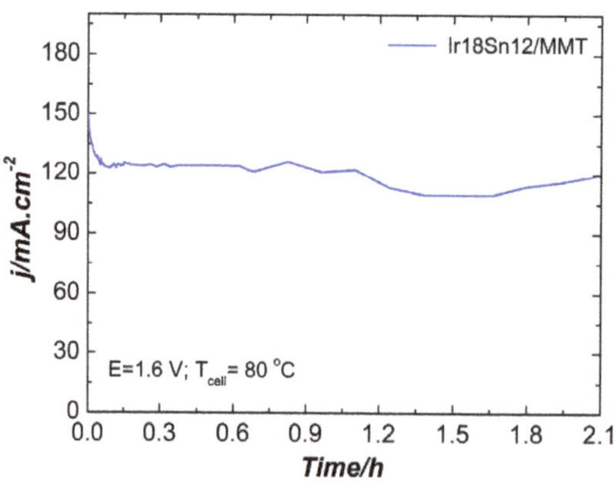

Figure 10. Potentiostatic polarization curve of Ir18Sn12/MMT at 1.8 V vs. RHE, T_{cell} = 80 °C for 2 h.

3. Materials and Methods

3.1. Materials

Tin (II) acetylacetonate (Sn-acac) (99,9% trace metals basis), isopropanol (99.5%) and sulfuric acid were purchased from Sigma-Aldrich (St. Louis, MO, USA). The Iridium (III) 2,4-pentanedionate (Ir-acac) (Ir 37.5% min), Toray paper (GP-H-30), and Nafion D250 solution (5 wt. %) were purchased from Alfa Aesar (Ward Hill, MA, USA). The Nafion 117 membrane was purchased from DuPont Engineering Polymers (Wilmington, DE, USA). The commercial gas diffusion electrode (GDE) with Pt/C catalyst (20 wt. %) was purchased from Johnson Matthey (London, UK). The MMT (Cloisites Na^+) was purchased from Southern Clay Products Inc. (Gonzales, TX, USA). All chemicals were used as received without further purification.

3.2. Synthesis of IrSn/MMT Electrocatalysts

Four different Ir:Sn/MMT samples with 30 wt. % total metal content and varying Ir:Sn ratio (denoted Ir20Sn10/MMT, Ir18Sn12/MMT, Ir15Sn15/MMT, and Ir10Sn20/MMT)) were synthesized through a modified sol–gel method. The calculated amounts of Ir-acac, Sn-acac, and MMT were mixed and dispersed in 20 mL isopropanol, followed by 2 h sonication at 60 °C until a fine gel was obtained. Then, the resulting gel was placed in a quartz mortar and thermally treated in a tube furnace under a hydrogen atmosphere. The tube was first purified by purging with Ar for 10 min, and then a reducing atmosphere was established by introducing H_2 gas for 30 min. Afterward, the system was sealed, and the temperature was gradually increased with a heating rate of 20 °C min^{-1} to 240 °C and kept constant for 2 h to remove the organics and obtain MMT-supported metal nanoparticles.

3.3. Physicochemical Characterization of the Electrocatalysts

The morphology and distribution of Ir and Sn particles on the substrate were observed by High Resolution Transmission Electron Microscopy (HRTEM) and Energy Dispersive X-ray Spectrometry (EDX). HRTEM images were recorded using a JEOL JEM-2100 transmission electron microscope (JEOL Ltd., Tokyo, Japan) equipped with an XEDS system (X-MAXN 80T) fromOxford Instruments (Abingdon, Oxfordshire, UK) and a CCD Camera (ORIUS 1000, 11 Mp) from Gatan Inc. (Pleasanton, CA, USA) operated at an accelerating voltage of 200 kV. Image analysis was performed using Digital Micrograph software. (Version 3.32, Gatan Inc., USA)

X-ray diffractor (XRD) measurements were carried out using a Philips ADP15 diffractometer (Philips Analytical, Eindhoven, Netherlands) with Cu-Kα radiation (λ = 1.54178 Å) at a constant rate of $0.02°s^{-1}$ over an angle range 2θ = 4° ÷ 80°.

3.4. Electrochemical Measurements

All electrochemical measurements were performed using Gamry Reference 3000 Potentiostat/Galvanostat (Gamry Instruments, Warminster, PA, USA). The activity and stability of the prepared IrSn/MMT catalysts toward OER were characterized firstly in a three-electrode glass cell using an aqueous solution of 0.5 M H_2SO_4 as electrolyte. Pt wire was used as a counter electrode and saturated Ag/AgCl electrode as a reference electrode. The metal loading in all samples was 0.5 $mg_{IrSn}cm^{-2}$, and the working area of the test electrodes—0.5 cm^{-2}. The anodic OER steady-state polarization curves in Ar saturated electrolyte were recorded to assess the OER activity of the catalysts. The potential was scanned from the open circuit (OC) potential to 1.8 V at a scan rate of 1 mV s^{-1}. Chronoamperometry experiments at a constant potential of 1.8 V were conducted to evaluate the long-term stability of the catalysts.

3.5. Assembly of MEA and PEMEC Tests

The bimetallic IrSn/MMT catalyst ink and membrane electrode assemblies (MEAs) were prepared following a previously reported procedure [7,19]. Briefly, the synthesized catalyst was mixed with a 5 wt. % Nafion® dispersion (used as an ionomer) and isopropanol as the solvent. The resulting ink was manually coated onto carbon paper gas diffusion layers (GDL). The metal loading was maintained at 0.5 mgIrSn·cm^{-2} for all samples.

The prepared electrodes were then integrated as anodes in MEAs using a Nafion® 117 membrane (Alfa Aesar) as the polymer electrolyte. The cathode and reference electrodes were cut from a commercial gas diffusion electrode (GDE) containing a carbon-supported platinum catalyst (20 wt. % Pt/C, 0.2 mgPt·cm^{-2}). Finally, these electrodes were hot-pressed onto the membrane to complete the assembly.

The catalytic performance was evaluated using a custom-made single-cell PEM electrolyzer with an active area of 1 cm^2. Experiments were performed at both room temperature and elevated temperatures ranging from 40 to 80 °C. Cyclic voltammograms (CVs) were recorded between the hydrogen and oxygen evolution regions at a scan rate of 100 mV·s^{-1}. Additionally, linear sweep voltammograms (LSVs) were collected at a scan rate of 1 mV·s^{-1}. Catalyst durability was assessed via chronoamperometry at a constant potential of 1.6 V vs. RHE.

4. Conclusions

Nanostructured bimetallic IrSn/MMT composites were successfully synthesized and evaluated as anode catalysts for the electrolysis of water in PEMEC. Structural characterization using XRD and HRTEM revealed homogeneous dispersion of both Ir and Sn nanoparticles on the surface of the support as well as intercalation in the interlayer space of the MMT.

The electrochemical investigations demonstrated that the samples with a balanced metal ratio, namely Ir18Sn12/MMT and Ir15Sn15/MMT, exhibit superior OER efficiency with lower overpotential and smaller Tafel slopes compared to the other compositions. The MEA with Ir18Sn12/MMT anode achieved the highest performance with a sustainable current density of around 120 mA.cm^{-2} at only 1.6 V in a long-term stability test in comparison with other presented samples.

Overall, the results indicate that optimizing the Ir:Sn ratio leads to a synergistic effect that enhances catalytic efficiency in PEM electrolysis. The superior performance of Ir18Sn12/MMT and Ir15Sn15/MMT positioned them as cost-effective alternatives to pure Ir catalysts for OER application in PEMECs. Although the exact mechanism behind the observed synergetic interactions between Ir and Sn and their intercalation in the complex layered structure of the catalytic support is not yet fully understood, the results proved enhanced catalyst utilization, efficiency, and stability under operating conditions in PEMEC.

Author Contributions: Conceptualization, I.B., E.L. and E.S.; Data curation, I.B.; Formal analysis, I.B., E.L. and D.P.; Investigation, I.B. and D.P.; Methodology, I.B., G.B. and D.P.; Supervision, E.L., G.B. and E.S.; Validation, G.B. and E.S.; Visualization, I.B.; Writing—original draft, I.B.; Writing—review and editing, E.L. and E.S. All authors have read and agreed to the published version of the manuscript.

Funding: The experiments were performed using equipment of project BG16RFPR002-1.014-0009 form 31 March 2025 "Development and Sustainability of Centre of Competence HITMOBIL", funded by Program "Research, Innovations and Digitalization for Smart Transformation" 2021–2027.

Institutional Review Board Statement: Not applicable.

Informed Consent Statement: Not applicable.

Data Availability Statement: The original contributions presented in this study are included in the article.

Conflicts of Interest: The authors declare no conflicts of interest.

Abbreviations

The following abbreviations are used in this manuscript:

PEMEC	Polymer electrolyte membrane electrolysis cell
HRTEM	High resolution transmission electron microscopy
EDX	Energy-dispersive X-ray spectrometry
XRD	X-ray diffraction patterns
MEA	Membrane electrode assembly
OER	Oxygen evolution reaction
RHE	Reference hydrogen electrode
PGM	Platinum group metal
MMT	Montmorillonite
DFT	Density functional theory
PAW	Projector augmented wave
PBE	Perdew-Burke-Ernzerhof
GDE	Gas diffusion electrode
OC	Open circuit
OCV	Open-circuit voltage
LSV	Linear sweep voltammetry
CV	Cyclic voltammograms

References

1. Li, G.; Yu, H.; Wang, X.; Sun, S.; Li, Y.; Shao, Z.; Yi, B. Highly effective $Ir_xSn_{1-x}O_2$ electrocatalysts for oxygen evolution reaction in the solid polymer electrolyte water electrolyser. *Phys. Chem. Chem. Phys.* **2013**, *15*, 2858–2866. [CrossRef]
2. Wang, C.; Lan, F.; He, Z.; Xie, X.; Zhao, Y.; Hou, H.; Guo, L.; Vignesh, M.; Shao, Q.; Liu, H.; et al. Iridium-based Catalysts for Solid Polymer Electrolyte Electrocatalytic Water Splitting. *ChemSusChem* **2019**, *12*, 1576–1590. [CrossRef] [PubMed]
3. Sun, K.; Mao, W.; Jin, L.; Shi, W.; Niu, W.; Wei, C.; He, Y.; Yan, Q.; Wang, R.; Li, Y.; et al. Enhancing heterointerface coupling for durable industrial-level proton exchange membrane water electrolysis. *Angew. Chem.* **2025**, *137*, e202502250. [CrossRef]
4. Wu, Y.; Yu, Z.; Tong, Y.; Lin, C.; Zhang, N.; Chen, P. Interfacial bridge bonds induced strong electronic coupling of Co@V-WOx catalyst for enhanced concurrent co-electrolysis performance. *Adv. Funct. Mater.* **2025**, 2502104. [CrossRef]
5. Lin, H.Y.; Li, W.J.; Lin, M.Y.; Xu, H.G.; Fang, S.R.; Lv, Y.; Li, W.; Guo, J.; Fu, H.Q.; Yuan, H.Y.; et al. Leaching-induced Ti trapping stabilizes amorphous IrOx for proton exchange membrane water electrolysis. *Angew. Chem.* **2025**, *137*, e202504212.
6. Song, S.; Zhang, H.; Ma, X.; Shao, Z.; Baker, R.T.; Yi, B. Electrochemical Investigation of Electrocatalysts for the Oxygen Evolution Reactionin PEM Water Electrolyzers. *Int. J. Hydrogen Energy* **2018**, *33*, 4955–4961. [CrossRef]
7. Slavcheva, E.; Radev, I.; Bliznakov, S.; Topalov, G.; Andreev, P.; Budevski, E. Sputtered Iridium Oxide Films as Electrocatalysts for Water Splitting Via PEM Electrolysis. *Electrochim. Acta* **2007**, *52*, 3889–3894. [CrossRef]
8. Khan, M.A.; Zhao, H.; Zou, W.; Chen, Z.; Cao, W.; Fang, J.; Xu, J.; Zhang, L.; Zhang, J. Recent Progresses in Electrocatalysts for Water Electrolysis. *Electrochem. Energy Rev.* **2018**, *4*, 483–530. [CrossRef]
9. Ortel, E.; Reier, T.; Strasser, P.; Kraehnert, R. Mesoporous IrO_2 Films Templated by PEO-PB-PEO Block-Copolymers: Self-Assembly, Crystallization Behavior, and Electrocatalytic Performance. *Chem. Mater.* **2011**, *23*, 3201–3209. [CrossRef]
10. Li, G.; Yu, H.; Wang, X.; Yang, D.; Li, Y.; Shao, Z.; Yi, B. Triblock Polymer Mediated Synthesis of Ir-Sn Oxide Electrocatalysts for Oxygen Evolution Reaction. *J. Power Sources* **2014**, *249*, 175–184. [CrossRef]
11. Ardizzone, S.; Trasatti, S. Interfacial Properties of Oxides with Technological Impact in Electrochemistry. *Adv. Colloid Interface Sci.* **1996**, *64*, 173–251. [CrossRef]
12. Mazur, P.; Polonsky, J.; Paidar, M.; Bouzek, K. Non-conductive TiO_2 as the Anode Catalyst Support for PEM Water Electrolysis. *Int. J. Hydrogen Energy* **2012**, *37*, 12081–12088. [CrossRef]
13. Wu, X.; Scott, K. RuO_2 Supported on Sb-doped SnO_2 Nanoparticles for Polymer Electrolyte Membrane Water Electrolysers. *Int. J. Hydrogen Energy* **2011**, *36*, 5806–5810. [CrossRef]
14. Marshall, A.; Børresen, B.; Hagen, H.; Tsypkin, M.; Tunold, R. Electrochemical Characterisation of $Ir_xSn_{1-x}O_2$ Powders as Oxygen Evolution Electrocatalysts. *Electrochim. Acta* **2006**, *51*, 3161–3167. [CrossRef]
15. Chen, G.H.; Chen, X.M.; Yue, P.L. Electrochemical Behavior of Novel Ti/IrOx$-Sb_2O_5-SnO_2$ Anodes. *J. Phys. Chem. B* **2002**, *106*, 4364–4369. [CrossRef]
16. Marshall, A.; Haverkamp, R. Electrocatalytic Activity of IrO_2–RuO_2 Supported on Sb-Doped SnO_2 Nanoparticles. *Electrochim. Acta* **2010**, *55*, 1978–1984. [CrossRef]

17. Di Blasi, A.; D'Urso, C.; Baglio, V.; Antonucci, V.; Arico, A.; Ornelas, R.; Matteucci, F.; Orozco, G.; Beltran, D.; Meas, Y.; et al. Preparation and Evaluation of RuO_2–IrO_2, IrO_2–Pt and IrO_2–Ta_2O_5 Catalysts for the Oxygen Evolution Reaction in an SPE Electrolyzer. *J. Appl. Electrochem.* **2009**, *39*, 191–196. [CrossRef]
18. Jiang, G.; Yu, H.; Hao, J.; Chi, J.; Fan, Z.; Yao, D.; Qin, B.; Shao, Z. An Effective Oxygen Electrode Based on $Ir_{0.6}Sn_{0.4}O_2$ for PEM Water Electrolyzers. *J. Energy Chem.* **2019**, *39*, 23–28. [CrossRef]
19. Felix, C.; Maiyalagan, T.; Pasupathi, S.; Bladergroen, B.; Linkov, V. Synthesis. Characterisation and Evaluation of IrO_2 Based Binary Metal Oxide Electrocatalysts for Oxygen Evolution Reaction. *Int. J. Electrochem. Sci.* **2012**, *7*, 12064–12077. [CrossRef]
20. Boshnakova, I.; Lefterova, E.; Slavcheva, E. Investigation of Montmorillonite as Carrier for OER. *Int. J. Hydrogen Energy* **2018**, *43*, 16897–16904. [CrossRef]
21. Jiang, B.; Dou, B.; Wang, K.; Zhang, C.; Song, Y.; Chen, H.; Xu, Y. Hydrogen Production By Chemical Looping Steam Reforming of Ethanol Using NiO/Montmorillonite Oxygen Carriers in a Fixed-Bed Reactor. *Chem. Eng. J.* **2016**, *298*, 96–106. [CrossRef]
22. Martín, Z.; Jiménez, I.; Gómez, M.A.; Ade, H.W.; Kilcoyne, D.A.; Hernádez-Cruz, D. Spectro microscopy Study of Intercalation and Exfoliation in Polypropylene/Montmorillonite Nanocomposites. *J. Phys. Chem. B* **2009**, *113*, 11160–11165. [CrossRef] [PubMed]
23. Bokov, D.; Jalil, A.T.; Chupradit, S.; Suksata, W.; Ansari, M.J.; Shewael, I.H.; Valiev, G.H.; Kianfar, E. Nanomaterial by sol gel method: Synthesis and application. *Adv. Mater. Sci. Eng.* **2021**, *2021*, 5102014. [CrossRef]
24. Slavcheva, E.; Borisov, G.; Lefterova, E.; Petkucheva, E.; Boshnakova, I. Ebonex Supported Iridiumas Anode Catalystfor PEM Water Electrolysis. *Int. J. Hydrogen Energy* **2015**, *40*, 11356–11361. [CrossRef]
25. Fukushima, Y.; Inagaki, S. Synthesis of an Intercalated Compound of Montmorillonite and 6-Polyamide. *J. Incl. Phenom.* **1987**, *5*, 473–482. [CrossRef]
26. Li, Y.; Wang, X.; Wang, J. Cation Exchange, Interlayer Spacing, and Thermalanalysis of Na/Ca-Montmorillonite Modified with Alkaline and Alkalineearth Metal Ions. *J. Therm. Anal. Calorim.* **2012**, *110*, 1199–1206. [CrossRef]
27. Lanfgord, J.; Wilson, A.A. Survey and some New results in the determination of crystallite size. *J. Appl. Crystallogr.* **1978**, *11*, 102–113.
28. Burke, L.; Whelan, D.A. Voltammetric Investigation of the Charge Storage Reactions of Hydrous Iridium Oxide Layers. *J. Electroanal. Chem. Interfacial Electrochem.* **1984**, *162*, 121–141. [CrossRef]
29. Lee, W.H.; Kim, H. Oxidized Iridium Nanodendrites as Catalysts for Oxygen Evolution Reactions. *Catal. Commun.* **2011**, *12*, 408–411. [CrossRef]
30. Bonanno, M.; Müller, K.; Bensmann, B.; Hanke-Rauschenbach, R.; Aili, D.; Franken, T.; Chromik, A.; Peach, R.; Freiberg, A.T.S.; Thiele, S.; et al. Review and Prospects of PEM Water Electrolysis at Elevated Temperature Operation. *Adv. Mater. Technol.* **2024**, *9*, 2300281. [CrossRef]

inorganics

MDPI

Communication

$NiCo(OH)_2/NiCo_2O_4$ as a Heterogeneous Catalyst for the Electrooxidation of 5-Hydroxymethylfurfural

Wen Li, Di Yin, Wanxin Liu, Yi Li and Yijin Wu *

Hunan Engineering Research Center for Monitoring and Treatment of Heavy Metals Pollution in the Upper Reaches of XiangJiang River, Key Laboratory of Functional Metal-Organic Compounds of Hunan Province, College of Chemistry and Material Science, Hengyang Normal University, Hengyang 421001, China; wenliw2025@126.com (W.L.); z3102685535@163.com (D.Y.); 13875879102@163.com (W.L.); liyi12345678901@foxmail.com (Y.L.)

* Correspondence: wuyj313978@163.com or yjwu@hynu.edu.cn

Abstract: The electrochemical oxidation of biomass-derived 5-hydroxymethylfurfural (HMF) coupled with water electrolysis for green hydrogen production is a promising strategy to address energy crises and environmental pollution. Despite the suitable adsorption energy for HMF due to their partially filled d-band electronic structures, Ni- or Co-based oxides/hydroxides still face challenges in insufficient activity and stability. In this study, a porous heterogeneous nickel cobalt oxide/hydroxide growth on nickel foam (NF), which is defined as NF@NiCo-H/O, was developed via immersion in concentrated alkali solution. Compared with the single-component NiCo oxides, the NF@NiCo-H/O catalyst exhibits a lower application potential of only 1.317 V, 1.395 V, and 1.443 V to achieve current densities of 20, 50, and 100 mA cm^{-2}, respectively, in an alkaline solution containing HMF. Additionally, it demonstrates rapid reaction kinetics with a Tafel slope of 27.6 mV dec^{-1} and excellent cycling stability. Importantly, the presence of more high-valent Ni^{3+}-O species on the catalyst surface contributes to its exceptional selectivity for 2,5-furandicarboxylic acid (86.7%), Faradaic efficiency (93.1%), and conversion rate (94.4%). This catalyst provides some theoretical guidance for the development of biomass electrooxidation catalysts for sustainable energy and chemical production.

Keywords: 5-hydroxymethylfurfural; heterogeneous catalyst; nickel–cobalt compounds; electrochemical performance; active species

1. Introduction

In the context of dwindling global fossil fuels and the transition to green energy, the oxidation reaction of 5-hydroxymethylfurfural (HMF) has emerged as a sustainable biomass conversion technology of significant research importance [1]. As a renewable biomass platform molecule, HMF can be oxidized to produce 2,5-furandicarboxylic acid (FDCA), a high-value chemical that serves as an alternative to traditional petrochemical feedstocks [2]. The electrochemical oxidation of HMF (HMFOR) to FDCA not only enables the efficient utilization of biomass resources but also reduces energy consumption when coupled with water electrolysis for hydrogen production [3]. Moreover, compared to traditional water electrolysis, the HMF electro-oxidation process features lower oxidation potentials and faster reaction kinetics, which enhance hydrogen production efficiency and mitigate potential safety risks [1,2]. Therefore, the in-depth investigation of the surface and interfacial structures of HMFOR catalysts, as well as their reaction mechanisms, is of

https://doi.org/10.3390/inorganics13070211

significant importance for promoting the electrochemical upgrading of biomass and the development of sustainable energy.

At present, the large-scale commercialization of electrochemical oxidation of HMF to produce FDCA faces many severe challenges. The construction of high-performance and cost-effective electrocatalysts is one of the key challenges to address the above-mentioned challenges. A variety of non-noble metal electrocatalysts, including metal oxides (oxides [4,5], hydroxides [6,7], oxyhydroxides [8,9]), sulfides [10,11], phosphides [12,13], carbides [14,15], and metal complexes [16,17], have been developed and their structure–activity relationships have been investigated. In particular, NiCo-based electrocatalysts with abundant 3d electrons and easily tunable e_g orbitals can effectively manage the covalency of metal–oxygen bonding, thereby optimizing the adsorption/desorption behavior of HMF and other intermediates, reducing the protonation/deprotonation energy barriers of active Ni sites, and significantly improving the performance of HMFOR [18]. For example, Liu et al. developed benzoyl ligand-doped $NiCo(OH)_X$ nanowires (BZ-$NiCo(OH)_X$), which have abundant electron-deficient Ni/Co sites for high-performance HMFOR [19]. Li et al. found that $NiCo_2O_4$ nanosheets with low Co/Ni ratios can provide more active sites, which are kinetically favorable for enhancing the intrinsic catalytic activity of HMFOR and the selectivity of FDCA products (98.6%) [20]. Yang et al. found that the introduction of F dopants into $NiCo_2O_4$ to form oxygen vacancies led to crystal structure expansion, increased electron density, and weakened metal–oxygen bonds, ultimately promoting HMF adsorption [21]. Liu et al. introduced Pd nanoclusters into the NiCo surface to obtain Pd/NiCo catalysts; the Pd atoms promoted the formation rate of active Ni^{3+}-O species, thereby enhancing the hybridization adsorption of Pd's d orbitals with the furan ring of HMF [22]. The above reports indicate that the current research on HMF catalysts is primarily focused on the structural and performance regulation of single NiCo-based oxides or hydroxides. However, the impact of NiCo dual-species hybrids or heterogeneous compounds on the performance and mechanism of HMFOR remains underexplored.

In this work, the partial transformation of $NiCo_2O_4$ grown on nickel foam (NF) into $NiCo(OH)_2$ was achieved via immersion in concentrated alkali solution, resulting in a heterogeneous structure with partial conversion from $NiCo_2O_4$ to $NiCo(OH)_2$. This process was confirmed to introduce a greater number of oxygen vacancies in the catalyst compared to the pristine $NiCo_2O_4$. The partial conversion of $NiCo_2O_4 \| NiCo(OH)_2$ was verified by TEM, SEM, and XPS. Further electrochemical performance evaluations revealed that the NF@NiCo-H/O catalyst exhibited rapid HMFOR reaction kinetics, enhanced catalytic activity, stability, and superior FDCA selectivity in an alkaline electrolyte containing HMF. Raman spectroscopy of the NF@NiCo-H/O catalyst before and after the HMFOR reaction demonstrated that the resulting heterogeneous catalyst possessed a higher concentration of high-valent and HMFOR-active Ni^{3+}-O species compared to the $NiCo_2O_4$ catalyst, which is conducive to achieving high HMFOR performance.

2. Results and Discussion

2.1. Characterization of Structure

Figure 1a illustrates the synthetic scheme of the NF@$NiCo_2O_4$ and NF@NiCo-H/O catalysts. The partial substitution of oxygen atoms in $NiCo_2O_4$ with OH^- ions from a concentrated alkaline solution to form $NiCo(OH)_2$ is based on the fact that solids at the nanoscale can undergo flexible anion or cation exchange processes. Consequently, as shown in Equation (1), when metal cations in a metal oxide are in a solution with a high concentration of ligand molecules (L, i.e., OH^-), the O^{2-} ions will exchange with L to form hydroxides. By controlling the reaction time, the heterogeneous NF@NiCo-H/O catalyst can be formed.

$$M_xO_y + OH^- \rightarrow xMOH^{(2y-1)+} + yO^{2-} \quad (1)$$

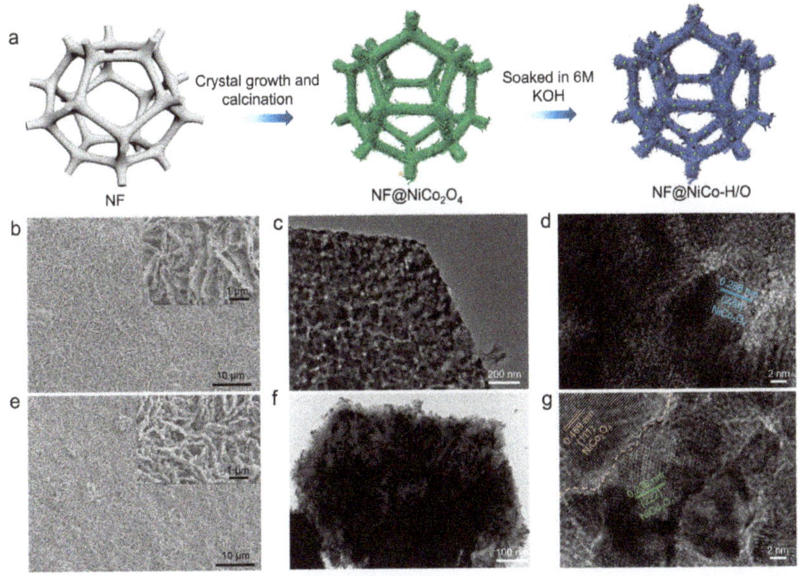

Figure 1. (**a**) Schematic illustration of the synthesis of the NF@NiCo-H/O. SEM images of the (**b**) NF@$NiCo_2O_4$ and (**e**) NF@NiCo-H/O, with insets showing the corresponding magnified images. TEM images for NF@$NiCo_2O_4$ with (**c**) low- and (**d**) high-magnification. TEM images for NF@NiCo-H/O with (**f**) low- and (**g**) high-magnification TEM.

As shown in the SEM image in Figure 1b, NF@$NiCo_2O_4$ is densely grown on the NF substrate, and the initial $NiCo_2O_4$ exhibits a 2D porous sheet morphology at the micrometer scale. This porous sheet structure is expected to expose a greater number of accessibly active sites. In sharp contrast, after the concentrated alkali reaction, the O^{2-} ions in the $NiCo_2O_4$ crystals spontaneously exchange with OH^- ions, leading to the formation of $NiCo(OH)_2$ on the surface of the $NiCo_2O_4$ crystals and ultimately resulting in the heterogeneous NF@NiCo-H/O. The SEM image in Figure 1e displays the 2D sheet morphology of NF@NiCo-H/O. Compared with NF@$NiCo_2O_4$, the surface of NF@NiCo-H/O appears rougher overall. The TEM image of NF@$NiCo_2O_4$ shown in Figure 1c reveals that the micrometer-scale NF@$NiCo_2O_4$ is composed of many small $NiCo_2O_4$ particles that are piled up, with numerous gaps between the particles, which is beneficial for the rapid mass transfer of electrolytes and other reactive substances. The high-magnification HRTEM image shows a lattice spacing of 0.288 nm, corresponding to the (220) crystal plane of $NiCo_2O_4$, indicating that the prepared NF@$NiCo_2O_4$ has a good crystalline structure (Figure 1d). In stark contrast to the formation of NF@$NiCo_2O_4$, the TEM image of NF@NiCo-H/O shows many crystals with lower contrast (Figure 1f), attributed to the formation of $NiCo(OH)_2$. In addition, many small nanoparticles are present on the surface of $NiCo(OH)_2$. The HRTEM image further indicates that these small nanoparticles are composed of $NiCo_2O_4$ with lattice spacings of 0.469 nm and 0.236 nm (Figure 1g). These characterizations demonstrate that after concentrated alkali treatment, the $NiCo_2O_4$ catalyst is partially transformed into $NiCo(OH)_2$, that is, the formation of the heterogeneous $NiCo(OH)_2||NiCo_2O_4$ catalyst.

X-ray diffraction (XRD) was employed to conduct the crystalline analysis of the synthesized catalysts. Owing to the growth of $NiCo_2O_4$ and $NiCo(OH)_2/NiCo_2O_4$ catalysts on the highly conductive NF surface, severe Ni characteristic diffraction peaks from the NF substrate were generated during the PXRD test, thereby masking the characteristic diffraction peak signals of the $NiCo_2O_4$ and $NiCo(OH)_2/NiCo_2O_4$ catalysts grown on its surface. To obtain better PXRD signals, NF@$NiCo_2O_4$ and NF@NiCo-H/O were first subjected to thorough ultrasonication in ethanol, followed by centrifugation, drying, and collection to obtain the corresponding powders. As shown in Figure 2a, a series of diffraction peaks appeared for NF@$NiCo_2O_4$ powder at 18.9°, 31.1°, 36.7°, 38.4°, 44.6°, 55.4°, 59.1°, 64.9°, 68.3°, and 77.5°, which correspond to $NiCo_2O_4$ of PDF#20-0781, indicating that $NiCo_2O_4$ was well grown on the NF surface. For the NF@NiCo-H/O, a series of $NiCo_2O_4$ diffraction peaks of PDF#20-0781 still existed, demonstrating the presence of $NiCo_2O_4$ on its surface. However, the intensity of the $NiCo_2O_4$ diffraction peaks decreased, suggesting a reduction in the amount of $NiCo_2O_4$. Moreover, a sharp peak appeared at 10.2°, corresponding to the (001) plane of $NiCo(OH)_2$, which further indicated that the decreased $NiCo_2O_4$ diffraction peak signals were due to the successful partial transformation of $NiCo_2O_4$ into $NiCo(OH)_2$.

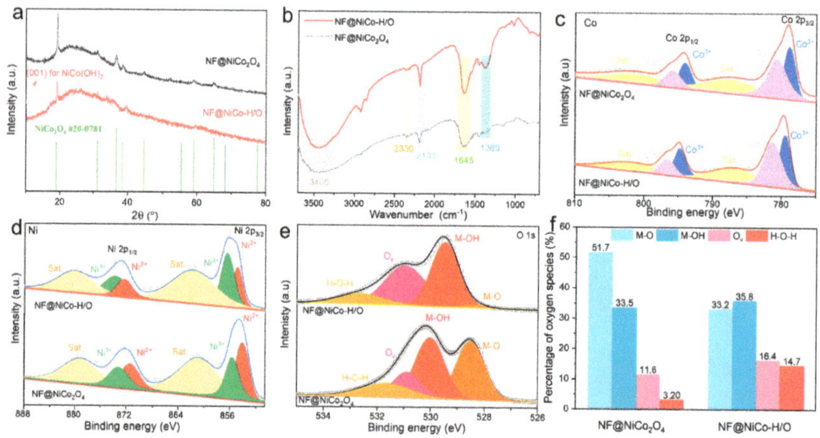

Figure 2. NF@$NiCo_2O_4$ catalyst and the NF@NiCo-H/O of (**a**) PXRD patterns, (**b**) FT-IR spectra. (**c**–**f**) XPS spectrum of (**c**) Co 2p, (**d**) Ni 2p, (**e**) O 1s, and (**f**) the proportion of different oxygen species.

As shown in Figure 2c, FT-IR spectra of NF@$NiCo_2O_4$ and NF@NiCo-H/O are revealed. The broad peaks appearing at 3250–3600 cm^{-1} are attributed to the stretching vibrations of interlayer bonded OH^- in hydroxides or O–H vibrations of surface-adsorbed water molecules [23]. The strong peak at 2330 cm^{-1} corresponds to the stretching vibrations of CO_2 in the air. The strong peak at 2193 cm^{-1} is associated with water molecules adsorbed on the catalyst surface, while the bending vibration signal of adsorbed water is observed at 1645 cm^{-1} [24]. The small peak at 1380 cm^{-1} is attributed to the bending vibration of nitrate anions intercalated in the hydroxide layers [25]. Compared with the NF@$NiCo_2O_4$, the signal intensities of the peaks at 3460, 1645, and 1380 cm^{-1} for NF@NiCo-H/O are enhanced, which, once again, confirms the successful preparation of the heterogeneous catalyst. The surface composition and elemental valence states of the catalysts were analyzed using XPS spectra. As shown in Figure 2c, both catalysts exhibit Co^{3+} (778.9 eV), Co^{2+} (780.7 eV), and corresponding satellite peaks (787.9 eV). However, NF@NiCo-H/O shows a lower Co^{3+}/Co^{2+} peak ratio (1.02) compared with the NF@$NiCo_2O_4$ (1.17), supporting the reduction of high-valent Co^{3+} species in NF@NiCo-H/O. For the metal Ni element (Figure 2d), both catalysts exhibit significant Ni^{3+} (854.1 eV), Ni^{2+} (855.7 eV), and satellite peaks (860.9 eV). Importantly, the Ni^{3+}/Ni^{2+} peak ratio significantly increases from 0.74 (NF@$NiCo_2O_4$) to 1.26 (NF@NiCo-H/O). This decrease in the valence state of Co and increase in that of Ni indicate that partial electron transfer (PET) occurs within the NF@NiCo-H/O crystal, that is, from Co to Ni. The occurrence of this PET also promotes the formation of high-valent active Ni^{3+} species, supporting high-performance HMFOR.

2.2. *Performance of HMFOR*

The HMFOR and OER performance of NF@$NiCo_2O_4$ and NF@NiCo-H/O were evaluated in 1.0 M KOH electrolyte with and without 50 mM HMF using a three-electrode system. Figure 3a shows the linear sweep voltammetry (LSV) curves, with the solution without HMF serving as the control. As shown in Figure 3b, NF@NiCo-H/O required only 1.317 V, 1.395 V, and 1.443 V to achieve current densities of 20 mA cm^{-2}, 50 mA cm^{-2}, and 100 mA cm^{-2}, respectively, which are significantly lower than those of NF@$NiCo_2O_4$ (1.413 V, 1.458 V, and 1.514 V). Moreover, NF@NiCo-H/O achieved a limiting current density of 380 mA cm^{-2} at a voltage of 1.7 V, which is notably higher than that of the NF@$NiCo_2O_4$ catalyst (250 mA cm^{-2}). Compared with the OER performance in 1.0 M KOH electrolyte without 50 mM HMF, the HMFOR performance of both NF@NiCo-H/O and NF@$NiCo_2O_4$ catalysts was significantly superior to their OER performance. Specifically, the potential difference between HMFOR and OER for the NF@NiCo-H/O catalyst at 100 mA cm^{-2} was 197 mV, while for the NF@$NiCo_2O_4$ catalyst at 20 mA cm^{-2}, the HMFOR and OER potential difference was 236 mV. In addition, the NF@NiCo-H/O catalyst outperforms the recently reported non-noble metal catalysts (Figure 3f) [26–32], which also demonstrates the great potential of heterogeneous NF@NiCo-H/O catalysts for HMFOR.

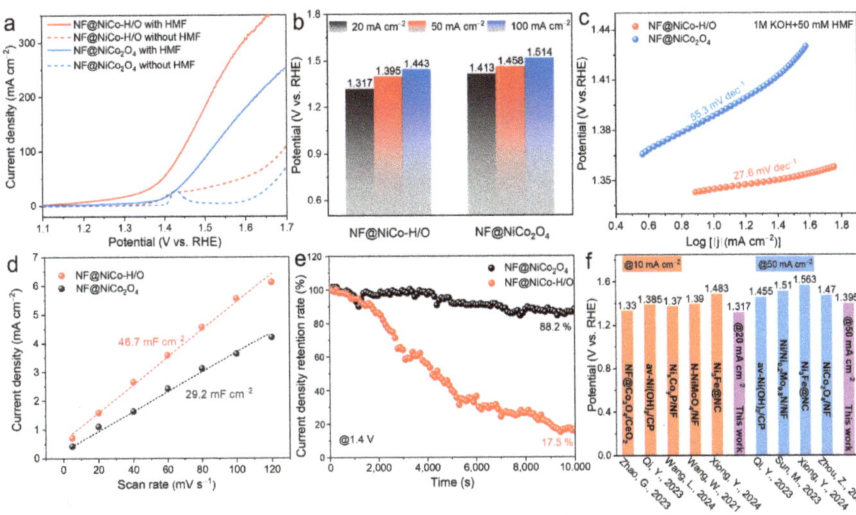

Figure 3. The NF@$NiCo_2O_4$ and NF@NiCo-H/O of (**a**) the LSV profiles, (**b**) the corresponding potentials value for current density of 20, 50, 100 mA cm^{-2}, (**c**) Tafel plot, (**d**) capacitive currents as a function of scan rates, and (**e**) chronoamperometry measurement at 1.4 V. (**f**) Comparison of HMFOR activity in this work [26–32].

The LSV curves and Tafel equation ($\eta = a + b \cdot \log j$) are employed to obtain the Tafel slope, where η represents the overpotential, j denotes the current density, and the slope b corresponds to the Tafel slope. A lower Tafel slope reflects a faster electron transfer or a more facile kinetic nature of the chemical steps in the electrocatalytic reaction. The corresponding Tafel plot in Figure 3c reveals that NF@NiCo-H/O possesses an extremely low Tafel slope

(27.6 mV dec^{-1}), which, compared with NF@$NiCo_2O_4$ (36.7 mV dec^{-1}), supports the notion that the heterogeneous NF@NiCo-H/O exhibits superior HMFOR catalytic kinetics than that of single-component $NiCo_2O_4$. Furthermore, the electrochemical active surface area (ECSA) of the as-prepared catalysts, an important parameter for assessing activity, was estimated through double-layer capacitance (C_{dl}), which was determined by the cyclic voltammetry curves with the different scan rates shown in Figure 3d. As shown in Figure 3d, NF@NiCo-H/O exhibits a higher C_{dl} value (46.7 mF cm^{-2}) than NF@$NiCo_2O_4$ (29.2 mF cm^{-2}), supporting the presence of a heterogeneous interface and a large number of high-valent Ni^{3+} species in NF@NiCo-H/O, which may be the reasons for the high ECSA. The long-term durability of the two catalysts was also investigated. During a chronoamperometry measurement at 1.4 V (in 1.0 M KOH electrolyte containing 50 mM HMF, without replacing the electrolyte during the test). After a reaction process of 10,000 s, the current density of NF@NiCo-H/O showed a decrease of 82.5% (Figure 3e). In sharp contrast, NF@$NiCo_2O_4$ exhibited only an 11.8% decrease in current. HPLC analysis of the electrolyte after the durability test revealed that the concentration of HMF in the electrolyte of NF@NiCo-H/O was essentially close to 0 mM, while that of NF@$NiCo_2O_4$ was still around 44 mM. The above durability results indicate that NF@NiCo-H/O can rapidly and continuously oxidation HMF molecules. The above results on HMFOR activity and stability suggest that the formation of a heterogeneous hybrid of $NiCo(OH)_2$ and $NiCo_2O_4$ can enhance the catalytic activity and stability of HMFOR than the single-component $NiCo_2O_4$.

The HMFOR was performed using NF@NiCo-H/O as the working electrode in an H-type electrolysis cell with an anion exchange membrane to analyze the oxidation products in detail. As shown in Figure 4a, the oxidation of HMF to FDCA typically involves two potential pathways due to the coexistence of hydroxyl and aldehyde groups [1]. For Pathway 1, the aldehyde group is preferentially oxidized to a carboxyl group to form HMFCA [2]. In Pathway 2, the hydroxyl group is oxidized to an aldehyde group to generate DFF [3]. Figure 4b,c shows that, as the reaction time progresses, the peak intensity of HMF in the HPLC spectrum decreases (15 C), while the peak intensity of FDCA increases. Meanwhile, the concentration of the product intermediate HMFCA first rises (not exceeding 1 mM) and then declines, eventually approaching zero. Additionally, a small amount of FFCA intermediate was detected, while the signal for DFF was essentially zero. Therefore, it is that this may be due to the rate of conversion of HMFCA to FFCA being slower than the rate of conversion of FFCA to FDCA, leading to the accumulation of the HMFCA intermediate until most of the HMF is consumed. This indicates that the oxidation of HMFCA to FFCA may be the rate-determining step (RDS) in the HMFOR process.

To confirm the rate-determining step in the conversion of HMF to FDCA, we investigated the oxidation rates of HMF and its intermediate products. Assuming that the oxidation process of HMF, HMFCA, and FFCA follows pseudo-first-order kinetics, the plots of $\ln(C_0/C)$ versus reaction time should be linear (Figure S3a–c). It was found that both the NF@NiCo-H/O and NF@$NiCo_2O_4$ catalysts exhibited lower rate constants for the oxidation of HMFCA compared to those for HMF and FFCA, further confirming that the conversion of HMFCA to FFCA is the RDS in both catalysts. Interestingly, the NF@NiCo-H/O catalyst showed significantly higher oxidation rates for HMF, HMFCA, and FFCA compared to NF@$NiCo_2O_4$, with rate constants that were 2.42, 2.79, and 2.86 times higher, respectively. This suggests that the reaction kinetics and RDS process of HMFOR are significantly improved through the synergistic interaction between the heterojunction phases of $NiCo_2O_4$ and $NiCo(OH)_2$. The above results indicate that the HMFOR process of NF@NiCo-H/O belongs to Pathway I, that is, HMF→HMFCA→FFCA→FDCA. Notably, NF@NiCo-H/O exhibits outstanding HMFOR performance at 1.424 V, with a high HMF conversion rate of 94.4%, an FDCA yield of 86.7%, a selectivity of 92.2%, and a Faradaic efficiency (F.E.) of

93.1%. To investigate the cyclic durability of the catalyst, twelve consecutive cycles were conducted at 1.424 V to assess the durability of the NF@NiCo-H/O catalyst. As shown in Figure 4c and Figure S4, the selectivity of FDCA, the Faradaic efficiency, and the conversion rate of HMF all remained above 85.8% after twelve cycles, demonstrating the excellent stability of the NF@NiCo-H/O.

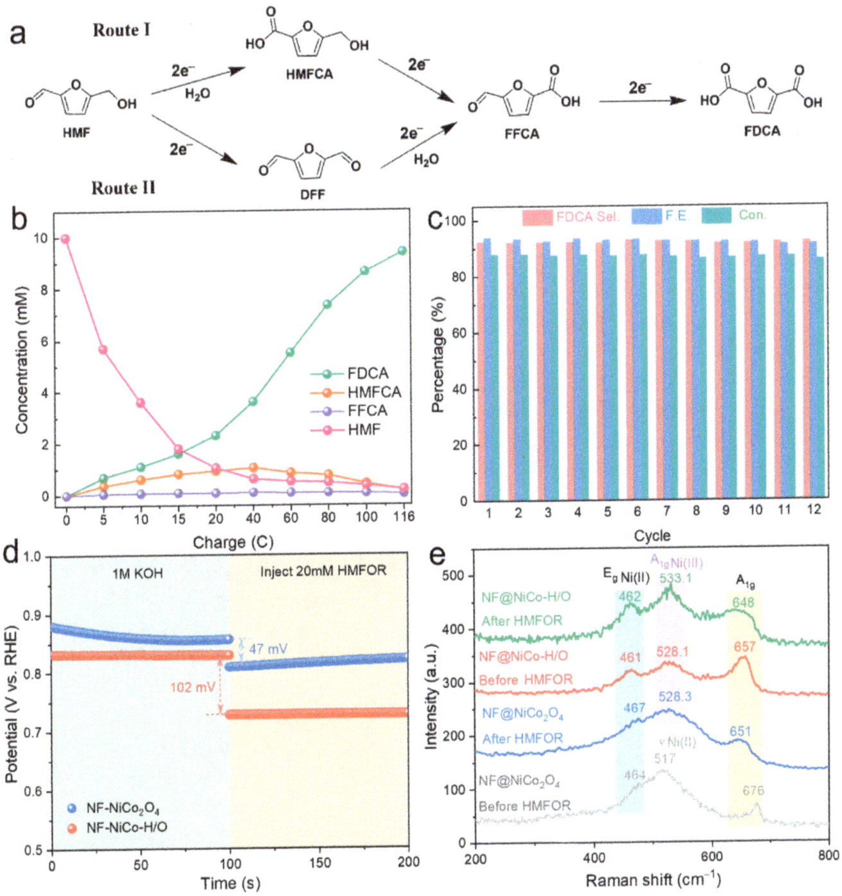

Figure 4. (**a**) Two pathways for the electrooxidation of HMF to FDCA. NF@NiCo-H/O catalyst of (**b**) concentration changes of various intermediate products under different charge quantities, (**c**) conversion rate of HMF, selectivity of FDCA, and Faradaic efficiency during twelve consecutive HMF electrooxidation reactions at 1.424 V. (**d**) OCP measurement of NF@$NiCo_2O_4$ and NF@NiCo-H/O. (**e**) Raman spectra of the NF@$NiCo_2O_4$ and NF@NiCo-H/O before and after chronoamperometry measurement.

In electrochemical reactions of biomass, the electrocatalytic activity is closely related to the concentration, accessibility of active sites, and the adsorption capacity for organic molecules [33]. During the HMFOR process, the adsorption of HMF molecules on the electrode surface plays a crucial role in constructing an efficient HMFOR. Therefore, it is of great importance to regulate the adsorption of HMF molecules. The adsorption of HMF molecules on NF@$NiCo_2O_4$ and NF@NiCo-H/O was investigated using the open-circuit potential (OCP) [4]. When 50 mM HMF was injected into a 1 M KOH solution, the OCP drop of NF@NiCo-H/O (102 mV) was larger than that of NF@$NiCo_2O_4$ (47 mV), indicating that NF@NiCo-H/O has a stronger adsorption capacity toward HMF molecules than NF@$NiCo_2O_4$ (Figure 4d). To further explore the catalytic mechanism of NF@NiCo-H/O for HMF, Raman spectroscopy was employed to analyze the catalysts before and after the durability test. For the NF@$NiCo_2O_4$ catalyst, the Raman signals at 464, 517, and 676 cm^{-1} before the reaction correspond to the E_g (Ni^{II}-O), Ni^{II}-O, and A_{1g} vibrations of $NiCo_2O_4$ [34]. However, after the HMFOR stability test, the NF@$NiCo_2O_4$ catalyst exhibited a Raman signal at 528.3 cm^{-1}, corresponding to the A_{1g} (Ni^{III}-O) vibration of nickel-based hydroxides, and the A_{1g} vibration at 676 cm^{-1} red-shifted to 651 cm^{-1} with a decrease in intensity, indicating partial transformation of the NF@$NiCo_2O_4$ catalyst's surface, resulting in the formation of more high-valent and HMFOR-active Ni^{3+}-O species [35]. In contrast, the NF@NiCo-H/O catalyst exhibited the A_{1g} (Ni^{III}-O) vibration signal of nickel-based hydroxides before the reaction. After the HMFOR stability test, the A_{1g} (Ni^{III}-O) vibration signal of nickel-based hydroxides blue-shifted by 5 cm^{-1} and increased in intensity, while the A_{1g} vibration signal at 657 cm^{-1} decreased, further indicating that more $NiCo_2O_4$ structures were transformed into HMFOR-active nickel-based hydroxides, that is, more Ni^{3+}-O species were generated.

Furthermore, ICP-OES analysis was performed on the NF@NiCo-H/O catalysts before and after the HMFOR stability test. Notably, the Co/Ni ratio of the NF@NiCo-H/O decreased from 2.3 before testing to 1.9 after testing, indicating partial dissolution of metallic Co species and structural evolution of the catalyst. High-resolution XPS characterization was further employed to investigate the changes in valence states and electronic structures of the catalysts before and after HMFOR stability testing. The electronic structural changes of metallic Ni and Co species were analyzed by XPS spectra. The atomic percentages obtained by XPS reveal that the Ni/Co atomic ratio on the surface of the NF@NiCo-H/O catalyst has increased from 0.70 before testing to 1.16 after testing (Figure S5a). As shown in Figure S5b,d, compared to the Ni^{3+}/Ni^{2+} ratio of 1.26 in the NF@NiCo-H/O-before testing, the high-resolution Ni 2p spectra of the NF@NiCo-H/O-after testing show only the presence of Ni^{3+} species. Additionally, the high-resolution Co 2p spectra show a lower Co^{3+}/Co^{2+} ratio (0.44) for NF@NiCo-H/O-after testing compared to the value of 1.17 toward NF@NiCo-H/O-before testing (Figure S5c). Additionally, significant differences were observed in the high-resolution O 1s spectra. As shown in Figure S5e,f, the M-O bonds disappeared in the NF@NiCo-H/O-after testing, and the proportion of oxygen vacancies increased significantly. These changes in the high-resolution XPS spectra indicate that partial dissolution of metallic Co species occurred on the surface of the NF@NiCo-H/O-after testing, with some electrons from Ni being transferred to Co (inset of Figure S5d), leading to partial structural evolution on the catalyst surface, forming a heterogeneous interface $NiCo_2O_4||NiCo(OH)_2$ and generating more Ni^{3+}-O species and oxygen vacancies. The SEM image of the NF@NiCo-H/O-after testing also supports this (Figure S6). These findings are consistent with the Raman results presented in the revised manuscript (Figure 4e).

Previous literature has indicated that nickel/cobalt species can work synergistically to enhance the catalytic activity and stability of HMFOR [2,3]. Consequently, we delved further into the synergistic catalytic mechanism of nickel and cobalt in the NF@NiCo-H/O catalyst. To this end, we synthesized monometallic heterojunction catalysts containing only nickel or cobalt (denoted as NF@Ni-H/O and NF@Co-H/O catalysts), using the same synthesis procedure. As depicted in Figure S7a,b, the LSV curves and potential values at 10, 50, and 100 mA cm^{-2} demonstrate the following activity order: NF@NiCo-H/O > NF@$NiCo_2O_4$ > NF@Ni-H/O > NF@Co-H/O. Specifically, the NF@Ni-H/O catalyst performs markedly better than the NF@Co-H/O catalyst. Moreover, the NF@NiCo-H/O catalyst, which incorporates both nickel and cobalt, exhibits the highest catalytic activity.

Hence, in this study, it is proposed that cobalt and nickel in the NF@NiCo-H/O catalyst synergistically catalyze HMF conversion, leading to high HMFOR activity.

Previous literature has also indicated that different metal species can selectively oxidize different functional groups of HMF [8]. Thus, to analyze the role of individual Ni, Co, and bimetallic Ni/Co catalysts in the selective oxidation of HMF, we compared their cyclic voltammetry (CV) curves on 2-furaldehyde and furfuryl alcohol. As shown in Figure S8, all single-metal (Co and Ni) and bimetallic catalysts (NF@NiCo-H/O and NF@$NiCo_2O_4$) exhibited higher activity for furfuryl alcohol catalysis than for furaldehyde catalysis. At 1.4 V, the activity order for the four catalysts was NF@NiCo-H/O > NF@$NiCo_2O_4$ > NF@Ni-H/O > NF@Co-H/O, based on the current density. These results confirm that the synergistic effect between Ni/Co sites effectively accelerates the electrooxidation of both hydroxyl and aldehyde groups.

In summary, the above results indicate that the heterojunction interface formed between $NiCo_2O_4$ and $NiCo(OH)_2$, namely, $NiCo_2O_4||NiCo(OH)_2$, may facilitate the generation of high-activity Ni^{3+}-O species during the HMF oxidation process. In addition, through the synergistic action of cobalt and nickel metal species, the catalytic activity, selectivity, and long-term durability of the high-activity Ni^{3+}-O species towards HMF can be significantly enhanced.

3. Materials and Methods

3.1. Materials

Anhydrous ethanol (≥99.7%), acetone (≥99.5%), hydrochloric acid (36.0–38.0%), cobalt nitrate hexahydrate (99%), nickel nitrate hexahydrate (99%), urea (99%), ammonium chloride (≥99.5%), chromatographic-grade methanol, and chromatographic-grade formamide were purchased from Shanghai Guoyao Co., Ltd., Shanghai, China. 5-Hydroxymethylfurfural (95%) was obtained from Aladdin Reagents Co., Ltd., Riverside, CA, USA.

3.2. Preparation of NF@$NiCo_2O_4$

A piece of NF (2 × 0.5 cm) was subjected to ultrasonic cleaning for 15 min in a sequence of acetone, 1 M HCl, deionized water, and ethanol. The cleaned NF was then dried in an oven at 40 °C and kept for further use. Subsequently, a piece of clean NF was placed into a solution containing 2.0 mmol $Co(NO_3)_2{\cdot}6H_2O$, 1.0 mmol $Ni(NO_3)_2{\cdot}6H_2O$, 12 mmol urea, and 20 mmol ammonium chloride, which were mixed and dissolved in a mixture of 20 mL deionized water and 20 mL ethanol. This solution, along with the NF, was transferred into a 100 mL Teflon-lined stainless-steel autoclave and subjected to solvothermal reaction at 120 °C. After 2 h, the electrode was naturally cooled to room temperature and washed several times with deionized water and ethanol. The resulting electrode was heated to 300 °C at a heating rate of 5 $°C{\cdot}min^{-1}$ and maintained at this temperature for 2 h to obtain the NF@$NiCo_2O_4$.

3.3. Preparation of NF@NiCo-H/O

A piece of NF@$NiCo_2O_4$ was immersed in a 6 M KOH solution and stirred slowly for 12 h. The catalyst was then taken out and washed several times with deionized water and ethanol to obtain the NF@NiCo-H/O.

3.4. Physicochemical Characterization

The crystalline structure of the catalysts was determined using an X-ray diffractometer (XRD, Philips X-Pert Pro, Almere, The Netherlands) with Cu Kα radiation (1.5418 Å). The microscopic structure were analyzed by transmission electron microscopy (TEM, Tecnai G2

F20, Hillsboro, OR, USA, and JEM-2100, Tokyo, Japan) at 200 kV, in which the catalyst was first subjected to ultrasonication in an ethanol solution for 1 h, after which the catalyst was drop-cast onto a copper grid with an ultrathin carbon film. The morphology and structure of the catalyst were analyzed by Scanning electron microscopy (SEM, Hitachi SU8000, Tokyo, Japan) at 10 kV. The elemental composition and valence state were determined by X-ray photoelectron spectroscopy (XPS, ESCALAB 250Xi, Waltham, MA, USA) with a monochromatized Al Kα X-ray excitation source (1487 eV). ICP-OES was carried out on a NexION 2000-(A-10) (Waltham, MA, USA) to determine the Ni and Co concentrations. The surface structure and functional group information of the catalysts were identified using an Xplora confocal microprobe Raman system (HORIBA Jobin Yvon, Glasgow, UK) at 532 nm and Fourier-transform infrared spectroscopy (FT-IR, Nicolet Nexus, Shanghai, China).

3.5. Electrochemical Testing

Electrochemical tests were conducted using a dual potentiostat (Wuhan KEST CS2350M). The electrodes prepared as described above, a mercury oxide electrode and a carbon rod, were employed as the working electrode, reference electrode, and counter electrode, respectively. The exposed surface area of the working electrode toward as-prepared electrode is 0.5 cm^2 during the measurements. The OER or HMFOR performance was evaluated in 1 M KOH electrolyte with or without 50 mM HMF at room temperature. The catalyst was activated at a scan rate of 50 $mV \cdot s^{-1}$ until the curves overlapped in 1 M KOH with 50 mM HMF, after which other electrochemical performance tests could be carried out. The tested potentials were converted to potentials relative to the reversible hydrogen electrode (vs. RHE) through calculation. The electrochemical data were presented with 85% iR correction.

In order to quantitatively analyze the products of HMF oxidation and calculate the corresponding Faradaic efficiency, 10 μL of the electrolyte was extracted after electrolysis at a potential of 1.45 V (in a three-electrode configuration) for different amounts of charge (5C, 10C, 15C, 20C, 40C, 60C, 80C, 100C, 116C), and then neutralized with dilute sulfuric acid to neutrality. The as-solution were tested using high-performance liquid chromatography (HPLC, Waters 1525): the analysis was performed with a UV-visible detector at a detection wavelength of 265 nm; a C18 column (4.6 mm × 150 mm, Shim-pack GWS, 5 μm) was employed with mobile phase A (methanol) and B (5 mM ammonium formate aqueous solution, A:B = 3:7) at a flow rate of 0.6 $mL \cdot min^{-1}$.

The external standard method was employed for qualitative and quantitative analysis of production (HMF/FDCA/HMFCA/FFCA). HMF conversion, FDCA yield, and FE value of FDCA were calculated using the following equations:

$$\text{HMF coversion} = \frac{mole\ of\ consumed\ HMF}{mole\ of\ initial\ HMF} \times 100\% \quad (2)$$

$$\text{FDCA yield} = \frac{mole\ of\ FDCA\ formed}{mole\ of\ initial\ HMF} \times 100\% \quad (3)$$

$$\text{FE of FDCA} = \frac{mole\ of\ FDCA\ formed}{total\ charge\ passed/(6 \times F)} \times 100\% \quad (4)$$

where F is the Faraday constant (96,485 C mol^{-1}).

4. Conclusions

In summary, the partial transformation of $NiCo_2O_4$ into $NiCo(OH)_2$ was achieved via immersion in concentrated alkali solution, resulting in the formation of the heterogeneous NF@NiCo-H/O catalyst. A variety of physicochemical characterizations supported the growth of $NiCo(OH)_2$ sub-nanoparticles on the porous micrometer-scale $NiCo_2O_4$ surface, creating abundant heterogeneous interfaces of $NiCo_2O_4||NiCo(OH)_2$. The HMFOR performance revealed that, compared with single-component NF@$NiCo_2O_4$, the heterogeneous NF@NiCo-H/O catalyst exhibited higher HMFOR activity and stability, as well as high HMF conversion rate (86.0%), FDCA yield (83.1%), and Faradaic efficiency (85.7%). More importantly, Raman spectra before and after the stability test indicated that the generation of more high-activity Ni^{3+}-O species on the NF@NiCo-H/O catalyst, compared with the single-component NF@$NiCo_2O_4$, during the HMFOR process, is the main reason for the higher activity and selectivity of HMFOR.

Supplementary Materials: The following supporting information can be downloaded at: https://www.mdpi.com/article/10.3390/inorganics13070211/s1, Figure S1: SEM images of low-magnification of (a) NF@$NiCo_2O_4$, and (b) NF@NiCo-H/O, and high-magnification of (c) NF@$NiCo_2O_4$, and (d) NF@NiCo-H/O; Figure S2: CV curves of (a) NF@NiCo-H/O, (b) NF@$NiCo_2O_4$ acquired at scan rates of 10, 20, 40, 60, 80, 100, and 120 mV s^{-1}; Figure S3: $\ln(C_0/C)$ based on (a) HMF, (b) HMFCA, and (c) FFCA vs the reaction time for the HMFOR process; Figure S4: CV curve of (a) NF@NiCo-H/O, (c) NF@$NiCo_2O_4$, (e) NF@Ni-H/O, and (g) NF@Co-H/O in 1 M KOH and 10 mM 2-furaldehyde solution at a scan rate of 5 mV s^{-1}. CV curve of (b) NF@NiCo-H/O, (d) NF@$NiCo_2O_4$, (f) NF@Ni-H/O, and (h) NF@Co-H/O in 1 M KOH and 10 mM furfuryl alcohol solution at a scan rate of 5 mV s^{-1}; Figure S5: XPS characterization of NF@NiCo-H/O catalysts before and after HMFOR stability testing in 1 M KOH + 50 mM HMF for 12 h. (a) The atomic ratios of the NF@NiCo-H/O catalyst before and after the HMFOR stability testing. High-resolution XPS spectra of (b) Ni 2p and (c) Co 2p, (d) the ratio of different oxidation state species of metals, (e) O 1s, and (f) proportion of different oxygen species for NF@NiCo-H/O before and after HMFOR stability testing. Inset of (d) is schematic diagram of partial electron transfer between Ni-O and Co-O; Figure S6: SEM images of the catalyst at different magnifications after HMFOR stability testing in 1 M KOH + 50mM HMF for 12 h; Figure S7: Catalysts of (a) LSV profiles in solution of 1M KOH with 50mM HMF, (b) the corresponding potentials value for current density of 10, 50, 100 mA cm^{-2}. Catalysts of (c) LSV profiles in solution of 1M KOH without 50mM HMF; Figure S8: CV curve of (a) NF@NiCo-H/O, (c) NF@$NiCo_2O_4$, (e) NF@Ni-H/O, and (g) NF@Co-H/O in 1 M KOH and 10 mM 2-furaldehyde solution at a scan rate of 5 mV s^{-1}. CV curve of (b) NF@NiCo-H/O, (d) NF@$NiCo_2O_4$, (f) NF@Ni-H/O, and (h) NF@Co-H/O in 1 M KOH and 10 mM furfuryl alcohol solution at a scan rate of 5 mV s^{-1}; Table S1: Comparison of HMFOR activity between NF@NiCo-H/L catalyst and other well-developed electrocatalysts. References [26–32] are cited in the supplementary materials.

Author Contributions: Conceptualization, Y.W.; methodology, Y.W.; software, D.Y. and W.L. (Wen Li); validation, W.L. (Wen Li), Y.L. and D.Y.; formal analysis, D.Y.; investigation, W.L. (Wanxin Liu), W.L. (Wen Li) and Y.L.; resources, Y.W.; data curation, Y.L.; writing—original draft preparation, W.L. (Wen Li) and D.Y.; writing—review and editing, W.L. (Wen Li) and Y.W.; visualization, W.L. (Wanxin Liu); supervision, Y.W.; project administration, Y.W.; funding acquisition, Y.W., W.L. (Wanxin Liu) and D.Y. contributed equally to this work; all authors analyzed the data; all authors discussed the results of this paper. All authors have read and agreed to the published version of the manuscript.

Funding: This research was supported by the Academic College Student Innovation and Entrepreneurship Training Program (No. S202210546010), the National Natural Science Foundation Program of China (No. 52301271), and the Natural Science Foundation of Hunan Province of China (No. 2024JJ6087).

Institutional Review Board Statement: Not applicable.

Informed Consent Statement: Not applicable.

Data Availability Statement: The raw data supporting the conclusions of this article will be made available by the authors upon request.

Conflicts of Interest: The authors declare no conflicts of interest.

References

1. Wu, X.; Wang, Y.; Wu, Z.-S. Design principle of electrocatalysts for the electrooxidation of organics. *Chem* **2022**, *8*, 2594–2629. [CrossRef]
2. Chen, C.; Lv, M.; Hu, H.; Huai, L.; Zhu, B.; Fan, S.; Wang, Q.; Zhang, J. 5-Hydroxymethylfurfural and its downstream chemicals: A review of catalytic routes. *Adv. Mater.* **2024**, *36*, 2311464.
3. Simoska, O.; Rhodes, Z.; Weliwatte, S.; Cabrera-Pardo, J.R.; Gaffney, E.M.; Lim, K.; Minteer, S.D. Advances in electrochemical modification strategies of 5-hydroxymethylfurfural. *ChemSusChem* **2021**, *14*, 1674–1686.
4. Pham, M.T.H.; Ang, A.W.Y.; Vo, T.G.; Hayashi, T.; Chiang, C.Y. Unlocking the potential of mixed-valence silver oxide for electrochemical valorization of 5-hydroxymethylfurfural into valuable products. *Mater. Today Sustain.* **2024**, *28*, 100992. [CrossRef]
5. Kubota, S.R.; Choi, K.S. Electrochemical oxidation of 5-hydroxymethylfurfural to 2, 5-furandicarboxylic acid (FDCA) in acidic media enabling spontaneous FDCA separation. *ChemSusChem* **2018**, *11*, 2138–2145. [CrossRef] [PubMed]
6. Zubair, M.; Usov, P.M.; Ohtsu, H.; Yuwono, J.A.; Gerke, C.S.; Foley, G.D.; Hackbarth, H.; Webster, R.F.; Yang, Y.; Lie, W.H.; et al. Vacancy mediated electrooxidation of 5-hydroxymethyl furfuryl using defect engineered layered double hydroxide electrocatalysts. *Adv. Energy Mater.* **2024**, *14*, 2400676. [CrossRef]
7. Woo, J.; Moon, B.C.; Lee, U.; Oh, H.S.; Chae, K.H.; Jun, Y.; Lee, D.K. Collaborative electrochemical oxidation of the alcohol and aldehyde groups of 5-hydroxymethylfurfural by NiOOH and $Cu(OH)_2$ for superior 2, 5-furandicarboxylic acid production. *ACS Catal.* **2022**, *12*, 4078–4091. [CrossRef]
8. Taitt, B.; Nam, D.; Choi, K. A comparative study of nickel, cobalt, and iron oxyhydroxide anodes for the electrochemical oxidation of 5-hydroxymethylfurfural to 2, 5-furandicarboxylic acid. *ACS Catal.* **2018**, *9*, 660–670. [CrossRef]
9. Bender, M.T.; Choi, K.S. Electrochemical oxidation of HMF via hydrogen atom transfer and hydride transfer on NiOOH and the impact of NiOOH composition. *ChemSusChem* **2022**, *15*, e202200675. [CrossRef]
10. Han, W.; Tang, M.; Li, J.; Li, X.; Wang, J.; Zhou, L.; Yang, Y.; Wang, Y.; Ge, H. Selective hydrogenolysis of 5-hydroxymethylfurfural to 2,5-dimethylfuran catalyzed by ordered mesoporous alumina supported nickel-molybdenum sulfide catalysts. *Appl. Catal. B* **2020**, *268*, 118748. [CrossRef]
11. Sun, Y.; Wang, J.; Qi, Y.; Li, W.; Wang, C. Efficient electrooxidation of 5-hydroxymethylfurfural using Co-doped Ni_3S_2 catalyst: Promising for H_2 production under industrial-level current density. *Adv. Sci.* **2022**, *9*, 2200957.
12. Xu, X.; Song, X.; Liu, X.; Wang, H.; Hu, Y.; Xia, J.; Chen, J.; Shakouri, M.; Guo, Y.; Wang, Y. A highly efficient nickel phosphate electrocatalyst for the oxidation of 5-hydroxymethylfurfural to 2,5-furandicarboxylic acid. *ACS Sustain. Chem. Eng.* **2022**, *10*, 5538–5547. [CrossRef]
13. Fu, J.; Yang, G.; Jiao, Y.; Tian, C.; Yan, H.; Fu, H. Ni-Cu-based phosphide heterojunction for 5-hydroxymethylfurfural electrooxidation-assisted hydrogen production at large current density. *Nano Energy* **2024**, *127*, 109727.
14. Huang, R.; Jiang, J.; Liang, J.; Wang, S.; Chen, Y.; Zeng, X.; Wang, K. Selective hydrogenation of 5-hydroxymethylfurfural triggered by a high Lewis acidic Ni-based transition metal carbide catalyst. *Green Energy Environ.* **2025**, *10*, 573–584. [CrossRef]
15. Li, Y.; Alorku, K.; Shen, C.; Yan, L.; Li, Q.; Tian, X.; Li, W.; Xu, Y.; Wang, C.; Li, C.; et al. In-situ redispersion of Ni@C catalyst boosts 5-hydroxymethylfurfural electrooxidation by increasing Ni^{4+} sites. *Appl. Catal. B* **2024**, *357*, 124250.
16. Huang, N.; Chu, B.; Chen, D.; Shao, B.; Zheng, Y.; Li, L.; Xiao, X.; Xu, Q. Rational design of a quasi-metal–organic framework by ligand engineering for efficient biomass upgrading. *J. Am. Chem. Soc.* **2025**, *147*, 8832–8840. [CrossRef]
17. Yang, S.; Guo, Y.; Zhao, P.; Jiang, H.; Shen, H.; Chen, Z.; Jiang, L.; Xue, X.; Zhang, Q.; Zhang, H. Unraveling the electrooxidation mechanism of 5-(hydroxymethyl)furfural at a molecular level via nickel-based two-dimensional metal–organic frameworks catalysts. *ACS Catal.* **2024**, *14*, 449–462. [CrossRef]
18. Prabhu, P.; Wan, Y.; Lee, J.M. Electrochemical conversion of biomass derived products into high-value chemicals. *Matter* **2020**, *3*, 1162–1177. [CrossRef]
19. Liu, X.; Wang, X.; Mao, C.; Qiu, J.; Wang, R.; Liu, Y.; Chen, Y.; Wang, D. Ligand-hybridization activates lattice-hydroxyl-groups of $NiCo(OH)_x$ nanowires for efficient electrosynthesis. *Angew. Chem. Int. Ed.* **2024**, *63*, e202408109. [CrossRef]
20. Li, K.; Lu, H.; Shi, H. Engineering the morphology and electronic structure of $NiCo_2O_4$ to boost the electrocatalytic oxidation of 5-hydroxymethylfurfural. *ChemSusChem* **2025**, *18*, e202402605. [CrossRef]
21. Yang, S.; Xiang, X.; He, Z.; Zhong, W.; Jia, C.; Gong, Z.; Zhang, N.; Zhao, S.; Ya, C. Anionic defects engineering of $NiCo_2O_4$ for 5-hydroxymethylfurfural electrooxidation. *Chem. Eng. J.* **2023**, *457*, 141344. [CrossRef]
22. Liu, G.; Nie, T.; Song, Z.; Sun, X.; Shen, T.; Bai, S.; Zheng, L.; Song, Y. Pd Loaded NiCo hydroxides for biomass electrooxidation: Understanding the synergistic effect of proton deintercalation and adsorption kinetics. *Angew. Chem. Int. Ed.* **2023**, *135*, e202311696. [CrossRef]
23. Jadhav, H.; Roy, A.; Desalegan, B.; Seo, J. An advanced and highly efficient Ce assisted NiFe-LDH electrocatalyst for overall water splitting. *Sustain. Energy Fuels* **2020**, *4*, 312–323. [CrossRef]

24. Nagappan, S.; Karmakar, A.; Madhu, R.; Sankar, S.; Kumaravel, S.; Bera, K.; Dhandapani, H.; Sarkar, D.; Yusuf, S.M.; Kundu, S. 2D CoFe-LDH nanosheet-incorporated 1D microfibers as a high-performance OER electrocatalyst in neutral and alkaline media. *ACS Appl. Energy Mater.* **2022**, *5*, 11483–11497. [CrossRef]
25. Nyongombe, G.E.; Kabongo, G.L.; Noto, L.L.; Dhlamini, M.S. Up-scalable synthesis of highly crystalline electroactive Ni-Co LDH nanosheets for supercapacitor applications. *Int. J. Electrochem. Sci.* **2020**, *15*, 4494–4502. [CrossRef]
26. Zhao, G.; Hai, G.; Zhou, P.; Liu, Z.; Zhang, Y.; Peng, B.; Xia, W.; Huang, X.; Wang, G. Electrochemical oxidation of 5-hydroxymethylfurfural on CeO_2-modified Co_3O_4 with regulated intermediate adsorption and promoted charge transfer. *Adv. Funct. Mater.* **2023**, *33*, 2213170. [CrossRef]
27. Qi, Y.F.; Wang, K.Y.; Zhou, Y.; Sun, Y.; Wang, C. Effects of different vacancies in nickel hydroxides on the electrooxidation towards 5-hydroxymethylfurfural. *Chem. Eng. J.* **2023**, *477*, 146917. [CrossRef]
28. Wang, L.; Wang, Z.; Chu, L.; Huang, Z.; Yang, M.; Wang, G. Electronic structure modulation and morphological engineering of trifunctional Ni_xCo_yP electrode for ultrastable overall water splitting at 1 A cm^{-2} and efficient biomass oxidation valorization. *Int. J. Hydrogen Energy* **2024**, *64*, 830–841. [CrossRef]
29. Wang, W.; Wang, M. Nitrogen modulated $NiMoO_4$ with enhanced activity for the electrochemical oxidation of 5-hydroxymethylfurfural to 2, 5-furandicarboxylic acid. *Catal. Sci. Technol.* **2021**, *11*, 7326–7330. [CrossRef]
30. Xiong, Y.; Hu, S.; Jiang, J.; Liu, Y.; Zhao, W.; Ji, X.; Chen, C.; Fan, M.; Wang, K. Enhancing Ni oxidation reconstruction in Ni_3Fe nanoalloy for efficient electro-oxidation of 5-Hydroxymethylfurfural. *Chem. Eng. J.* **2024**, *499*, 156320. [CrossRef]
31. Sun, M.; Yang, J.; Huang, J.; Wang, Y.; Liu, X.; Qi, Y.; Zhang, L. Interfacial engineering of $Ni/Ni_{0.2}Mo_{0.8}N$ heterostructured nanorods realizes efficient 5-hydroxymethylfurfural electrooxidation and hydrogen evolution. *Langmuir* **2023**, *39*, 3762–3769. [CrossRef] [PubMed]
32. Zhou, Z.; Xie, Y.; Sun, L.; Wang, Z.; Wang, W.; Jiang, L.; Tao, X.; Li, L.; Li, X.; Zhao, G. Strain-induced in situ formation of NiOOH species on CoCo bond for selective electrooxidation of 5-hydroxymethylfurfural and efficient hydrogen production. *Appl. Catal. B* **2022**, *305*, 121072. [CrossRef]
33. Ghosh, S.; Bagchi, D.; Mondal, I.; Sontheimer, T.; Jagadeesh, R.; Menezes, P. Deciphering the role of nickel in electrochemical organic oxidation reactions. *Adv. Energy Mater.* **2024**, *14*, 2400696. [CrossRef]
34. Iliev, M.N.; Silwal, P.; Loukya, B.; Datta, R.; Kim, D.H.; Todorov, N.D.; Pachauri, N.; Gupta, A. Raman studies of cation distribution and thermal stability of epitaxial spinel $NiCo_2O_4$ films. *J. Appl. Phy.* **2013**, *114*, 033514. [CrossRef]
35. Wu, Y.J.; Yang, J.; Tu, T.X.; Li, W.Q.; Zhang, P.F.; Zhou, Y.; Li, J.F.; Li, J.T.; Sun, S.G. Evolution of cationic vacancy defects: A motif for surface restructuration of OER precatalyst. *Angew. Chem. Int. Ed.* **2021**, *60*, 26829–26836. [CrossRef]

MDPI AG
Grosspeteranlage 5
4052 Basel
Switzerland
Tel.: +41 61 683 77 34

Inorganics Editorial Office
E-mail: inorganics@mdpi.com
www.mdpi.com/journal/inorganics

www.ingramcontent.com/pod-product-compliance
Lightning Source LLC
Chambersburg PA
CBHW041019270926
51850CB00024B/2056
* 9 7 8 3 7 2 5 8 7 3 3 6 4 *